I0054709

Ken Kurosaki, Yoshiki Takagiwa and Xun Shi (Eds.)
**Thermoelectric Materials**

## Also of interest

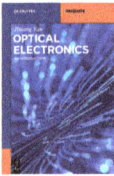

*Optical Electronics.*
*An Introduction*
Jixiang Yan, 2019
ISBN 978-3-11-050049-3, e-ISBN 978-3-11-050060-8

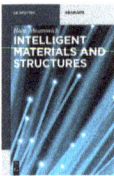

*Intelligent Materials and Structures*
Haim Abramovich, 2016
ISBN 978-3-11-033801-0, e-ISBN 978-3-11-033802-7

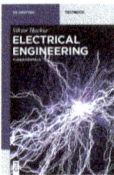

*Electrical Engineering.*
*Fundamentals*
Viktor Hacker, Christof Sumereder, 2020
ISBN 978-3-11-052102-3, e-ISBN 978-3-11-052111-5

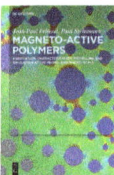

*Magneto-Active Polymers.*
*Fabrication, characterisation, modelling and simulation at the*
*micro- and macro-scale*
Jean-Paul Pelteret, Paul Steinmann, 2019
ISBN 978-3-11-041951-1, e-ISBN 978-3-11-041857-6

# Thermoelectric Materials

Principles and Concepts for Enhanced Properties

Edited by
Ken Kurosaki, Yoshiki Takagiwa and Xun Shi

**DE GRUYTER**

**Editors**

Dr. Ken Kurosaki
Kyoto University
Institute for Integrated Radiation and
Nuclear Science
2, Asashiro-Nishi, Kumatori-cho, Sennan-gun
Osaka 590-0494
Japan
kurosaki.ken.6n@kyoto-u.ac.jp

Dr. Yoshiki Takagiwa
National Institute for Materials Science
1-2-1 Sengen Tsukuba
Ibaraki 305-0047
Japan
takagiwa.yoshiki@nims.go.jp

Dr. Xun Shi
Shanghai Institute of Ceramics, CAS
1295 Dingxi Road
200050 Shanghai
China
xshi@mail.sic.ac.cn

ISBN 978-3-11-059648-9
e-ISBN (PDF) 978-3-11-059652-6
e-ISBN (EPUB) 978-3-11-059453-9

**Library of Congress Control Number: 2020932187**

**Bibliographic information published by the Deutsche Nationalbibliothek**
The Deutsche Nationalbibliothek lists this publication in the Deutsche Nationalbibliografie;
detailed bibliographic data are available on the Internet at http://dnb.dnb.de.

© 2020 Walter de Gruyter GmbH, Berlin/Boston
Cover image: Science Photo Library/Giphotostock
Typesetting: Integra Software Services Pvt. Ltd.
Printing and binding: CPI books GmbH, Leck

www.degruyter.com

# Contents

# List of contributors

**Ikumi Ando**
Department of Applied Physics
Graduate School of Engineering
Tohoku University
Sendai
Miyagi 980-8579, Japan

**Raju Chetty**
Research Institute for Energy
Conservation
National Institute of Advanced Industrial
Science and Technology (AIST)
Tsukuba
Ibaraki 305-8568, Japan

**Nader Farahi**
Institute of Materials Research
Linder Höhe
German Aerospace Center (DLR)
51147 Köln, Germany
Nader.Farahi@dlr.de

**Jerrold A. Floro**
Department of Materials Science and
Engineering
University of Virginia
Charlottesville, VA, USA

**Kei Hayashi**
Department of Applied Physics
Graduate School of Engineering
Tohoku University
Sendai
Miyagi 980-8579, Japan
hayashik@crystal.apph.tohoku.ac.jp

**Yuta Hayashibara**
Department of Applied Physics
Graduate School of Engineering
Tohoku University
Sendai
Miyagi 980-8579, Japan

**Teruyuki Ikeda**
Ibaraki University
College of Engineering
4-12-1 Nakanarusawa
Hitachi
Ibaraki 316-8511, Japan

**Feng Jiang**
Department of Materials Science and
Engineering
Southern University of Science and Technology
ShenZhen 518055, China

**Priyanka Jood**
Research Institute for Energy Conservation
National Institute of Advanced Industrial
Science and Technology (AIST)
Tsukuba
Ibaraki 305-8568, Japan
p.jood@aist.go.jp

**Holger Kleinke**
Department of Chemistry and Waterloo
Institute for Nanotechnology
University of Waterloo
Waterloo N2L 3G1, ON, Canada

**Masataka Kubouchi**
Department of Applied Physics
Graduate School of Engineering
Tohoku University
Sendai
Miyagi 980-8579, Japan

**Ken Kurosaki**
Kyoto University
Institute for Integrated Radiation and
Nuclear Science
2, Asashiro-Nishi, Kumatori-cho, Sennan-gun
Osaka 590-0494, Japan
kurosaki.ken.6n@kyoto-u.ac.jp

https://doi.org/10.1515/9783110596526-203

**Naiming Liu**
Department of Materials Science and
Engineering
University of Virginia
Charlottesville, VA, USA

**Weishu Liu**
Department of Materials Science and
Engineering
Southern University of Science and Technology
ShenZhen 518055, China
liuws@sustc.edu.cn

**Yuzuru Miyazaki**
Department of Applied Physics
Graduate School of Engineering
Tohoku University
Sendai
Miyagi 980-8579, Japan

**Eckhard Müller**
Institute of Inorganic and Analytical Chemistry
Justus-Liebig University Giessen
35392 Gießen, Germany

**Yohei Ogawa**
Department of Applied Physics
Graduate School of Engineering
Tohoku University
Sendai
Miyagi 980-8579, Japan

**Yuji Ohishi**
Graduate School of Engineering
Osaka University
Suita 565-0871, Japan
ohishi@see.eng.osaka-u.ac.jp

**Michihiro Ohta**
Research Institute for Energy Conservation
National Institute of Advanced Industrial
Science and Technology (AIST)
Tsukuba
Ibaraki 305-8568, Japan

**Supree Pinitsoontorn**
Department of Physics
Faculty of Science
Khon Kaen University
Khon Kaen 40002, Thailand
psupree@kku.ac.th

**Wataru Saito**
Department of Applied Physics
Graduate School of Engineering
Tohoku University
Sendai
Miyagi 980-8579, Japan

**Dr. Xun Shi**
Shanghai Institute of Ceramics, CAS
1295 Dingxi Road
200050 Shanghai, China

**Christian Stiewe**
Institute of Materials Research
Linder Höhe
German Aerospace Center (DLR)
51147 Köln, Germany

**Shogo Suzuki**
Department of Applied Physics
Graduate School of Engineering
Tohoku University
Sendai
Miyagi 980-8579, Japan

**Dr. Yoshiki Takagiwa**
National Institute for Materials Science
1-2-1 Sengen Tsukuba
Ibaraki 305-0047, Japan
takagiwa.yoshiki@nims.go.jp

**Dao Y Nhi Truong**
Institute of Materials Research
Linder Höhe
German Aerospace Center (DLR)
51147 Köln, Germany

**Jun Xie**
Graduate School of Engineering
Osaka University
Suita 565-0871, Japan

**Mona Zebarjadi**
Department of Electrical and Computer
Engineering
University of Virginia
Charlottesville, VA, USA
mz6g@virginia.edu

**Yongbin Zhu**
Department of Materials Science and
Engineering
Southern University of Science and Technology
ShenZhen 518055, China

Ken Kurosaki, Xun Shi, and Yoshiki Takagiwa

# 1 Thermoelectric materials and applications: a brief review

**Abstract:** Attempts to improve the performance of thermoelectric materials called dimensionless figure of merit $zT$, have been made over the past several decades. Owing to the efforts, recently, materials having a high $zT$ enough to achieve a conversion efficiency of 10% or more, which is a standard criterion for practical use of thermoelectric power generation, has been reported. This chapter reviews recent trends in thermoelectric materials and thermoelectric power generation applications. Regarding the former, we will describe conventional methods for improving $zT$, and their limitations, and then introduce examples of new materials design guidelines that have the potential to break through those limitations. Regarding the latter, here are some examples of thermoelectric power generation applications that have made remarkable progress in recent years.

**Keywords:** Thermoelectric materials, Thermoelectric power generation, Dimensionless figure of merit, Thermoelectric module, Application of thermoelectric power generation

## 1.1 Thermoelectrics and thermoelectric materials

When the temperature gradient is built in a material, electricity is generated. This phenomenon is called the Seebeck effect. Conversely, when an electric current is applied in a material, a temperature gradient occurs. This phenomenon is called the Peltier effect. In recent years, thermoelectric power generation by utilizing the Seebeck effect has paid much attention for efficient use of unused heat and development of energy harvesting technology. For the thermoelectric power generation, thermoelectric modules are used, in which $p$- and $n$-type semiconductors called thermoelectric materials are continuously connected thermally- in parallel but electrically in series and then sandwiched by two ceramic substrates. The energy conversion efficiency of this module is determined by the temperature difference added across the device as well as the performance of the thermoelectric materials. Therefore, a subject in the thermoelectric research area is to enhance the performance of thermoelectric materials.

**Ken Kurosaki,** Institute for Integrated Radiation and Nuclear Science, Kyoto University, Osaka, Japan
**Xun Shi,** Shanghai Institute of Ceramics, CAS, Shanghai, China
**Yoshiki Takagiwa,** National Institute for Materials Science, Ibaraki, Japan

https://doi.org/10.1515/9783110596526-001

The performance of thermoelectric materials is determined by a material's dimensionless figure of merit, $zT$. The $zT$ is expressed as $zT = S^2\sigma T/\kappa$, where $S$, $\sigma$, $T$, $\kappa$ are the Seebeck coefficient, the electrical conductivity, the absolute temperature, and the thermal conductivity, respectively. Figure 1.1 shows the carrier concentration dependences on these thermoelectric properties. To enhance the $zT$, not only an increase in $S$ and $\sigma$ but also a decrease in $\kappa$ is required. On the other hand, $\kappa$ is composed of two components: one is lattice thermal conductivity $\kappa_{lat}$ in which phonons carry heat and the other is electronic thermal conductivity $\kappa_{el}$ in which charge carriers carry heat. Further, $\kappa_{el}$ is correlated with $\sigma$ via the Wiedemann–Franz law, that is, $\kappa_{el} = L\sigma T$, where $L$ is the Lorenz number. In this situation, $S$, $\sigma$, and $\kappa_{el}$ are interrelated to each other based on the carrier concentration and have a trade-off relationship, that is, by increasing the carrier concentration, $\sigma$ and $\kappa_{el}$ increase but $S$ decreases. Therefore, the conventional guideline for enhancement of $zT$ is to optimize the carrier concentration to maximize the $S^2\sigma$ then to decrease the $\kappa_{lat}$, which is not dependent on the carrier concentration. This is a traditional strategy to enhance the thermoelectric properties. Based on this guideline, various thermoelectric materials have been developed [1]. Typical examples are $Bi_2Te_3$, $PbTe$, and $Zn_4Sb_3$, in which $Bi_2Te_3$, $PbTe$ are composed of heavy elements and $Bi_2Te_3$, $Zn_4Sb_3$ have complicated crystal structure, and their $\kappa_{lat}$ are intrinsically low. Therefore, their $zT$ can be enhanced through optimizing their carrier concentration. Besides, the rattling effect [2] observed in filled skutterudites and nanostructuring [3] have been studied as a method to reduce the $\kappa_{lat}$ with no influences on the $S$ and $\sigma$.

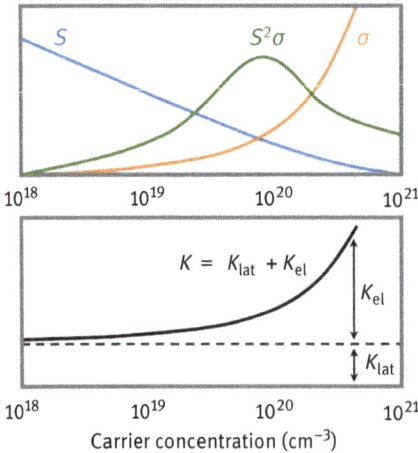

**Figure 1.1:** Carrier concentration dependences on the thermoelectric properties. $S$: Seebeck coefficient; $\sigma$, electrical conductivity; $\kappa$, thermal conductivity; $\kappa_{lat}$, lattice thermal conductivity; and $\kappa_{el}$, electronic thermal conductivity.

However, there exists a critical upper limit for the $zT$ enhanced based on this guideline. For example, the $zT$ value of a nanostructured $p$-type $(Bi,Sb)_2Te_3$ showing the best performance at around room temperature is 1.4 using $S = 210\ \mu V\ K^{-1}$, $\sigma = 8.4 \times 10^4\ S\ m^{-1}$, $\kappa = 1.0\ W\ m^{-1}\ K^{-1}$, and $T = 373\ K$ [4]. Here, we consider an unrealistic assumption, that is, the $\kappa_{lat}$ can be reduced to be 0 with no influences on the $S$, $\sigma$, and $\kappa_{el}$, which leads to $zT = 2.3$ at the largest. On the other hand, $zT > 4$ is required to realize consumer applications of thermoelectric power generation like exhaust heat recovery in automobiles. Thus, a $zT$ value that cannot be achieved from the basic guideline is required. Therefore, a new strategy, in other words, a method to solve a trade-off relationship between $S$ and $\sigma$ as well as a method to reduce the $\kappa_{lat}$ not to be zero but as much as possible is needed. In the next part, such methods are briefly reviewed by giving three representative examples.

## 1.1.1 Enhancement of thermoelectric properties by creating a resonant state

In 2018, an epoch-making method to solve a trade-off relationship between $S$ and $\sigma$ has been experimentally demonstrated in a small amount of Tl-doped PbTe [5]. In the Tl-doped PbTe, Tl not only dopes holes into PbTe but also creates a resonant state around the Fermi level. The resonant state distorts the density of states (DOS) of PbTe (Figure 1.2(a)). According to the Mott theory, $S$ is proportional to the slope of DOS with energy at the Fermi level [6], which means that PbTe doped with Tl makes a resonant state showing larger $S$ than that of PbTe doped with Na, which does not make it even though these two PbTe have the same carrier concentration (Figure 1.2(b)). This result indicates that the trade-off relationship between $S$ and $\sigma$ is solved. As a result, the $zT$ of Tl-doped PbTe is enhanced to be around 1.5, approximately two times higher than that of conventional PbTe.

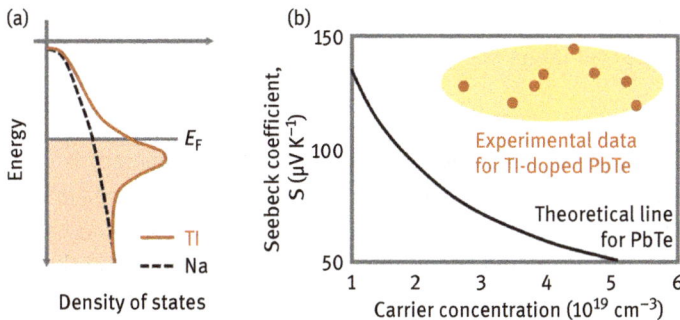

**Figure 1.2:** (a) Schematic representation of the density of states of the valence band of Na-doped (dashed line) and Tl-doped (solid line) PbTe and (b) theoretical line for the Seebeck coefficient at 300 K versus carrier concentration for $p$-type PbTe (solid line) compared to the experimental data for Tl-doped PbTe [5].

## 1.1.2 Enhancement of the Seebeck coefficient by band convergence

One of the strategies to break through the trade-off relationship between $S$ and $\sigma$ is band convergence. Convergence of the valence bands or the conduction bands will increase the effective mass of carriers with keeping high carrier mobility $\mu$, which results in the enhancement of $S$. Here, the mechanism is briefly described.

For simplification, a single parabolic band is assumed. Here, the energy dispersion relation is expressed as follows [7]:

$$E = \frac{\hbar^2 k^2}{2m_b{}^*},$$

(1.1)

where $\hbar$ is the reduced Plank constant, $\kappa$ is the wave number, and $m_b{}^*$ is the band mass. Thus, as indicated by the following equation, $m_b{}^*$ is related to a curvature of the band edge:

$$m_b{}^* = \frac{\hbar^2}{d^2 E / dk^2}.$$

(1.2)

On the other hand, $m_b{}^*$ is inversely proportional to $\mu$ [8]:

$$\mu \propto \left(\frac{1}{m_b{}^*}\right)^{\frac{5}{2}}.$$

(1.3)

Namely, focusing on one band, a broader band edge with larger $m_b{}^*$ but lower $\mu$ is called the heavy band, while a sharper band edge with smaller $m_b{}^*$ but higher $\mu$ is called the light band. Here, a trade-off relationship exists between $S$ and the band convergence [1, 8]. Assume that a parabolic band, $S$, can be expressed as follows:

$$S = \frac{8\pi^2 k_B{}^2}{3eh^2} m^* T \left(\frac{\pi}{3n}\right)^{2/3},$$

(1.4)

where $m^*$ is the DOS effective mass and $n$ is the carrier concentration. It can be confirmed from this equation that larger $m^*$, larger $S$. Furthermore, $m^*$ is given by

$$m^* = N_V{}^{2/3} \times m_b{}^*,$$

(1.5)

where $N_V$ is the valley degeneracy. It is clear that a large valley degeneracy is good for the thermoelectric material. Even though each band is a light band, large $m^*$ can be obtained by increasing $N_V$, which results in the enhanced $S$. Thus, $S$ can be enhanced without degeneration of $\mu$, and herein the trade-off relationship is broken.

This strategy has been demonstrated in various materials such as $PbTe_{1-x}Se_x$ [9] and $CoSb_3$ [10] (band convergence associated with temperature change; Figure 1.3(a) and (b)), $CuGaTe_2$ (band convergence associated with the crystal symmetry; Figure 1.3(c)) [11, 12], and $Mg_2(Si_{1-x}Sn_x)$ (band convergence associated with the composition change; Figure 1.3(d)) [13].

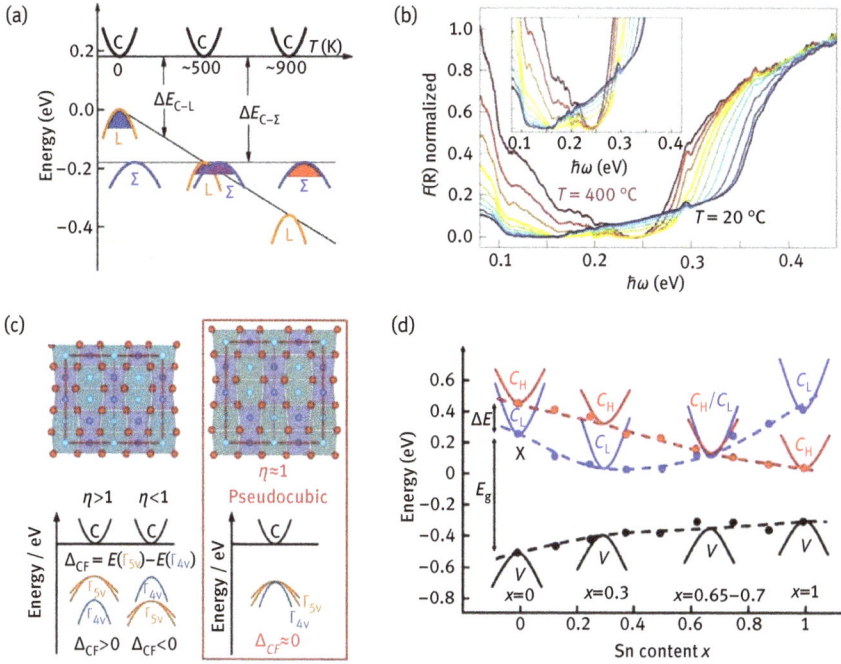

**Figure 1.3:** Band convergence observed in various thermoelectric materials: (a) $PbTe_{1-x}Se_x$ [9], (b) $CoSb_3$ [10], (c) $CuGaTe_2$ [11, 12], and $Mg_2(Si_{1-x}Sn_x)$ [13].

### 1.1.3 Reduction of lattice thermal conductivity by lone-pair electrons

General methods to reduce the $\kappa$ of solids are introduction of phonon impurity scattering by alloying, enhancement of phonon grain boundary scattering by structure size control, or utilization of special phonon characteristics such as rattling. Recently, in addition to these, the introduction of strong phonon–phonon scattering due to the anharmonicity of interatomic bonds has attracted attention [14]. Since the anharmonicity of the interatomic bond is material specific, it is not affected by the degree of randomness or the size of the structure. Therefore, in developing a material having a substantially lower $\kappa$, this becomes an attractive strategy.

The $\kappa$ of $AgInTe_2$ decreases as the temperature rises, while that of $AgSbTe_2$ is almost constant regardless of the temperature, and its value is close to the minimum limit (Figure 1.4) [15]. In $AgSbTe_2$, there exist lone pair of electrons consisting of 5s electrons of Sb, but In in $AgInTe_2$ does not have such lone pair of electrons. The existence of this lone pair of electrons is considered to enhance the anharmonicity of interatomic bonds. In fact, the Grüneisen parameter (a measure of anharmonicity) of $AgSbTe_2$ is about 2, a very high value among the relevant materials.

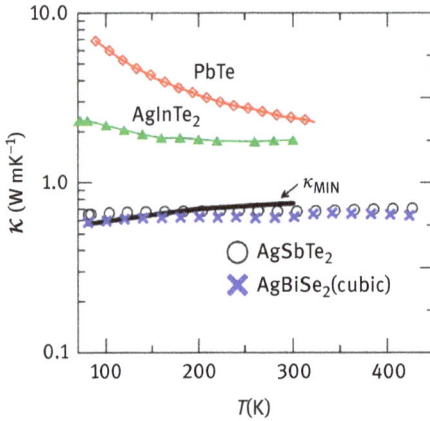

**Figure 1.4:** Temperature dependence of experimentally determined thermal conductivity ($\kappa$) of AgInTe$_2$ and AgSbTe$_2$ [15]. The comparison of data for AgBiSe$_2$ and single-crystal PbTe is shown. The electronic contributions to the $\kappa$ are negligible in all samples, which mean that the data shown represent the lattice thermal conductivity.

It has been considered that the reliable $zT$ values of bulk thermoelectric materials are around unity in maximum. However, new innovative concepts that are not in the traditional guideline have been proposed recently, and materials developed based on the new concepts have been reported. Apart from the above three concepts, several concepts including energy filtering [16, 17] and modulation doping [18] to enhance the $S$ and $\sigma$ by tuning the electronic state and the band structure of materials have been proposed. The energy filtering enhances the $S$ by scattering low-energy carriers at the interfaces between thermoelectric materials as the matrix phase and precipitates as the secondary phase (Figure 1.5). The modulation doping achieves high $\mu$ by cutting off the influence on the ionized impurity scattering through separating the carrier supply phase and the carrier mobile phase. Regarding the $\kappa_{lat}$ reduction, concepts for full-spectrum phonon scattering such as an introduction of dense dislocation arrays (Figure 1.6) [19], and all-scale hierarchical architectures (Figure 1.7) [20] have been proposed. Some of these new concepts are described in other chapters of this book. The development of thermoelectric materials advances to the next stage during the past decade, and further progress is expected in the future.

## 1.2 Thermoelectric modules and applications

The development of thermoelectric modules is also a relevant research field toward practical usage as well as basic thermoelectric materials research. Great progress has been achieved in developing various high-performance TE devices and modules in a

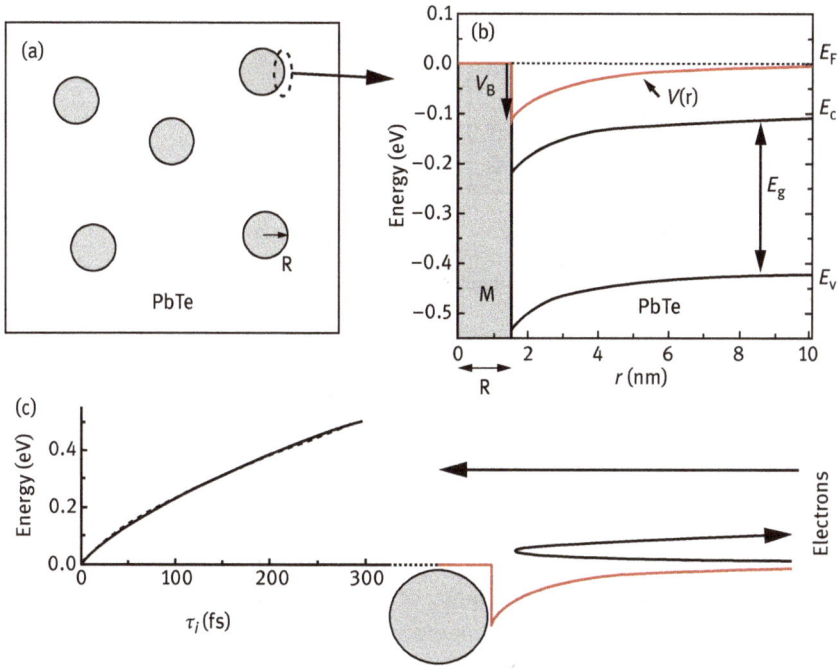

**Figure 1.5:** Images of energy filtering [16]. (a) Schematic of the PbTe matrix with nanoscale inclusions, (b) potential barrier for PbTe and metal under a typical condition, (c) and (d) the concept of energy filtering where low-energy electrons scatter strongly with the potential, but high-energy electrons are unaffected.

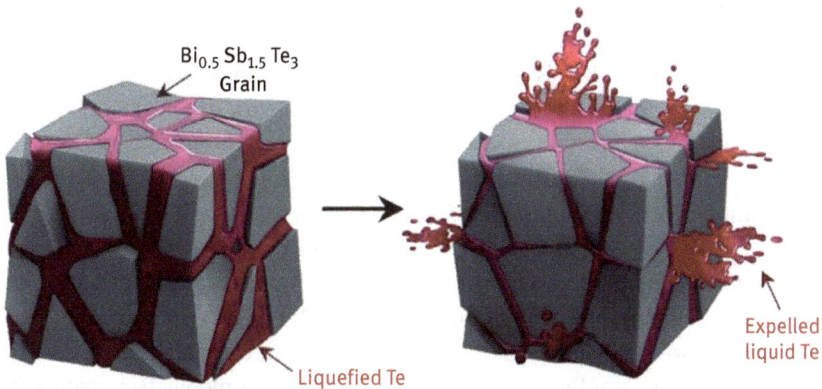

**Figure 1.6:** Generation of dislocation arrays during the liquid-phase compaction process, where the liquid existed at grain boundaries flows out during the compacting process and facilitates the formation of dislocation arrays [19].

$ZT \approx 1.1$   $ZT \approx 1.7$   $ZT \approx 2.2$

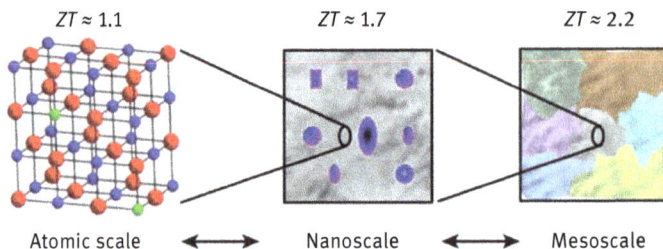

Atomic scale  ⟷  Nanoscale  ⟷  Mesoscale

**Figure 1.7:** All-length-scale hierarchy in thermoelectric material, PbTe–SrTe doped with Na, synthesized by spark plasma sintering [20].

wide temperate range. $Bi_2Te_3$-based TE materials are extensively used for TE generation modules working below 473 K and room temperature TE cooling modules. The conversion efficiency of $Bi_2Te_3$-based power generation modules (e.g., those reported by Hi-z in the USA and KELK in Japan) ranges from 4% to 7% [21, 22]. For mid-temperature range (523–873 K), the TE power generation devices based on n-type PbTe and p-type TAGS have been widely used in the radioisotope thermoelectric generator for space missions by NASA. The efficiency of such modules is ~7% [23]. Recently, as the best candidates used in mid-temperature range, the TE modules based on skutterudites have been successfully manufactured in the institutions including Jet Propulsion Laboratory [24], GM Global Research & Development, Marlow Industries, Fraunhofer Institute for Physical Measurement Techniques [25], and Shanghai Institute of Ceramics, Chinese Academy of Sciences (SICCAS) [26] with a maximum efficiency over 8%. In SICCAS, a high conversion efficiency of 9.3% at a temperature difference of 831 K [27] for skutterudites is achieved. Furthermore, efficiency is pushed to 12% for $Bi_2Te_3$-skutterudite segmented modules under a temperature difference of 814 K [28]. Above 873 K, SiGe and oxide-based materials are mostly used. In fact, SiGe is still the sole material used at high temperatures currently as a power source in space mission with an efficiency of ~7% [29]. The half-Heusler compounds have been recently used for high-temperature module fabrication with the advantages of good TE performance, mechanical robustness, and good stability. The Fraunhofer Institute reported the half-Heusler-based TE modules with a high power density of 3.2 W cm$^{-2}$ and a conversion efficiency of 5% [30]. SICCAS and Zhejiang University exhibit a high efficiency of 6.2% and a power density of 2.2 W cm$^{-2}$ at a temperature difference of 928 K [31] for the half-Heusler-based module.

Furthermore, we will mention the selected commercialized thermoelectric modules and possible applications using traditional thermoelectric materials, that is, iron disilicide ($\beta$-FeSi$_2$) and bismuth telluride (Bi-Te) alloys. Also, we will briefly review the recent progress of module development using novel materials such as half-Heusler and skutterudite alloys, mainly in the case of Japanese private companies [32].

The first application of thermoelectric power generation in Japan was a candle radio. The used U-shaped thermoelectric power generators were composed of $\beta$-FeSi$_2$ materials (Figure 1.8(a)), that is, p-type Mn-doped FeSi$_2$ and n-type Co-doped one. The low-temperature phase ($\beta$-FeSi$_2$) possesses a semiconducting transport property up to 1259 K. Above the peritectoid temperature, the metallic eutectic alloys of tetragonal $\alpha$-Fe$_2$Si$_5$ and cubic $\varepsilon$-FeSi are stable. Therefore, the $\beta$-FeSi$_2$ cannot be synthesized from a simple melting method. Extensive studies on processing to obtain a single phase of $\beta$-FeSi$_2$ have been performed [33]. Also, the purity of source materials affects the resulting thermoelectric properties: the nondoped $\beta$-FeSi$_2$ prepared with 99.99% purity of source materials exhibits p-type conduction, whereas nondoped single crystals prepared with 99.999% purity exhibit n-type [34]. Therefore, the conduction type of $\beta$-FeSi$_2$ should be tuned by chemical doping: Mn, Cr V, Ti, Al, Zr as p-type and Co, Ni, Pd, Pt, Cu as n-type dopants [35].

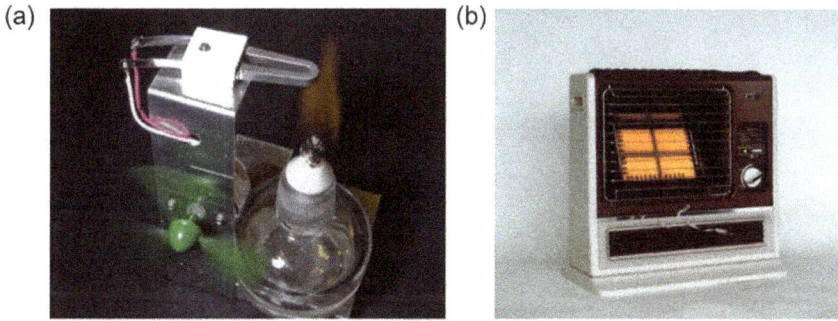

**Figure 1.8:** (a) U-shaped thermoelectric power generator using $\beta$-FeSi$_2$ thermoelectric materials, and (b) a prototype of codeless gas fan heater using FeSi$_2$-based thermoelectric modules [36].

The prototype of the electrode-less U-shaped module with each leg size of about 20 mm and 2 mm thick was developed by the National Institute for Materials Science in Japan [36]. The significant advantage of using $\beta$-FeSi$_2$ thermoelectric materials that consist of low-cost and nontoxic constituent elements is high oxidation resistance at higher temperatures even in the air. The open voltage of each pair is 0.49 V, mean resistivity is 0.051 $\Omega$ cm, and the maximum effective power $P_{max}$ is 110 W m$^{-1}$ (n-type) and 80 W m$^{-1}$ (p-type) at $\Delta T = 800$ K. A prototype of a codeless gas fan heater was fabricated using 30 pairs of U-shaped modules (Figure 1.8(b)). Very recently, Iwatani Co. commercially released a codeless gas fan heater (KAZEDAN) using Bi–Te-based thermoelectric modules [37].

Commercially available candle radio (YOUTES Co., CR-100) released in 1990 installed five couples of U-shaped thermoelectric power generator [36]. These p–n junctions are heated by candle's flame: the high-temperature side and low-temperature one is 680 and 330 K, respectively, and the resulting power output is approximately 30 mW (rated voltage of 1.1 V and current of 20 mA). A candle radio can provide over 12 h of stable power generation and can receive AM broadcasts. The superior high oxidation resistance at the mid-temperature region for β-FeSi$_2$ can enable us to use such practical applications. However, the remaining issue is to improve its mechanical strength because the contact between p- and n-type thermoelectric materials is intrinsically brittle.

About ten years later of the first commercialized application of thermoelectric power generators, a thermoelectric wristwatch was successfully developed by CITIZEN WATCH Co., Ltd [38]. Very finely fabricated Bi–Te thermoelectric module (Figure 1.9) could provide the power output of 14 μW at $\Delta T = 1$ K using body temperature. The number of p–n pairs was 1,242 in only 7.5 × 7.5 mm dimensions module. This application was the first attempt as low-grade thermal energy harvesting technique is considered recently toward the Internet of things society. Recently, this type of wristwatch (PowerWatch) is commercially released by Matrix Industries, USA [39].

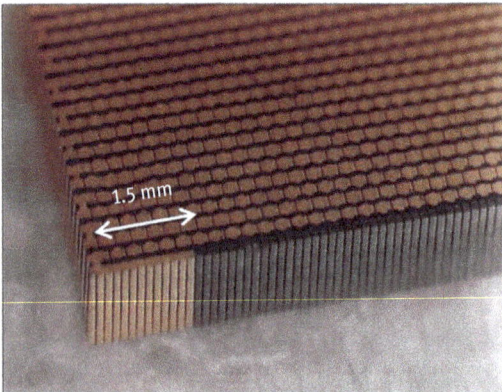

**Figure 1.9:** Bi–Te-based thermoelectric modules installed in the wristwatch by CITIZEN WATCH Co., Ltd [32].

Related to Bi–Te-based thermoelectric modules, Yamaha Corporation developed an independent power supply for a liquid crystal projector in 2006 [40]. The maximum power density is 0.5 W cm$^{-2}$ at $\Delta T = 100$ K between 298 and 398 K. Yamaha Co. is developing the sensing system of human body information using the modules as an energy harvester; the temperature sensing system for a human body and humidity sensing system for industries were developed in 2013.

In 2009, KELK Ltd., which is a leading company of Bi–Te-based thermoelectric modules in Japan, commercialized a Bi–Te module for power generation named "Thermogeneration module" (Figure 1.10) [41]. There are three types of thermoelectric modules available: high-performance type, multipurpose type, and a microgeneration module. The maximum power output of a high-performance module (Figure 1.10(a)) is 24 W at $\Delta T = 250$ K between 303 and 553 K, resulting in the power density of 1 W cm$^{-2}$ and efficiency of 7.2%. The dimensions of the module are $51.5 \times 54 \times 4.4$ mm, and the weight is 47 g. On the other hand, a multipurpose module (Figure 1.10(b)) that is not only economical but highly durable product produces a power output of 9 W at $\Delta T = 170$ K between 303 and 473 K. The dimensions of the module are $48 \times 56.5 \times 1.36$ mm. The last one is a microgeneration module with dimensions of $7.57 \times 7.16 \times 1.32$ mm (Figure 1.10(c)). This module provides the power output of 2.5 mW at $\Delta T = 10$ K, which can be combined with the low-voltage DC–DC converter and capacitor as a power manager.

**Figure 1.10:** Bi–Te-based thermoelectric modules and units produced by KELK Ltd [41]: (a) high-performance type, (b) multipurpose-type module, (c) micropower generation module, (d) thermoelectric energy harvesting device, (e) thermoelectric autonomous power supply unit, and (f) thermoelectric exhausted heat recovery unit.

Recently, KELK Ltd. developed three types of Bi–Te thermoelectric units: thermoelectric energy harvesting device, thermoelectric autonomous power supply unit, and thermoelectric exhausted heat recovery unit, as shown in Figure 1.10(d)–(f). The thermoelectric energy harvesting device installed temperature monitoring sensor with dimensions of $W60 \times D42 \times H25$ mm and $W100 \times D100 \times H30$ mm and the weight of 0.1 and 0.3 kg, respectively. This device can transmit the temperature data at 1–100 s interval.

When putting at the hot side of 473 K, the thermoelectric autonomous power supply units produce the maximum power output of 2.5 and 4 W for supplying USB output with the different dimensions of $W95 \times D115 \times H90$ mm and $W150 \times D148 \times H81$ mm, respectively. On the other hand, the thermoelectric exhausted heat recovery unit provides the more powerful output of 240 W under the temperature condition between 523 K and cooling water.

KELK company tested the power generation system of 10 kW class at the iron continuous casting process in cooperation with JFE Steel Co [42]. A newly launched NEDO project on the demonstration of this system by KELK Ltd. and JFE Steel Corporation started in 2017 [43].

The thermoelectric power generation modules using half-Heusler (Ti, Zr, Hf)-Ni-(Sn, Sb) with a high power density of 1 W cm$^{-2}$ at $\Delta T = 480$ K between 293 and 773 K were commercialized by Toshiba Co. in 2004 [44]. The maximum power output is 15 W with 14.4 cm$^2$ size module (37 mm × 39 mm) (Figure 1.11). The module power density was enhanced to 4.1 W cm$^{-2}$ at $\Delta T = 644$ K between 339 and 984 K in 2012.

**Figure 1.11:** (Ti, Zr, Hf)-Ni-(Sn,Sb) half-Heusler-based thermoelectric modules produced by Toshiba Co [44].

ATSUMITEC Co., Ltd. in cooperation with the Nagoya Institute of Technology developed a thermoelectric module using Fe$_2$VAl Heusler materials. This module produces the power density of 0.25 W cm$^{-2}$ at $\Delta T = 280$ K between 293 and 573 K, which is applied to the subpower generation from the exhausted heat of a motorbike. Recently, this company succeeded in the development of "Exhaust Gas Power Generating System" in cooperation with the National Institute of Advanced Industrial Science and Technology [45]. The central part consists of a solid oxide fuel cell (SOFC) cell using unused fuel and thermoelectric module using exhaust heat energy, as shown in Figure 1.12. The newly developed power generation unit produces 200 W with a unit size of 340 cm$^3$.

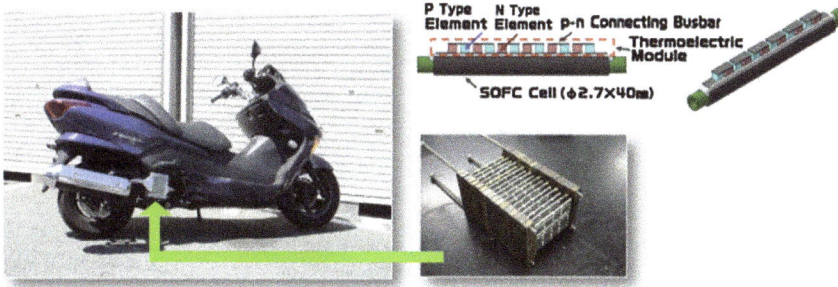

**Figure 1.12:** Exhaust gas power generating system developed by ATSUMITEC Co., Ltd. [45].

Finally, we briefly introduce the research and development of thermoelectric modules using clathrate compounds by Furukawa Electric Co., Ltd. [46]. They developed a sintered body U-shaped element for thermoelectric power generation in mid- and high-temperature range, and succeeded in generating electricity with a small flame such as a candle, which is similar to the $\beta$-FeSi$_2$ thermoelectric module, as shown in Figure 1.13(a). The electrode-less U-shaped thermoelectric module integrated with p-type and n-type sintered materials of environmentally friendly silicon clathrate compounds. This module over a flame such as a candle produces a voltage of 50 mV. Figure 1.13(b) shows a proposed compact power generation system installed in a stove. In combination with a portable stove that is a recommended disaster preparedness stockpile item, it is also expected that the technology can be used as a means to charge small electronic devices such as smartphones [46].

The following fundamental and technical issues should be overcome to develop the high-performance and reliable thermoelectric modules: (1) material development with high power output, (2) excellent mechanical properties of materials, (3) high chemical stability and oxidation resistance, (4) lowering thermal conductivity, (5) matching the coefficient of thermal expansion, and (6) contact design including low contact resistivity. In addition, the cost of materials and nontoxicity are other critical factors for practical usage.

(a)

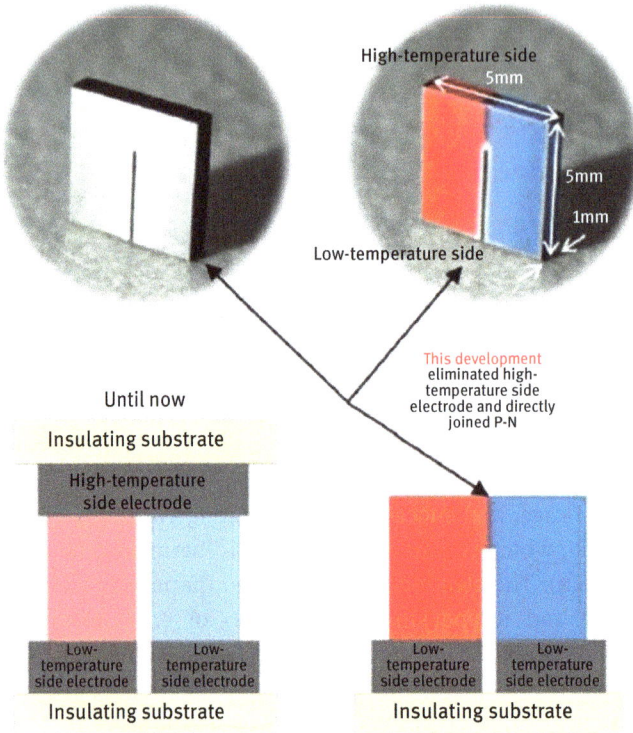

(b)

U-shaped elements are arranged circumferentially at high density around the outer edge of a stove flame.

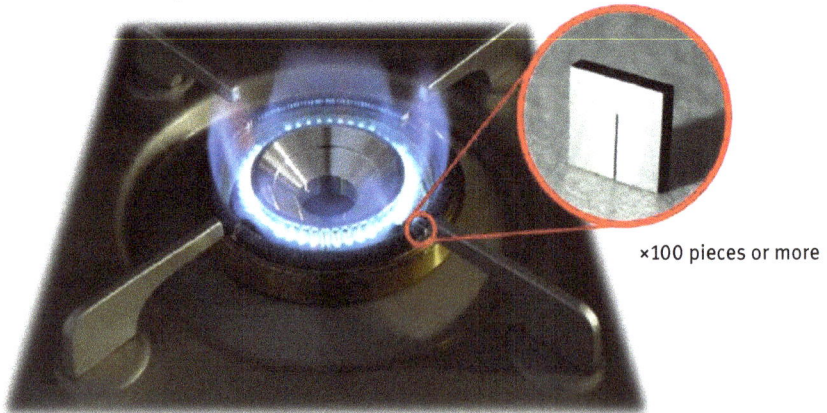

×100 pieces or more

**Figure 1.13:** (a) U-shaped thermoelectric power generation using clathrate compounds by Furukawa Electric Co., Ltd [46]. (b) Example of an application using stove flame.

# References

[1]    Snyder G J, Toberer, E S. Complex thermoelectric materials, Nat. Mater. 2008, 7, 105–114.
[2]    Sales B C, Mandrus, D, Williams, R K. Filled Skutterudite Antimonides: A New Class of
       Thermoelectric Materials, Science 1996, 272, 1325–1328.
[3]    Dresselhaus M S, Chen, G, Tang, M Y, Yang, R G, Lee, H, Wang, D Z, Ren, Z F, Fleurial, J P,
       Gogna, P. New Directions for Low-Dimensional Thermoelectric Materials, Adv. Mater. 2007,
       19, 1043–1053.
[4]    Poudel B, Hao, Q, Ma, Y, Lan, Y, Minnich, A, Yu, B, Yan, X, Wang, D, Muto, A, Vashaee, D,
       Chen, X, Liu, J, Dresselhaus, M S, Chen, G, Ren, Z. High-Thermoelectric Performance of
       Nanostructured Bismuth Antimony Telluride Bulk Alloys, Science 2008, 320, 634–638.
[5]    Heremans J P, Jovovic, V, Toberer, E S, Saramat, A, Kurosaki, K, Charoenphakdee, A,
       Yamanaka, S, Snyder, G J. Enhancement of Thermoelectric Efficiency in PbTe by Distortion of
       the Electronic Density of States, Science 2008, 321, 554–557.
[6]    Cutler M, Mott, N F. Observation of Anderson Localization in an Electron Gas, Phys. Rev. 1969,
       181, 1336–1340.
[7]    Kane E. J. Band Structure of Indium Antimonide, Phys. Chem. Solids 1957, 1, 249–261.
[8]    Pei Y, LaLonde, A D, Wang, H, Snyder, G J. Low effective mass leading to high thermoelectric
       performance, Energy Environ. Sci. 2012, 5, 7963–7969.
[9]    Pei Y, Shi, X, LaLonde, A, Wang, H, Chen, L, Snyder, G J. Convergence of electronic bands for
       high performance bulk thermoelectrics, Nature 2011, 473, 66–69.
[10]   Tang Y, Gibbs, Z M, Agapito, L A, Li, G, Kim, H-S, Nardelli, M B, Curtarolo, S, Snyder, G J.
       Convergence of multi-valley bands as the electronic origin of high thermoelectric
       performance in $CoSb_3$ skutterudites, Nat. Mater. 2015, 14, 1223–1228.
[11]   Plirdpring T, Kurosaki, K, Kosuga, A, Day, T, Firdosy, S, Ravi, V, Snyder, G J,
       Harnwunggmoung, A, Sugahara, T, Ohishi, Y, Muta, H, Yamanaka, S. Chalcopyrite CuGaTe 2:
       A high-efficiency bulk thermoelectric material, Adv. Mater. 2012, 24, 3622–3626.
[12]   Zhang J, Liu, R, Cheng, N, Zhang, Y, Yang, J, Uher, C, Shi, X, Chen, L, Zhang, W.
       High-Performance Pseudocubic Thermoelectric Materials from Non-cubic Chalcopyrite
       Compounds, Adv. Mater. 2014, 26, 3848–3853.
[13]   Liu W, Tan, X, Yin, K, Liu, H, Tang, X, Shi, J, Zhang, Q, Uher, C. Convergence of Conduction
       Bands as a Means of Enhancing Thermoelectric Performance of n-Type $Mg_2Si_{1-}xSnx$ Solid
       Solutions, Phys. Rev. Lett 2012, 108, 166601-1–166601-5.
[14]   Nielsen M D, Ozolins, V, Heremans, J P. Lone pair electrons minimize lattice thermal
       conductivity, Energy Environ. Sci. 2013, 6, 570–578.
[15]   Morelli D T, Jovovic, V, Heremans, J P. Intrinsically Minimal Thermal Conductivity in Cubic I–V–VI2
       Semiconductors, Phys. Rev. Lett 2008, 101, 035901-1–035901-4.
[16]   Faleev S V, Leonard, F. Theory of enhancement of thermoelectric properties of materials with
       nanoinclusions, Phys. Rev. B 2008, 77, 214304-1–214304-9.
[17]   Martin J, Wang, L, Chen, L, Nolas, G S. Enhanced Seebeck coefficient through energy-barrier
       scattering in PbTe nanocomposites, Phys. Rev. B 2009, 79, 115311-1–115311-5.
[18]   Dingle R, Störmer, H L, Gossard, A C, Wiegmann, W. Electron mobilities in modulation-doped
       semiconductor heterojunction superlattices, Appl. Phys. Lett. 1978, 33, 665–667.
[19]   Kim S, Lee, K H, Mun, H A, Kim, H S, Hwang, S W, Roh, J W, Yang, D J, Shin, W H, Li, X S,
       Lee, Y H, Snyder, G J, Kim, S W. Dense dislocation arrays embedded in grain boundaries for
       high-performance bulk thermoelectrics, Science 2015, 348, 109–114.
[20]   Biswas K, He, J, Blum, I D, Wu, C-I, Hogan, T P, Seidman, D N, Dravid, V P, Kanatzidis, M G.
       High-performance bulk thermoelectrics with all-scale hierarchical architectures, Nature 2012,
       489, 414–418.

[21] Kuroki T, Kabeya, K, Makino, K, Kajihara, T, Kaibe, H, Hachiuma, H, Matsuno, H, Fujibayashi, A. Thermoelectric generation using waste heat in steel works, J. Electron. Mater. 2014, 43, 2405–2410.

[22] Hao F, Qiu, P, Tang, Y, Bai, S, Xing, T, Chuc, H-S, Zhang, Q, Lu, P, Zhang, T, Ren, D, Chen, J, Shi, X, Chen, L. High efficiency $Bi_2Te_3$-based materials and devices for thermoelectric power generation between 100 and 300 °C, Energy Environ. Sci. 2016, 9, 3120–3127.

[23] Lange R G, Carroll, W P. Review of recent advances of radioisotope power systems, Energy Convers. Manage. 2008, 49, 393–401.

[24] Holgate T C, Bennett, R, Hammel, T, Caillat, T, Keyser, S, Sievers, B. Increasing the Efficiency of the Multi-mission Radioisotope Thermoelectric Generator, J. Electron. Mater. 2015, 44, 1814–1821.

[25] Salvador J R, Cho, J Y, Ye, Z, Moczygemba, J E, Thompson, A J, Sharp, J W, Koenig, J D, Maloney, R, Thompson, T, Sakamoto, J, Wang, H, Wereszczak, A A. Conversion efficiency of skutterudite-based thermoelectric modules, Phys. Chem. Chem. Phys. 2014, 16, 12510–12520.

[26] Shi X, Chen, L. Thermoelectric materials step up, Nat. Mater. 2016, 15, 691–692.

[27] Zhang Q, Zhou, Z, Dylla, M, Agne, M T, Pei, Y, Wang, L, Tang, Y, Liao, J, Li, J, Bai, S, Jiang, W, Chen, L, Snyder, G J. Realizing high-performance thermoelectric power generation through grain boundary engineering of skutterudite-based nanocomposites, Nano Energy 2017, 41, 501–510.

[28] Zhang Q, Liao, J, Tang, Y, Gu, M, Ming, C, Qiu, P, Bai, S, Shi, X, Uher, C, Chen, L. Realizing a thermoelectric conversion efficiency of 12% in bismuth telluride/skutterudite segmented modules through full-parameter optimization and energy-loss minimized integration, Energy Environ. Sci. 2017, 10, 956–963.

[29] Aswal D K, Basu, R, Singh, A. Key issues in development of thermoelectric power generators: High figure-of-merit materials and their highly conducting interfaces with metallic interconnects, Energy Convers. Manage. 2016, 114, 50–67.

[30] Bartholome K, Balke, B, Zuckermann, D, Kohne, M, Muller, M, Tarantik, K, Konig, J. Thermoelectric Modules Based on Half-Heusler Materials Produced in Large Quantities, J. Electron. Mater. 2014, 43, 1775–1781.

[31] Fu C, Bai, S, Liu, Y, Tang, Y, Chen, L, Zhao, X, Zhu, T. Realizing high figure of merit in heavy-band p-type half-Heusler thermoelectric materials, Nat. Commun. 2015, 6, 8144.

[32] Shinohara Y. Recent progress of thermoelectric devices or modules in Japan. Materials Today: Proceedings 2017, 4, 12333–12342.

[33] Kojima T. Semiconducting and Thermoelectric Properties of Sintered Iron Disilicide. Phys. Stat. Sol. (a) 1989, 111, 233–242.

[34] Isoda Y, Udono, H. Preparation and Thermoelectric Properties of Iron Disilicide. CRC Handbook Thermoelectrics and Its Energy Harvesting, (2012) 18-1-25.

[35] Nozariasbmarz A, Nozariasbmarz, A, Agarwal, A, Coutant, Z A, Hall, M J, Liu, J, Liu, R, Malhotra, A, Norouzzadeh, P, Öztürk, M C, Ramesh, V P, Sargolzaeiaval, Y, Suarez, F, Vashaee, D. Thermoelectric silicides: A review. Jpn. J. Appl. Phys 2017, 56, 05DA04-1-27.

[36] Nishida I A. Thermoelectrics-Principles and Applications, edited by M. Sakata, (2000) pp.199–209.

[37] http://www.i-cg.jp/product/stove/cb-gfh-2/.

[38] Watanabe, S, in: T. Kajikawa, S. Sano, J. Morimoto (Eds.), The New Edition of Thermoelectric Energy Conversion Systems, REALIZE Science & Engineering, Tokyo, (2004), pp. 299–304.

[39] https://www.powerwatch.com/pages/power-watch-japan.

[40] http://device.yamaha.com/ja/thermoelectric_cooler/generator/.

[41] http://www.kelk.co.jp/english/index.html.

[42] http://www.kelk.co.jp/news/120419.html (in Japanese).

[43] https://www.nedo.go.jp/koubo/DA3_100168.html (in Japanese).

[44] http://www.toshiba.co.jp/about/press/2004_03/pr_j2901.htm (in Japanese).

[45] http://www.atsumitec.co.jp/en/technology/basis.

[46] https://www.furukawa.co.jp/en/release/2017/kenkai_171117.html.

Yuji Ohishi and Jun Xie

# 2 Power factor optimization and lattice thermal conductivity reduction: a case study on Si

**Abstract:** Si is nontoxic and the second most abundant element in the Earth's crust, which makes it quite attractive as an environmentally friendly thermoelectric (TE) material. The TE performance of Si is, however, quite poor mainly due to its high thermal conductivity. This demerit might be overcome by nanostructuring since enhanced phonon scattering at nanostructures results in reduced thermal conductivity. From a practical point of view, self-assembled nanostructure is desirable. This chapter aims at introducing recent studies on Si with automatically formed nanostructures. To this end, the TE properties of Si are described first. Since the most reliable data can be obtained from single crystals, TE properties of P- and B-doped single crystal Si are shown. The carrier concentrations of these single crystal Si are at most $1 \times 10^{20}$ cm$^{-3}$, which are not enough for the evaluation of the optimum carrier concentration. Thus, the TE properties of Si are discussed based on simulations using models built based on data taken from single crystal Si. The calculation suggests that the best carrier concentration that maximizes the power factor is $5 \times 10^{20}$ cm$^{-3}$ for both p- and n-type Si. The TE properties of polycrystalline Si with the best carrier concentration are provided as the TE properties of optimized bulk Si. Next, the best nanostructure size is proposed to be 10 nm based on the calculated mean free paths of phonons and electrons. Then, three studies on Si with self-assembled nanostructures are introduced. The first one is Si/silicide nanocomposite films. Si and silicide nanocrystals are formed by the phase separation from amorphous phase. Significantly low thermal conductivity was achieved in this film. The second study is on nanometer-sized eutectic structures of Si/CrSi$_2$. Characteristic maze-like structure is formed by melt spinning of eutectic Si–CrSi$_2$ melt. The third study is about Si with nanoprecipitates and dislocations. P acts as an n-type dopant in Si, and when P is added to Si above its solution limit, the excess P atoms form nanoprecipitates in the Si matrix, improving the TE performance of Si. Dislocations are able to be formed in this excess P-doped Si through liquid phase sintering and reduce the thermal conductivity further by enhanced phonon scattering. Since the size of the nanostructures formed by these approaches is larger than the best size of 10 nm, there seems to be still space for further improvement of TE performance of Si.

**Keywords:** Si, Single crystal, Self-assembled nanostructure, Nanocomposite

**Yuji Ohishi and Jun Xie,** Graduate School of Engineering, Osaka University, Suita, Japan

https://doi.org/10.1515/9783110596526-002

# 2.1 Si as an environmentally friendly TE material

Although traditional thermoelectric (TE) materials such as $Bi_2Te_3$ and PbTe exhibit excellent TE performance, the toxicity and rarity of the constituent elements prevent the widespread use of these TE materials. On the contrary, Si is nontoxic and the second most abundant element in the Earth's crust, which makes it quite attractive as an environmentally friendly TE material. Pure Si is an intrinsic semiconductor with a band gap of 1.124 eV at 300 K [1]. By doping silicon with different elements, the conduction type and properties of Si can be modified. Boron-doped Si shows p-type conduction, whereas P-, As-, or Sb-doped Si show n-type conduction. In spite of these advantages in terms of rarity and toxicity, Si has not been used as a TE material since the TE performance of Si is poor due to its quite high thermal conductivity (140 W m$^{-1}$ K$^{-1}$ at 300 K [2]). However, if we look at the electrical properties of Si, we can notice that Si has good electrical property as other TE materials. The power factor (PF) of n-type Si at 300 K reaches 4.1 mW m$^{-1}$ K$^{-2}$ [3], which is almost identical to that of $Bi_2Te_3$. This fact suggests that the TE property of Si can be significantly improved by reducing the thermal conductivity.

It has been experimentally and theoretically proved that the reduction of the thermal conductivity is possible by introducing nanostructures that can act as scattering centers of phonons. Over the years, several groups have reported marked reduction in thermal conductivity and significant enhancement of $ZT$ in Si nanostructures such as nanocrystals (NCs) [4], nanowires [5, 6], and nanomeshes [7, 8]. Among them, Si nanowires fabricated using lithography technique have the highest $ZT$ value of around 1 at 200 K [5]. While this fact indicates that Si is a potential TE material, it is not realistic to use these nanostructured Si as a TE material since generally bulk size TE materials are required for practical use. But in other words, if such nanostructures can be automatically built in bulk Si, it will open the door to the practical use of Si-based TE materials.

In order to enhance the TE performance of nanostructured Si, it is quite important to understand the basic TE properties of Si. The physical properties of lightly doped Si have been keenly studied for decades [9–12], owing to the importance of Si in the semiconductor industry, whereas available information on the TE properties of heavily doped Si is limited. Therefore, the experimentally and theoretically evaluated basic TE properties of Si will be presented first. After that, recently developed Si-based TE materials with self-assembled nanostructures will be introduced.

## 2.1.1 Thermoelectric properties of single crystalline Si

Si nanostructures [4–8] have carrier concentration of around $1 \times 10^{20}$ cm$^{-3}$, which is the best carrier concentration for a lot of TE materials. However, in order to optimize the carrier concentration to maximize the TE performance of Si, it is essential to measure the TE property of Si with various carrier concentrations. In addition to that, information on the mean free paths (MFPs) of carriers and phonons in Si is also vital for designing nanostructure since the nanostructure should scatter phonons without affecting carriers in order to improve the TE performance.

The best way to get accurate physical properties is to conduct measurements with single crystals. Thus, the TE properties of Si measured using single crystal Si with various carrier concentration will be shown first. Second, the effects carrier concentration has on the TE properties of Si will be discussed based on calculated results. In this calculation, a model based on the Boltzmann transport equation (BTE) under the relaxation time approximation was employed in which parameters on relaxation time were set based on the measured TE properties of single crystal Si. Finally, the phonon and electron MFPs in Si will be calculated using the model.

### 2.1.1.1 Experimental study on the TE properties of single crystalline Si

Figure 2.1 shows the PF of heavily P- (n-type) and B- (p-type) doped single crystal Si measured by Osaka University group [13]. The values of PF are 2–3 mW m$^{-1}$ K$^{-2}$ at room temperature regardless of the carrier type and carrier concentration. The PF decreases with increasing temperature for Si with lower carrier concentration than $1 \times 10^{19}$ cm$^{-3}$ due to the reduction in the electrical conductivity caused by strong phonon–electron scattering. Thermal excitation of carriers becomes significant above 800 K, which also deteriorates PF. In terms of the carrier type, n-type shows higher PF than p-type owing to the higher carrier mobility of n-type. The maximum PF was 4.4 mW m$^{-1}$ K$^{-2}$ at 600–900 K for n-type Si with a carrier concentration of $6.8 \times 10^{19}$ cm$^{-3}$ and 2.6 mW m$^{-1}$ K$^{-2}$ at 600–1,000 K for p-type Si with a carrier concentration of $1.1 \times 10^{20}$ cm$^{-3}$.

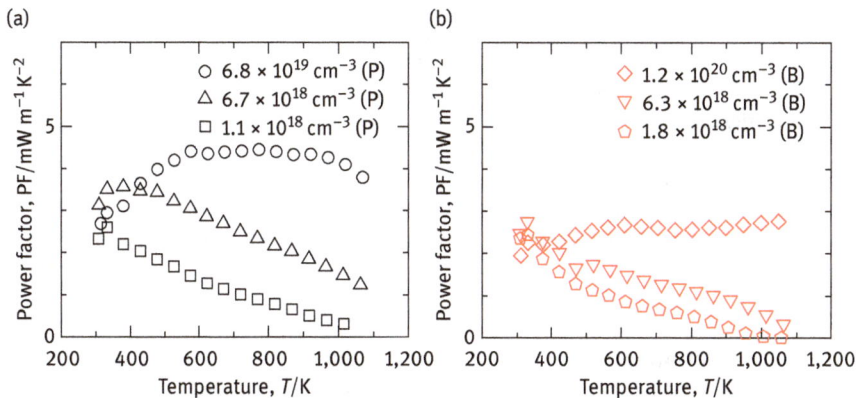

**Figure 2.1:** Temperature-dependent power factor for (a) single crystalline n-type Si and (b) single crystalline p-type Si, respectively [13]. The characters in the brackets stand for dopant elements.

Figure 2.2(a) and (b) gives the thermal conductivity of single crystal Si as a function of temperature. The electrical contribution to the thermal conductivity is negligible and phonon contribution is dominant across all samples. The thermal conductivity decreases with increasing temperature due to phonon–phonon scattering, and decreases with increasing carrier concentration because of phonon–impurity scattering. Rise in the thermal conductivity by thermally excited carriers is not observed even above 1,000 K.

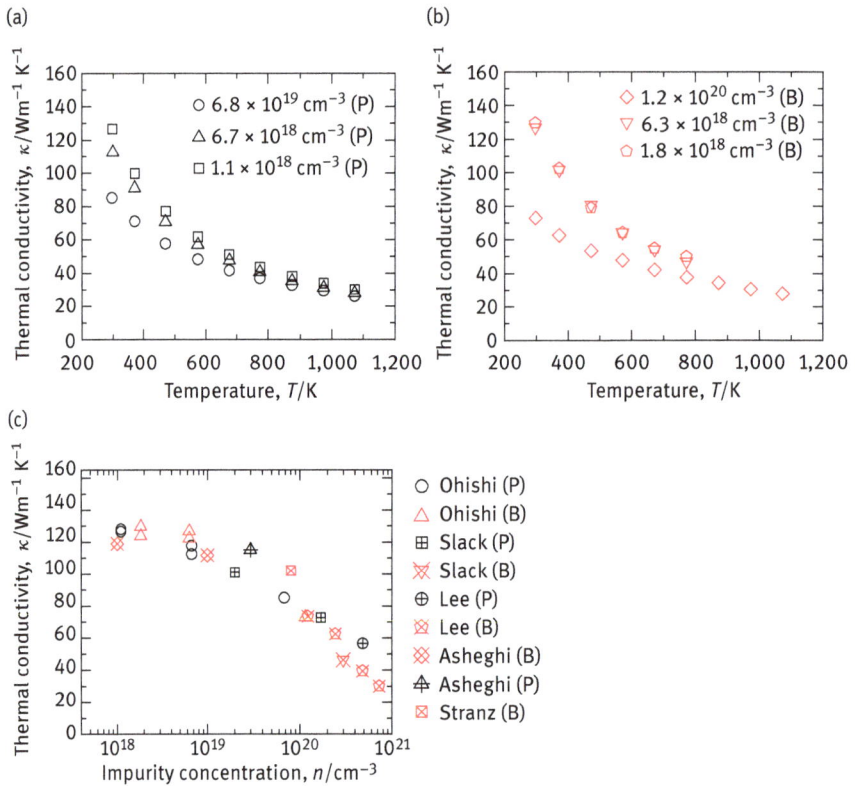

**Figure 2.2:** Temperature-dependent thermal conductivity of single crystalline Si for (a) n-type and (b) p-type, respectively. (c) Thermal conductivity of single crystalline Si at room temperature as a function of impurity concentration [9–13]. The characters in the brackets stand for dopant elements.

The carrier concentration-dependent thermal conductivity of Si at room temperature is shown in Figure 2.2(c). Although the thermal conductivity of pure Si is around 130 W m$^{-1}$ K$^{-1}$ [2], it can be suppressed significantly by heavy doping. While this reduction in the thermal conductivity would be mainly attributed to phonon–impurity scattering, other phonon scattering mechanisms such as phonon–electron scattering may also play an important role. This is because P and B doping equally suppress the thermal conductivity, which cannot be explained solely by phonon–impurity scattering since within the phonon–impurity scattering mechanism, B is expected to scatter phonons more effectively owing to its larger difference from Si in terms of atomic mass and covalent radius.

Figure 2.3 shows the $ZT$ values of single crystal Si with respect to temperature [13]. The higher PF of n-type Si (Figure 2.1) leads to higher $ZT$ values for n-type over p-type Si. The $ZT$ values increase with increasing carrier concentration for both n- and p-type Si, which indicates that the optimal carrier concentration of Si is higher than $6.8 \times 10^{19}$ cm$^{-3}$ for n-type and $1.2 \times 10^{20}$ cm$^{-3}$ for p-type Si. The maximum $ZT$ values are 0.17 and 0.10 for n-type and p-type, respectively.

(a)                                                    (b)

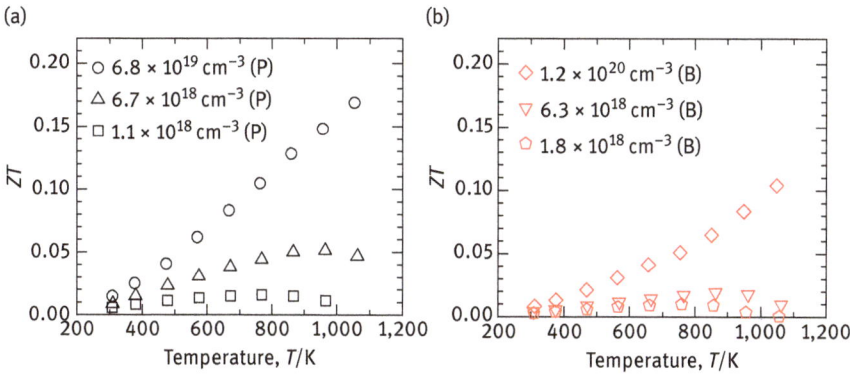

**Figure 2.3:** Temperature dependent $ZT$ of single crystalline Si for (a) n-type and (b) p-type, respectively. The characters in the brackets stand for dopant elements.

### 2.1.1.2 Relationship between carrier concentration and power factor

As shown earlier, experimental study on single crystal Si showed that the optimal carrier concentrations are higher than $6.8 \times 10^{19}$ cm$^{-3}$ and $1.2 \times 10^{20}$ cm$^{-3}$ for n-type and p-type Si, respectively. Although single crystal Si with higher carrier concentration will allow us to experimentally decide the optimal carrier concentration, unfortunately it is difficult to obtain such super heavily doped single crystal Si. Thus, theoretical calculation was performed to analyze the relationship between the carrier concentration and TE properties of Si.

The model used in this calculation is based on the BTE under the relaxation time approximation. The functional forms and parameters are given in ref. [13]. The parameters were set so that the model can reproduce the experimentally obtained TE properties of single crystal Si. Figure 2.4 shows the PFs of n-and p-type Si calculated at 300, 650, and 1,000 K as a function of the carrier concentration. This figure indicates that while the optimal carrier concentration changes depending on the temperature, it would be around $5 \times 10^{20}$ cm$^{-3}$ for both n-type and p-type Si. The PF of optimally doped Si is predicted to reach 7 and 3 mW m$^{-1}$ K$^{-2}$ at 650 K for n-type and p-type Si, respectively.

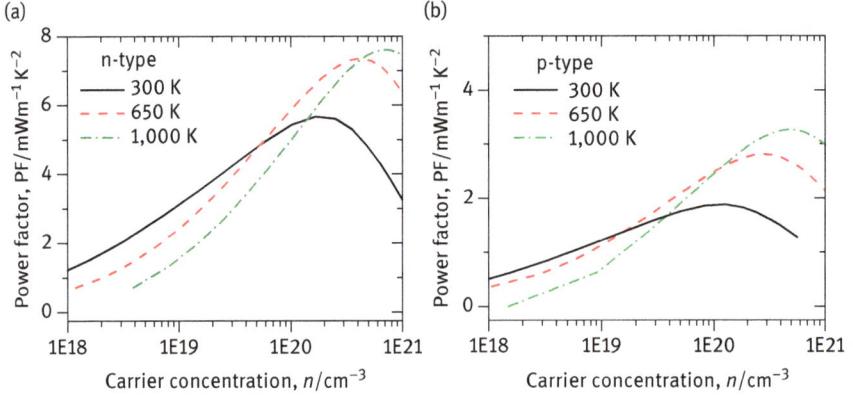

**Figure 2.4:** Calculated power factor as a function of carrier concentration for (a) n-type and (b) p-type single crystalline Si.

### 2.1.1.3 Accumulated thermal and electrical conductivity

It is recognized that nanostructures scatter phonons without affecting electrons, resulting in the reduced thermal conductivity and improved TE performance. This effect is attributed to the difference in the MFPs of phonons and electrons: generally phonons have longer MFP than electrons. Although this seems to be the case with Si since great improvements in TE properties have been achieved in nano-structured Si [4–8], it is necessary to get insight into the MFPs of phonons and electrons in Si in order to achieve maximum improvement in TE performance by nanostructuring.

The accumulated thermal conductivity of single crystal Si at 300 K as a function of phonon frequency is shown in Figure 2.5(a). This calculation was done using a BTE-based model that can reproduce the thermal conductivity of doped Si. In this model, the dispersion relation is approximated by a sine-type function. The parameters and functional forms of this model can be found in ref. [13]. The calculated result shows that phonons with wide frequency range contribute to heat transport, indicating that all-scale hierarchical architectures, which has been proven to be effective in improving the TE performance of PbTe [14], seem to work for Si as well.

(a)

(b)

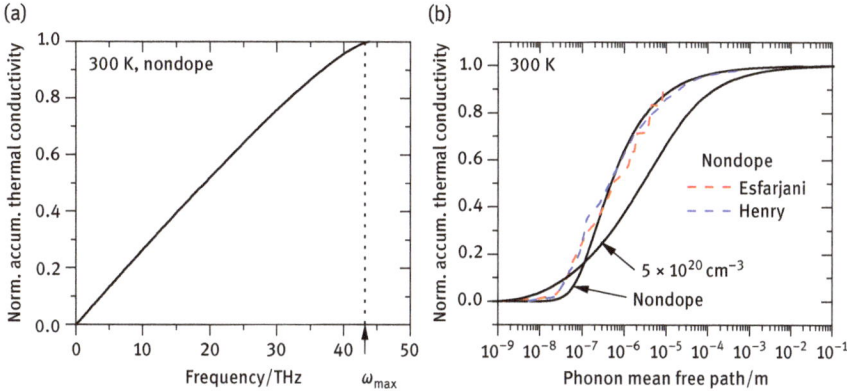

**Figure 2.5:** Accumulated thermal conductivity of single crystalline Si as a function of (a) phonon frequency and (b) phonon MFP.

The accumulated thermal conductivity was also calculated as a function of phonon MFP as shown in Figure 2.5(b). The calculated result for nondoped Si is in good agreement with those obtained by first-principles calculations [15, 16], suggesting that the model used here is roughly able to describe the heat transport process in Si in terms of phonon MFP. The calculation shows that 80–90% of heat in Si is transported by phonons with MFP longer than 100 nm. While the calculated result for heavily doped Si ($5 \times 10^{20}$ cm$^{-3}$) indicates that the contribution of phonons with MFPs smaller than 100 nm increases by heavy doping, contribution from phonons with long MFP remains dominant.

The accumulated electrical conductivity was also calculated in the same way for heavily doped Si as shown in Figure 2.6(a). While there is a slight difference between electrons and holes, electrical conduction of Si is dominated by carriers with MFPs shorter than 10 nm. The difference between phonons and electrons is clearly shown in Figure 2.6(b), where accumulated conductivities of electrons and phonons are shown together. This figure shows that nanostructures with 10 nm in size are the most effective to selectively scatter phonons without affecting electrons in Si.

## 2.1.2 TE properties of bulk polycrystalline Si at the optimized carrier concentration

The optimal carrier concentration in Si was estimated to be around $5 \times 10^{20}$ cm$^{-3}$ for both n-type and p-type Si. By assuming that all the dopant atoms are electrically active, the required amount of dopant to achieve a carrier concertation of $5 \times 10^{20}$ cm$^{-3}$ can be calculated and was found to be 1 at.%. Although it is difficult to obtain single

(a)

(b)

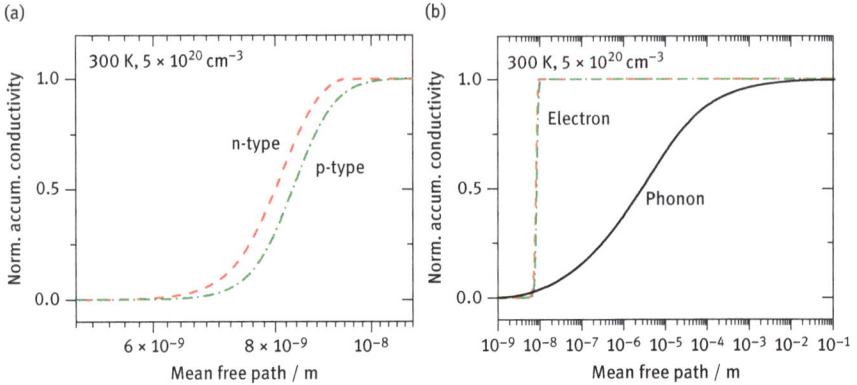

Figure 2.6: (a) Accumulated electrical conductivity of single crystalline Si as a function of carrier MFP and (b) accumulated conductivities of phonon and carriers are shown together.

crystal Si with such high doping concentration, it is possible to synthesize 1 at.% B or P added polycrystalline Si by arc-melting technique. The TE properties of arc-melted polycrystalline Si are shown in Figure 2.7. Although polycrystalline materials are likely to show lower $ZT$ values compared to single crystals because of pores and grain boundaries that can deteriorate electrical properties, this figure shows that polycrystalline Si has higher $ZT$ values than single crystal Si (Figure 2.3) owing to higher carrier concentration. The $ZT$ values of polycrystalline 1% P and B added Si reach 0.26 and 0.18 at 1,050 K for n- and p-types, respectively. These values are expected to be the TE performance of Si with optimized carrier concentration.

## 2.2 Lattice thermal conductivity reduction

Even if the carrier concentration is optimized, the TE performance of Si is still significantly lower than that of conventional TE materials. One of the promising ways to improve the TE performance is to reduce the lattice thermal conductivity with various kinds of nanostructures that can be constructed by ball milling, lithography, chemical etching, and so on. Although these techniques are effective in enhancing TE performance through the reduction of lattice thermal conductivity, these techniques are not so appropriate in the production of bulk size specimens. From the perspective of practical applications, it is desirable if such nanostructures could be built automatically. Some studies on self-assembled Si nanostructures are introduced later in this chapter.

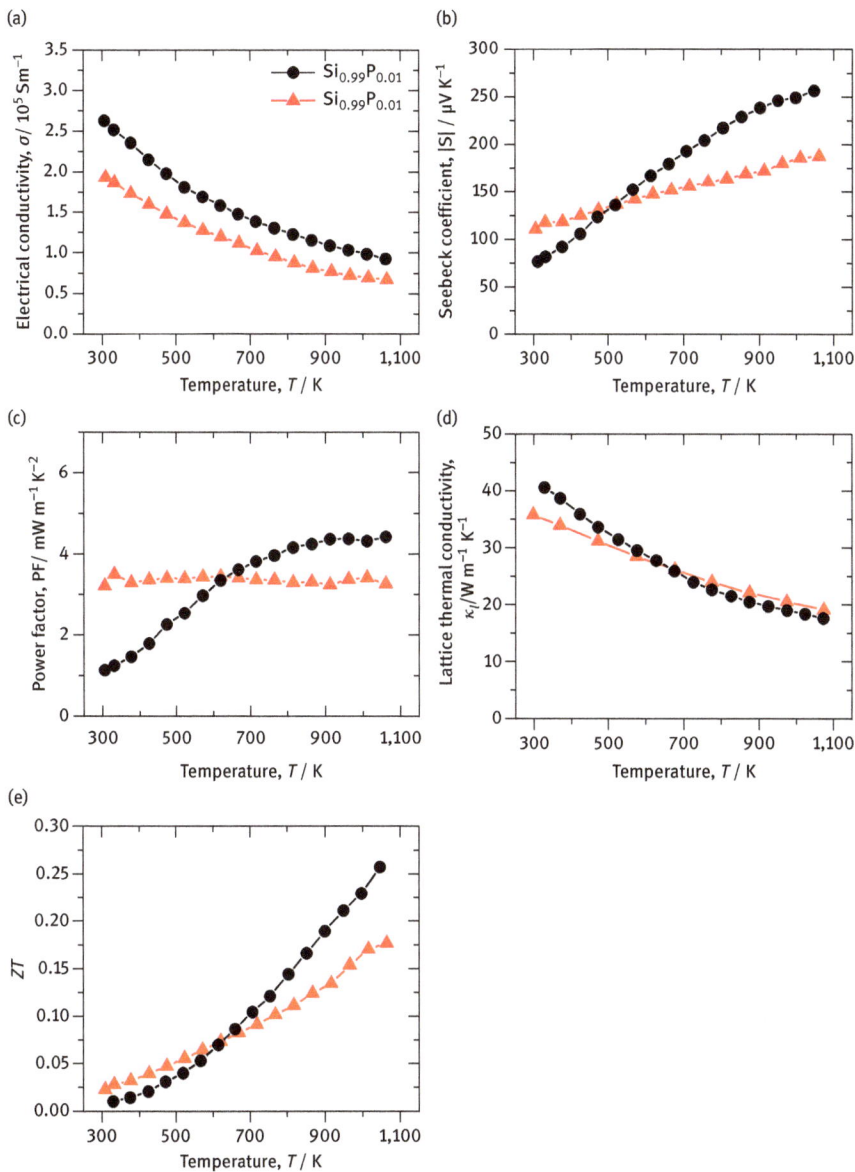

**Figure 2.7:** TE properties of Si-1 at.% P and Si-1 at.% B. (a) Electrical conductivity, (b) Seebeck coefficient, (c) lattice thermal conductivity, and (d) ZT.

## 2.2.1 Si/silicide nanocomposite films

NCs can be obtained by sintering fine powders that are synthesized by ball milling [4] or gas phase reaction process [17]. However, phase separation from amorphous phase is also an effective method to form dense NCs [18–21]. In this method, dense Si and silicide nanocomposite containing NCs can be obtained by thermal annealing of amorphous transition metal (M) and Si alloys deposited onto a substrate by sputtering. The microstructure of the formed NCs can be controlled by the transition metal, composition, annealing condition, and substrate. Among them, the substrate significantly affects the morphology of the NCs: while $SiO_2$ glass substrate results in randomly oriented NCs, SOQ (silicon on quartz) substrate leads to oriented NCs, which are expected to possess superior electrical properties than the randomly oriented ones owing to less electron scattering at the grain boundaries.

### 2.2.1.1 Randomly oriented Si/silicide nanocomposite films

Randomly oriented NCs can be obtained by using $SiO_2$ glass substrate [18, 19, 21]. Figure 2.8 shows the schematic of this process. First, amorphous film is deposited onto $SiO_2$ glass substrate by DC magnetron sputtering in Ar atmosphere. Si and transition metal mixture is used as a target. The deposited film thicknesses are 150–300 nm, which can be controlled by the sputtering time. The amorphous films are heat treated under Ar atmosphere to crystallize Si and silicide.

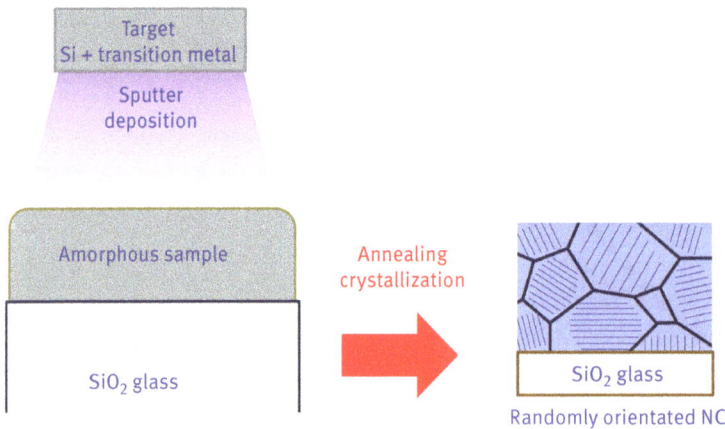

**Figure 2.8:** A schematic of the synthesis process of randomly oriented Si/silicide nanocomposite thin films.

Figure 2.9 shows the images of $MoSi_{12}$ composite film observed by in situ transmission electron microscope (TEM) heating technique. One can see from this figure that NCs are formed from amorphous phase during heat treatment. The dark region

**Figure 2.9:** In situ TEM images of Si/MoSi$_2$ nanocomposite (MoSi$_{12}$) film.

in the TEM corresponds to Si and the bright region corresponds to MoSi$_2$. While the NCs grow by the heat treatment at higher temperature for longer time, the diameters are typically 10–20 nm, which are the ideal size to reduce the lattice thermal conductivity by enhanced phonon scattering at grain boundaries, and improve TE performance. The thermal conductivity of Mo–Si nanocomposite film was measured by $2\omega$ method and found to be around 4 W m$^{-1}$ K$^{-1}$, which is significantly lower than that of single crystal Si as expected.

In order to achieve improved *ZT* values, what is important is not only reducing the lattice thermal conductivity but also keeping the electrical property as good as possible. In the case of Si/silicide nanocomposite films, the transition metal element affects the electrical property significantly. Various kinds of transition metal elements were examined, and it was found that Si and Ni composite films show the highest electrical property.

Figure 2.10 shows the temperature-dependent electrical properties (electrical conductivity and Seebeck coefficient) of p-type Si/MoSi$_2$ and Si/NiSi$_2$ composite films [21]. The added dopant was B, which was mixed with the sputtering targets, whose compositions are NiSi$_{20}$ and MoSi$_{20}$. The electrical property of single crystal Si is shown for comparison, of which carrier concentration is identical to the corresponding composite films. This figure shows that while the Seebeck coefficients of both films are in good agreement with those of single crystal Si, the electrical conductivities are much lower than that of single crystal Si.

This reduction in the electrical conductivity stems from the reduced carrier mobility, which is ascribed to carrier scattering at the grain boundaries. The reduction ratio from single crystal Si is however different between MoSi$_{20}$ and NiSi$_{20}$ films: the reduction ratio of NiSi$_{20}$ film is smaller than that of MoSi$_{20}$ film. This difference might be attributed to NiSi$_2$, which is known to form atomically flat interface with low shot key barrier height, resulting in less carrier scattering at Si/silicide interfaces.

The thermal conductivity at room temperature was measured by scanning thermal probe microimage (STPM) system and found to be 3.6 W m$^{-1}$ K$^{-1}$ for NiSi$_{20}$ film [18]. The *ZT* value of NiSi$_{20}$ film is calculated to be 0.13 at room temperature, and reaches 0.36 under the assumption that the thermal conductivity is constant.

(a)

(b)

(c)

(d)

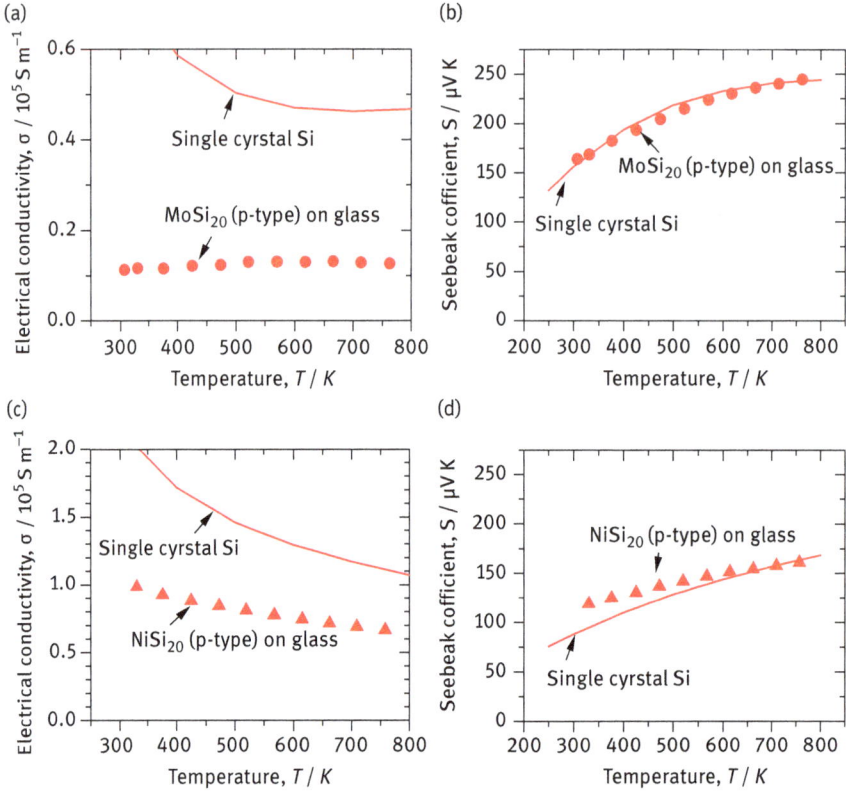

**Figure 2.10:** Electrical conductivity and Seebeck coefficient of $NiSi_{20}$ and $MoSi_{20}$ as a function of temperature. Solid lines represent the calculated results for single crystal Si with the same carrier concentration with $NiSi_{20}$ and $MoSi_{20}$.

Although significant improvement was achieved in $ZT$ values of Si/silicide nanocomposite films compared with single crystal Si owing to the reduced thermal conductivity, the $ZT$ values are still lower than that of conventional TE materials. This is because the electrical property (especially carrier mobility) was also significantly reduced by the nanostructure in Si/silicide nanocomposite films.

## 2.2.1.2 Oriented Si/NiSi$_2$ nanocomposite films

The oriented films were synthesized by using SOQ as substrate as shown in Figure 2.11 [20]. The thickness of the Si of the SOQ substrate was 75 nm with (100) crystal orientation, which was on the box $SiO_2$ substrate. Ni–Si amorphous film was deposited on the SOQ substrate by DC magnetron sputtering with the compositions of 2 at.% P or B added $NiSi_{20}$. The amorphous films were subsequently annealed at 800 °C for 5 min in

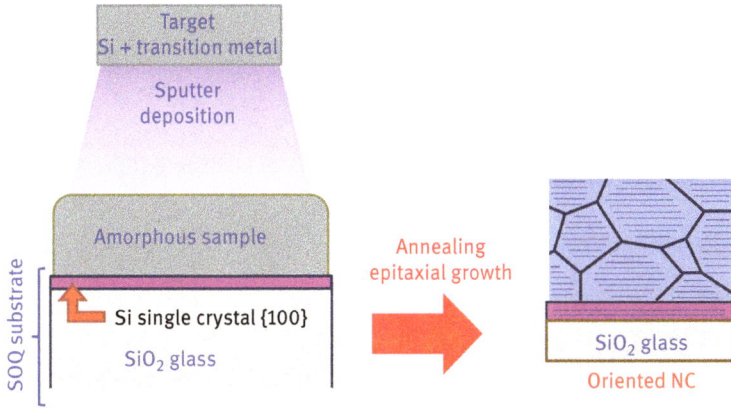

**Figure 2.11:** A schematic of the synthesis process of oriented Si/silicide nanocomposite thin films.

$N_2$ atmosphere. The film thickness was 750 nm to 1 μm. In order to activate the dopants, the films were subjected to rapid thermal annealing at 1,200 °C for n-type and 1,230 °C for p-type with infrared-lamp heating for 10 or 20 s in $N_2$ atmosphere.

Figure 2.12 shows the TE properties of p- and n-type $NiSi_{20}$ composite films on SOQ. The p-type $NiSi_{20}$ composite film show superior PF over that of n-type $NiSi_{20}$. The effect of the substrates can be seen by comparing PFs of $NiSi_{20}$ on SOQ and $NiSi_{20}$ on glass. $NiSi_{20}$ on SOQ shows improved PF than that formed on glass, suggesting that carrier scattering at grain boundaries was suppressed due to the oriented crystals. In addition to PF, the films formed on SOQ show lower thermal conductivity (2.3 W m$^{-1}$ K$^{-1}$ for p-type and 3.3 W m$^{-1}$ K$^{-1}$ for n-type measured by STPM at room temperature) than those formed on glass, possibly because of rapid thermal annealing which produced small voids and defects. The $ZT$ values were estimated assuming a constant thermal conductivity. The predicted $ZT$ values reach 0.65 for p-type and 0.40 for n-type at 770 K, which are significant improvements from single crystal Si.

## 2.2.2 Nanometer-sized eutectic structure of Si/CrSi$_2$

It is known that composites formed through eutectic solidification have eutectic structure characterized by fine microstructures with structural regularity. A lot of eutectic systems can be found between Si and metal silicides. For instance, Figure 2.13(a) shows a phase diagram between Si and CrSi$_2$, whose eutectic point is shown by an arrow. The microstructure of the Si–CrSi$_2$ eutectic alloy synthesized by arc melting is shown in Figure 2.13(b). This figure shows the fine lamella structure with characteristic length of several micrometers formed by eutectic solidification.

The eutectic structure is formed directly from the melt, and the feature size decreases with increasing cooling rate during solidification. This means that self-

(a)

(b)

(c)

(d)

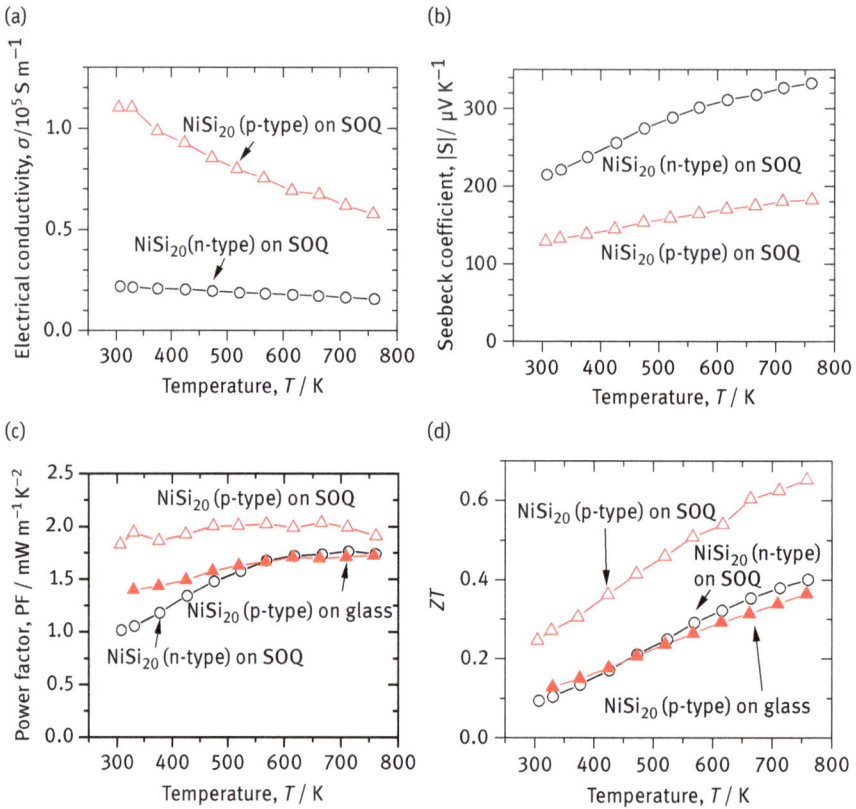

**Figure 2.12:** Temperature-dependent (a) electrical conductivity, (b) absolute value of Seebeck coefficient, (c) power factor, and (d) ZT of p-type and n-type $NiSi_{20}$ composite films on SOQ.

assembled nanostructure could possibly be formed from the rapid solidification of eutectic melts. Based on this idea, a melt-spinning (MS) technique was applied to the $Si–CrSi_2$ eutectic alloy [22, 23]. The MS technique is a commonly used quenching technique. As shown in Figure 2.14, a melt is ejected to a rotating copper wheel, resulting in ribbon-shaped quenched samples.

An image of the ribbon-shaped samples is shown in Figure 2.15(a). The cross-sectional image (Figure 2.15(b)) of the ribbon sample was observed by scanning electron microscope (SEM), which shows that the microstructure of the contact surface side is slightly different from that of the free surface side. The images parallel to the surface at different positions are observed by polishing the sample (Figure 2.15(c) and (d)). These figures show that maze-like patterns of Si and $CrSi_2$ are formed with a notably fine spacing of around 30 nm in free surface side. The contact surface structure is more complicated. It looks like fine $CrSi_2$ rods are

(a)

(b) Arc-melted Si-CrSi$_2$

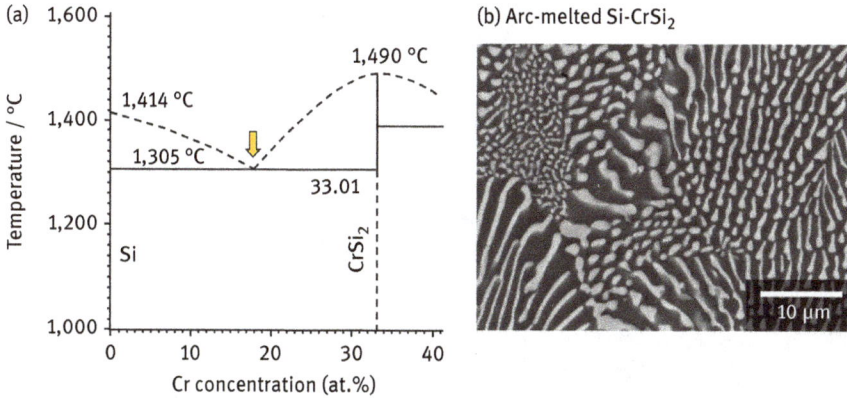

**Figure 2.13:** (a) Binary phase diagram of the Si–Cr system (Si-rich side only). An arrow shows the eutectic point. (b) SEM image of Cr$_{14.9}$(Si$_{0.99}$B$_{0.01}$)$_{85.1}$ eutectic alloy synthesized by arc melting. The bright and dark areas are CrSi$_2$ and Si, respectively.

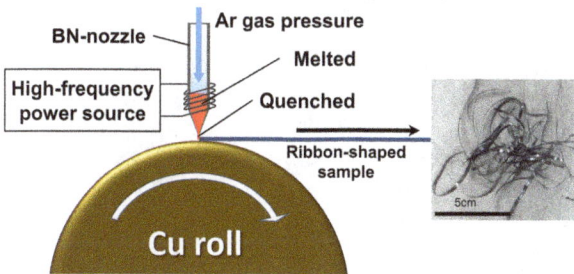

**Figure 2.14:** A schematic of the MS technique.

embedded in a Si matrix. Such structure might be ascribed to the faster cooling rate at the contact surface.

The TE properties of the melt spun Cr$_{14.9}$(Si$_{0.99}$B$_{0.01}$)$_{85.1}$ at room temperature are shown in Table 2.1 together with those of the arc melted sample [23]. The melt spun sample shows lower thermal conductivity than the arc melted one, suggesting that the fine structure formed by the MS technique effectively scatters phonons and reduces the thermal conductivity. Although the electrical property is deteriorated to some extent in the melt spun sample, the improved $ZT$ value is obtained in the melt spun sample owing to the reduction of the thermal conductivity. The size of the nanostructure formed by MS was at most 30 nm, which is much larger than the ideal size of 10 nm for TE property improvement. If finer structure is formed by improving the MS technique, higher $ZT$ value would be obtained.

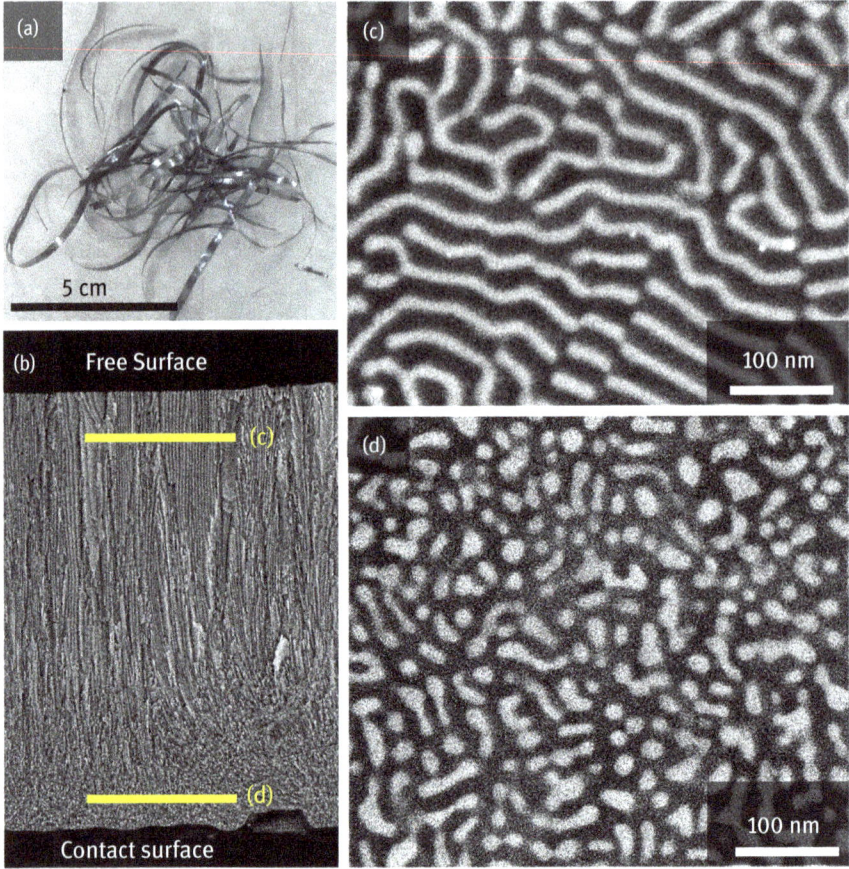

**Figure 2.15:** (a) Photo image of the ribbon-shaped samples with the Si–CrSi$_2$ eutectic composition synthesized by the MS technique and (b) cross-sectional SEM image of the ribbon. The dark and bright regions correspond to Si and CrSi$_2$, respectively; (c) and (d) are the SEM images parallel to the surface taken at different positions: free surface side and contact surface side, respectively.

**Table 2.1:** TE properties of arc melted (AM) and melt spun (MS) Cr$_{14.9}$(Si$_{0.99}$B$_{0.01}$)$_{85.1}$ samples at room temperature.

| Preparation method | $S$ ($\mu$V K$^{-1}$) | $r$ ($\mu\Omega$ cm$^{-1}$) | $\kappa$ (W m$^{-1}$ K$^{-1}$) | $PF$ (mW m$^{-1}$ K$^{-2}$) | $ZT$ |
|---|---|---|---|---|---|
| AM | 148 | 11.2 | 34 | 2.0 | 0.017 |
| MS at 52 m s$^{-1}$ | 71 | 4.4 | 12 | 1.2 | 0.030 |

## 2.2.3 Si with nanoprecipitates and dislocations

### 2.2.3.1 Si–P nanocomposite

Although the effects on carrier concentration and the kinetics of dopant precipita-
tions in Si–Ge alloy have been discussed by many authors [24–26], there are few studies that provide microscopic investigations on the morphology and size of the dopant
precipitates. P is the most widely used n-type dopant for Si-based TE material owing
to (1) high solubility that enables the optimization of carrier concentration (even
though the actual solubility of P in Si remains an ongoing debate [27, 28]); (2) precipitation of P-rich phase at elevated temperature that may induce additional phonon
scattering [29]. Herein, we demonstrate the effect of excess doping with 3 at.% P for
n-type Si on TE properties in this section [30].

The starting materials (Si and P) were arc melted, roughly crushed into micro-
scale powders, and finally transformed into the bulk state via Spark Plasma Sintering
(SPS) under appropriate conditions. The bright-field TEM images of the samples
$Si_{100}P_1$ and $Si_{100}P_3$ (depicted in Figure 2.16(a) and (b)) contain nanoscale precipitates
with various sizes below a few dozen nanometers, and the precipitate size decreases
near the grain boundaries, leading to areas where no precipitates are formed, as
shown in Figure 2.16(c). Two different types of precipitates exist: spherical precipitates with diameters of approximately 5 nm, and bar-shaped (plate-like) precipitates
with lengths of approximately 20 nm, as illustrated in Figure 2.16(d). The plate-like
precipitates interacted with the matrix phase in a semicoherent manner, while the
spherical precipitates interacted with the matrix phase in a coherent manner [for details on high-resolution TEM (HRTEM), please refer to ref. [30]], which are considered
to be important for controlling the transport properties of phonons and electrons.

Figure 2.17 shows the temperature dependences of TE properties, including $S$, $\sigma$,
$\kappa_{lat}$, and $ZT$ for the Si/SiP nanocomposites. Because P is an electron donor in Si, all
samples exhibited negative values of $S$ over the entire temperature range. As the temperature increased, the absolute value of S increased as well, while the magnitude of
$\sigma$ decreased, demonstrating the typical behavior of degenerate semiconductors. The
$\kappa_{lat}$ reduced significantly due to the enhanced phonon scattering by the formation of
nanoscale precipitates. It has been demonstrated for various materials that nanostructuring reduces $\kappa_{lat}$; however, in many cases, it also deteriorates the carrier mobility to a certain degree [31]. In this chapter, Si/SiP nanocomposites exhibited carrier
mobility values close to those of single crystalline Si and much higher than those of
Si bulk NCs [4], indicating that the nanoscale precipitates scattered phonons selectively due to the coherent or semicoherent interfaces. As a result, an enhanced $ZT$ of
0.35 was obtained for $Si_{100}P_3$ at 1,073 K (Figure 2.17(d)), which was higher than that
for crystalline Si. However, this value is still much lower than that of some best Si
bulk NCs [4, 32]. Further enhancements in $ZT$ values are expected if better control on
the size and distribution of the nanoscale features could be achieved.

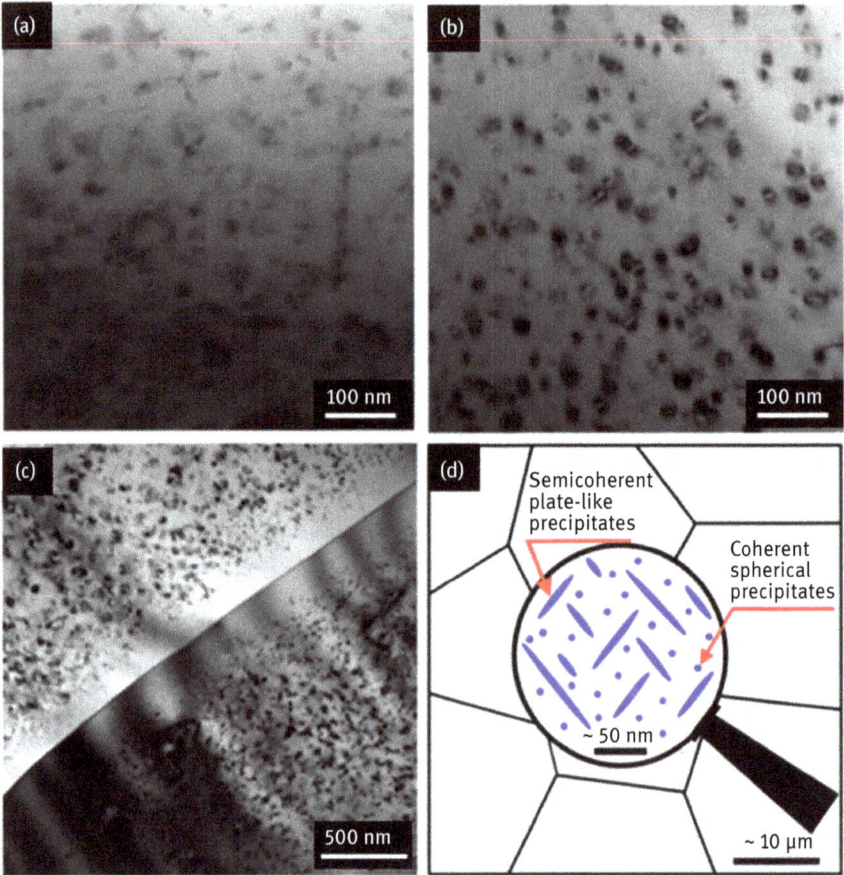

**Figure 2.16:** Bright-field TEM of the sintered pellets: (a) $Si_{100}P_1$ and (b) $Si_{100}P_3$. (c) low-magnification bright-field TEM showing a grain boundary. (d) Schematics of two types of the precipitates.

### 2.2.3.2 Si–CoSi$_2$ with dislocations

While a lot of studies on nanocomposites focused on thermal conductivity reduction effect through phonon scattering by small precipitates, we focused on phonon scattering by dislocations. Although the actual frequency dependence of dislocation scattering still remains controversial and requires further investigation, dislocations with a density over $10^{10}$ cm$^{-2}$ have been proved theoretically [33] and experimentally [34–37] to play an important role in lowering the $\kappa_{lat}$. This motivated us to develop a novel strategy to self-assembly by introducing dense dislocations into bulk Si [38].

(a)

(b)

(c)

(d)

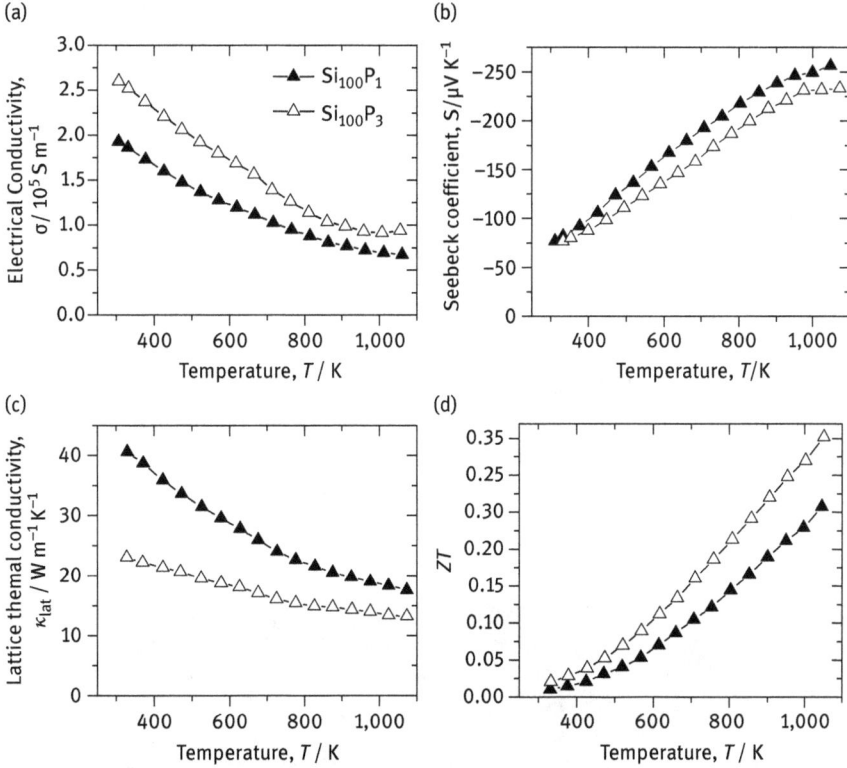

**Figure 2.17:** Temperature-dependent TE transport properties of Si/SiP composites: (a) electrical conductivity $\sigma$, (b) lattice thermal conductivity $\kappa_{lat}$, and (c) ZT.

If a composite where fine silicide particles are distributed in the Si matrix is heated to slightly higher than its eutectic temperature and cooled, a part of the composite will melt by eutectic reaction and solidify, which might result in the formation of dislocations as a result of nonequilibrium impurity trapped in the matrix. The Si and silicide composite can be synthesized by the MS technique. Sintering the MS ribbons by SPS will allow us to heat the sample at a slightly higher temperature than its eutectic point if the sintering temperature is carefully selected. In addition, ionized P atoms have a very strong pinning effect on dislocations due to the electrostatic interaction with acceptor sites on the dislocation in n-type Si [39], which means a high density of dislocations is expected by preventing the dislocations from disappearing at grain boundaries. Multiplication of the defects is also expected by the ionized P atoms. Motivated by the above-mentioned features, P doping is chosen for further investigation.

CoSi$_2$ was selected as the silicide because of its moderate eutectic point between Si. Although CoSi$_2$ is a metallic silicide, which is not favorable as a TE material, the similarity in crystal structures and lattice parameters (mismatch within 1.2% at RT and 0.5% above 973 K based on the thermal expansion [40]) may reduce the density of interfacial dislocations [41] and result in the formation of a connected network suitable for electron transport [42, 43].

Ingots with nominal composition $(Si_{0.97}P_{0.03})_{95}Co_5$ were synthesized by MS followed with high-temperature sintering by SPS, denoted as Si–CoSi$_2$. The sintering temperature was 1,080 °C. CoSi$_2$ was found to be expelled from the graphite die after SPS, which indicates that the temperature reached the eutectic point of Si–CoSi$_2$ during SPS.

In Si–CoSi$_2$, dense plate-like precipitates with a size of 50–100 nm were found in the matrix, as shown in bright-field TEM images in Figure 2.18(a). Such precipitates with a similar size and plate-like morphology were also found in P-supersaturated Si [30, 44], which were demonstrated to have semicoherent precipitates in Si. Note that high density of residual dislocations is distributed in Si–CoSi$_2$. The wavy, unstable orientations of short segments of partial dislocations reveal a strong pinning behavior at some of the defect sites like P solute atoms [45, 46].

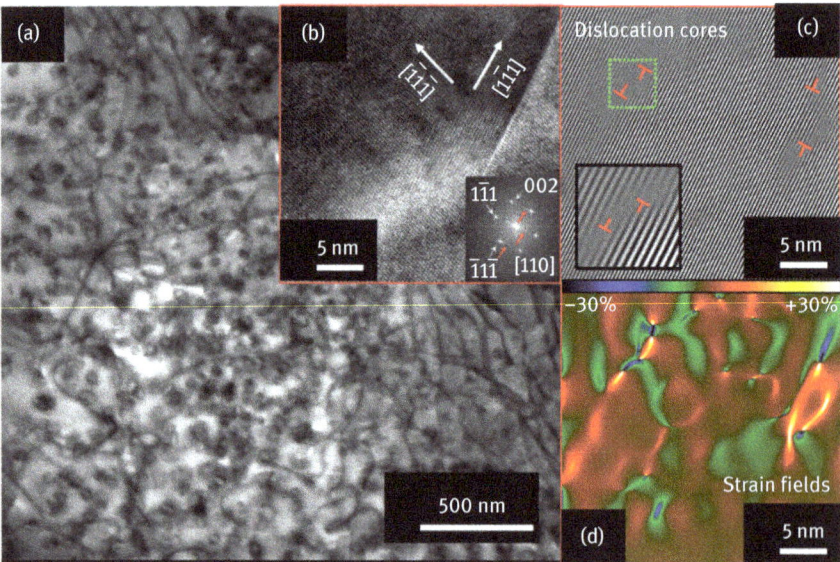

**Figure 2.18:** (a) Bright-field TEM image; (b) HRTEM images with inserted fast Fourier transform (FFT) pattern for the dislocation area; (c) inverse FFT (IFFT) images of (200) atomic planes of (b) with enlarged views of dot-boxed regions and (d) strain fields map of $\varepsilon_{xx}$ between (1-11) and (002) of (c) in Si–CoSi$_2$.

To investigate the detailed microstructure features of dislocation, an HRTEM image is shown in Figure 2.18(b). Interestingly, high-density and dislocation cores pervasively exist around the dislocations in all the three planes. The areal dislocation core density was found over $2.0 \times 10^{12}$ cm$^{-2}$ in the region shown in Figure 2.18(c), while the average core density along the dislocations is estimated about $0.7 \times 10^{12}$ cm$^{-2}$. Figure 2.18(d) shows the strain fields map of $\varepsilon_{xx}$ between (1-11) and (002) of (c) in Si–CoSi$_2$. Convergence regions of strains are found to exist around the dislocation core, which randomly distributed in all orientation.

Figure 2.19 shows the temperature dependences of $\kappa_{lat}$ and $ZT$ for the Si–MSi$_2$ (M = Co, Fe, Ni, and Cr). Over 15–65% reductions in $\kappa_{lat}$ (from 42.0 to 35.2–13.5 W m$^{-1}$ K$^{-1}$ at 300 K) were obtained in Si–MSi$_2$, especially for Si–CoSi$_2$ and Si–CrSi$_2$, which can be ascribed to the enhanced phonon scattering due to the complex microstructure including high-density SiP precipitates and dislocations. Thus, although reduction of carrier mobility by the higher concentration of grain boundary/lattice defects was considered to be responsible for lower $\sigma$ and $PF$ in all MS-SPS samples, owing to the large reduction in $\kappa_{lat}$, a high figure of merit $ZT$ of 0.37–0.39 at 1,073 K is achieved in Si–CoSi$_2$ and Si–CrSi$_2$.

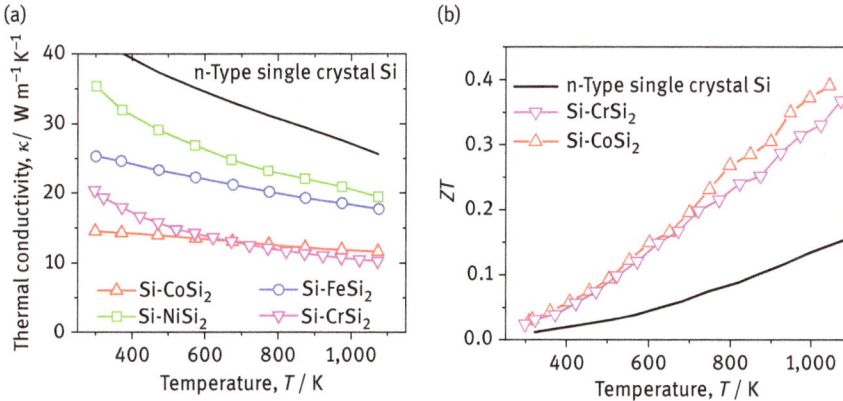

**Figure 2.19:** (a) Thermal conductivity and (b) $ZT$ of 3% P added Si–MSi$_2$ as a function of temperature.

## 2.3 Summary and outlook for the future work

In spite of its poor TE property, Si has attracted significant attention as an environmentally friendly TE material. We first showed the measured TE property of single and polycrystalline Si as well as the simulation results to clarify the intrinsic TE property of Si. These experimental and theoretical results revealed that the optimized carrier concentration of both p- and n-type Si is around $5 \times 10^{20}$ cm$^{-3}$, and the $ZT$ values of carrier optimized Si are 0.26 and 0.18 at 1,050 K for n- and p-types, respectively. These values are significantly lower than those of conventional TE materials. The reason of this poor performance is mainly ascribed to silicon's high thermal conductivity, which implies that nanostructuring is an effective approach to improve the TE performance of Si. Theoretical calculation suggests that while phonons with MFP longer than 10 nm mainly transport heat, electrons with MFP shorter than 10 nm play an important role in the electrical conduction, which means that 10 nm is the best characteristic length for improving the TE performance of silicon by nanostructuring.

While nanostructures can be formed by microfabrication techniques such as lithography and chemical etching, self-assembled nanostructure seems to be the key for the practical application. From this perspective, three approaches to automatically form nanostructures were introduced: Si/silicide nanocomposite films, nanometer-sized eutectic structure of Si/CrSi$_2$, and Si with nanoprecipitates and dislocations. While the $ZT$ values of these nanostructured Si-based TE materials are still below 1, significant improvement was achieved by nanostructuring. Figure 2.20 gives the relationship between $ZT$ and thermal conductivity of nanostructured Si at room temperature [4–6, 8, 17, 18, 22, 23, 30, 47–49]. This figure clearly shows that the TE performance of Si has been improved through the reduction of the thermal conductivity. Although the amorphous limit of the thermal conductivity of Si is estimated to be around 1 W m$^{-1}$ K$^{-1}$ [50], recent study on ultrafine coherent nanocrystalline Si [51] has shown that it is possible to defeat the amorphous limit while keeping the electrical conductivity high. Even though self-organized Si nanostructures have been extensively studied, there is still room for further improvement on the TE performance of Si.

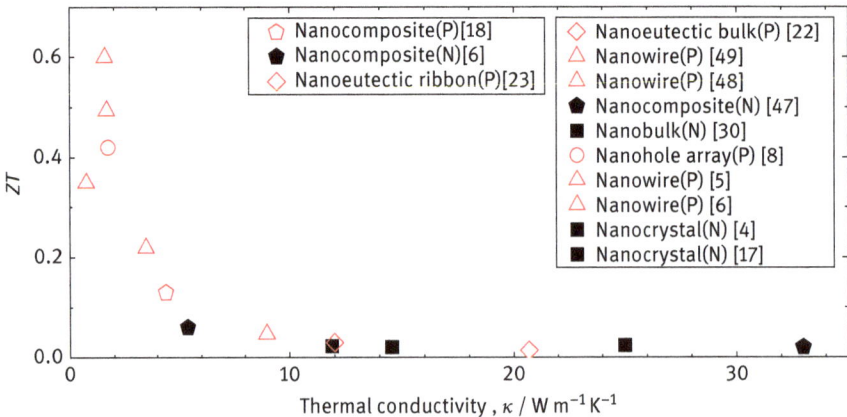

**Figure 2.20:** Relationship between $ZT$ and thermal conductivity of nanostructured Si at room temperature.

**Acknowledgments:** The authors thank Prof. M. Ishimaru for providing the in situ TEM images of Si/MoSi$_2$ nanocomposite (MoSi$_{12}$) film. The authors also thank Dr. Y. Miyazaki for providing the SEM images of the Si/CrSi$_2$ eutectic alloys.

# References

[1]  Bludau W, Onton, A, Heinke, W. Temperature dependence of the band gap of silicon, J. Appl. Phys. 1974, 45, 1846–1848. Doi: 10.1063/1.1663501.

[2]  Shanks H R, Maycock, P D, Sidles, P H, Danielson, G C. Thermal conductivity of silicon from 300 to 1400°K, Phys. Rev. 1963, 130, 1743–1748. Doi: 10.1103/PhysRev.130.1743.

[3]  Weber L, Gmelin, E. Transport properties of silicon, Appl. Phys. A. 1991, 53, 136–140.

[4]  Bux S K, Blair, R G, Gogna, P K, Lee, H, Chen, G, Dresselhaus, M S, Kaner, R B, Fleurial, J P. Nanostructured bulk silicon as an effective thermoelectric material, Adv. Funct. Mater. 2009, 19, 2445–2452. Doi: 10.1002/adfm.200900250.

[5]  Boukai A I, Bunimovich, Y, Tahir-Kheli, J, Yu, J-K, Goddard, W A, III, J R. Heath, Silicon nanowires as efficient thermoelectric materials, Nature 2008, 451, 168–171. Doi: 10.1038/nature06458.

[6]  Hochbaum A I, Chen, R, Delgado, R D, Liang, W, Garnett, E C, Najarian, M, Majumdar, A, Yang, P. Enhanced thermoelectric performance of rough silicon nanowires, Nature 2008, 451, 163–167. Doi: 10.1038/nature06381.

[7]  Yu J K, Mitrovic, S, Tham, D, Varghese, J, Heath, J R. Reduction of thermal conductivity in phononic nanomesh structures, Nat. Nanotechnol 2010, 5, 718–721. Doi: 10.1038/nnano.2010.149.

[8]  Tang J, Wang, H T, Lee, D H, Fardy, M, Huo, Z, Russell, T P, Yang, P. Holey silicon as an efficient thermoelectric material, Nano. Lett. 2010, 10, 4279–4283. Doi: 10.1021/nl102931z.

[9]  Slack G A. Thermal conductivity of pure and impure silicon, silicon carbide, and diamond, J. Appl. Phys. 1964, 35, 3460–3466. Doi: 10.1063/1.1713251.

[10]  Asheghi M, Kurabayashi, K, Kasnavi, R, Goodson, K E. Thermal conduction in doped single-crystal silicon films, J. Appl. Phys. 2002, 91, 5079–5088. Doi: 10.1063/1.1458057.

[11]  Lee Y, Hwang, G S. Mechanism of thermal conductivity suppression in doped silicon studied with nonequilibrium molecular dynamics, Phys. Rev. B. 2012, 86, 075202. Doi: 10.1103/PhysRevB.86.075202.

[12]  Stranz A, Kähler, J, Waag, A, Peiner, E. Thermoelectric properties of high-doped silicon from room temperature to 900 K, J. Electron. Mater. 2013, 42, 2381–2387. Doi: 10.1007/s11664-013-2508-0.

[13]  Ohishi Y, Xie, J, Miyazaki, Y, Aikebaier, Y, Muta, H, Kurosaki, K, Yamanaka, S, Uchida, N, Tada, T. Thermoelectric properties of heavily boron- and phosphorus-doped silicon, Japanese journal of applied physics 2015, 54, 071301. Doi: 10.7567/JJAP.54.071301.

[14]  Biswas K, He, J, Blum, I D, Wu, C-I, Hogan, T P, Seidman, D N, Dravid, V P, Kanatzidis, M G. High-performance bulk thermoelectrics with all-scale hierarchical architectures, Nature 2012, 489, 414–418. Doi: 10.1038/nature11439.

[15]  Ackbarow T, Buehler, M J. Hierarchical coexistence of universality and diversity controls robustness and multi-functionality in protein materials, J. Comput. Theor. Nanosci. 2008, 5, 1193–1204. Doi: 10.1166/jctn.2008.2554.

[16]  Esfarjani K, Chen, G, Stokes, H T. Heat transport in silicon from first-principles calculations, Phys. Rev. B – Condens. Matter Mater. Phys. 2011, 84, 1–11. Doi: 10.1103/PhysRevB.84.085204.

[17]  Schierning G, Theissmann, R, Stein, N, Petermann, N, Becker, A, Engenhorst, M, Kessler, V, Geller, M, Beckel, A, Wiggers, H, Schmechel, R. Role of oxygen on microstructure and

thermoelectric properties of silicon nanocomposites, J. Appl. Phys. 2011, 110. Doi: 10.1063/1.3658021.

[18] Uchida N, Tada, T, Ohishi, Y, Miyazaki, Y, Kurosaki, K, Yamanaka, S. Heavily doped silicon and nickel silicide nanocrystal composite films with enhanced thermoelectric efficiency, J. Appl. Phys. 2013, 114, 134311. Doi: 10.1063/1.4823814.

[19] Ohishi Y, Kurosaki, K, Suzuki, T, Muta, H, Yamanaka, S, Uchida, N, Tada, T, Kanayama, T. Synthesis of silicon and molybdenum–silicide nanocrystal composite films having low thermal conductivity, Thin Solid Films 2013, 534, 238–241. Doi: 10.1016/j.tsf.2013.02.127.

[20] Uchida N, Ohishi, Y, Miyazaki, Y, Kurosaki, K, Yamanaka, S, Tada, T. Thermoelectric Properties of (100) Oriented Silicon and Nickel Silicide Nanocomposite Films Grown on Si on Insulator and Si on Quartz Glass Substrates, Mater. Trans. 2016, 57, 1076–1081.

[21] Ohishi Y, Miyazaki, Y, Muta, H, Kurosaki, K, Yamanaka, S, Uchida, N, Tada, T. Carrier transport properties of p-type silicon–metal silicide nanocrystal composite films, J. Electron. Mater. 2015, 44, 2074–2079. Doi: 10.1007/s11664-015-3663-2.

[22] Norizan M N, Kurimoto, T, Miyazaki, Y, Ohishi, Y, Kurosaki, K, Muta, H. Fabrication and thermoelectric property of nanostructured, Mater. Res. Express. 2019, 6, 025702. Doi: 10.1088/2053-1591/aaee98.

[23] Norizan M N, Miyazaki, Y, Ohishi, Y, Muta, H, Kurosaki, K, Yamanaka, S. The nanometer-sized eutectic structure of Si/CrSi2 thermoelectric materials fabricated by rapid solidification, J. Electron. Mater. 2018, 47, 2330–2336. Doi: 10.1007/s11664-017-6060-1.

[24] Minnich J, Lee, H, Wang, W, Joshi, G, Dresselhaus, S, Ren, F, Chen, G, Vashaee, D. Modeling study of thermoelectric SiGe nanocomposites, Phys. Rev. B. 2009, 80, 155327. Doi: 10.1103/PhysRevB.80.155327.

[25] Nasby R, Burgess, E. Precipitation of dopants in silicon-germanium thermoelectric alloys, J. Appl. Phys. 1972, 43, 2908–2909. Doi: 10.1063/1.1661622.

[26] Savvides N, Rowe, D M. Precipitation of phosphorus from solid solutions in Si-Ge alloys and its effect on thermoelectric transport properties, J. Phys. D. Appl. Phys. 1981, 14, 723–732. Doi: 10.1088/0022-3727/14/4/025.

[27] Mostafa A, Medraj, M. Binary phase diagrams and thermodynamic properties of silicon and essential doping elements (Al, As, B, Bi, Ga, In, N, P, Sb and Tl), Materials 2017, 10, 676. Doi: 10.3390/ma10060676.

[28] Safarian J, Tangstad, M. Phase diagram study of the Si–P system in Si-rich region, J. Mater. Res. 2011, 26, 1494–1503. Doi: 10.1557/jmr.2011.130.

[29] Schierning G. Silicon nanostructures for thermoelectric devices: A review of the current state of the art, Phys, Status Solidi 2014, 211, 1235–1249. Doi: 10.1002/pssa.201300408.

[30] Yusufu A, Kurosaki, K, Miyazaki, Y, Ishimaru, M, Kosuga, A, Ohishi, Y, Muta, H, Yamanaka, S. Bottom-up nanostructured bulk silicon: A practical high-efficiency thermoelectric material, Nanoscale 2014, 6, 13921–13927. Doi: 10.1039/c4nr04470c.

[31] Androulakis J, Hsu, K F, Pcionek, R, Kong, H, Uher, C, D'Angelo, J J, Downey, A, Hogan, T, Kanatzidis, M G. Nanostructuring and high thermoelectric efficiency in p-type Ag(Pb 1-ySny) mSbTe2+m, Adv. Mater. 2006, 18, 1170–1173. Doi: 10.1002/adma.200502770.

[32] Shiomi J. Research Update: Phonon engineering of nanocrystalline silicon thermoelectrics, APL Mater 2016, 4, 104504. Doi: 10.1063/1.4962935.

[33] Wang T, Carrete, J, Van Roekeghem, A, Mingo, N, Madsen, G K H. Ab initio phonon scattering by dislocations, Phys. Rev. B. 2017, 95, 1–7. Doi: 10.1103/PhysRevB.95.245304.

[34] Chen Z, Ge, B, Li, W, Lin, S, Shen, J, Chang, Y, Hanus, R, Snyder, G J, Pei, Y. Vacancy-induced dislocations within grains for high-performance PbSe thermoelectrics., Nat. Commun 2017, 8, 13828. Doi: 10.1038/ncomms13828.

[35] Kim S, Lee, K, Mun, H, Kim, H, Hwang, S, Roh, J, Yang, D, Shin, W, Li, X, Lee, Y, Snyder, J, Kim, S. Dense dislocation arrays embedded in grain boundaries for high-performance bulk thermoelectrics, Science (80-.). 2015, 348, 109–114. Doi: 10.1126/science.aaa4166.

[36] Xin J, Wu, H, Liu, X, Zhu, T, Yu, G, Zhao, X. Mg vacancy and dislocation strains as strong phonon scatterers in Mg2Si1?xSbx thermoelectric materials, Nano Energy 2017, 34, 428–436. Doi: 10.1016/j.nanoen.2017.03.012.

[37] He J, Kanatzidis, M G, Dravid, V P. High performance bulk thermoelectrics via a panoscopic approach, Mater. Today 2013, 16, 166–176. Doi: 10.1016/j.mattod.2013.05.004.

[38] Xie J, Ohishi, Y, Ichikawa, S, Muta, H, Kurosaki, K, Yamanaka, S. Naturally decorated dislocations capable of enhancing multiple-phonon scattering in Si-based thermoelectric composites, J. Appl. Phys. 2018, 123, 115114. Doi: 10.1063/1.5019614.

[39] Sumino K, Imai, M. Interaction of dislocations with impurities in silicon crystals studied by in situ X-ray topography, Philos. Mag. A Phys. Condens. Matter, Struct. Defects Mech. Prop. 1983, 47, 37–41. Doi: 10.1080/01418618308245262.

[40] Smeets D, Vantomme, A, De Keyser, K, Detavernier, C, Lavoie, C. The role of lattice mismatch and kinetics in texture development: Co 1-xNixSi2 thin films on Si(100), J. Appl. Phys. 2008, 103, 1–11. Doi: 10.1063/1.2888554.

[41] Stringfellow G. The importance of lattice mismatch in the growth of Ga x In 1– x P epitaxial crystals, J. Appl. Phys. 1972, 43, 3455–3460. Doi: 10.1063/1.1661737.

[42] Zhao L, Dravid, V, Kanatzidis, M. The panoscopic approach to high performance thermoelectrics, Energy Environ. Sci 2014, 7, 251–268. Doi: 10.1039/C3EE43099E.

[43] Biswas K, He, J, Zhang, Q, Wang, G, Uher, C, Dravid, V P, Kanatzidis, M G. Strained endotaxial nanostructures with high thermoelectric figure of merit., Nat. Chem. 2011, 3, 160–166. Doi: 10.1038/nchem.955.

[44] Armigliato A, Werner, P. High resolution electron microscopy observations of precipitates in p-supersaturated silicon, Ultramicroscopy 1984, 15, 61–69. Doi: 10.1016/0304-3991(84)90075-5.

[45] Imai M, Sumino, K. In situ X-ray topographic study of the dislocation mobility in high-purity and impurity-doped silicon crystals, Philos. Mag. A. 1983, 47, 599–621. Doi: 10.1080/01418618308245248.

[46] Taylor P, Heggie, M, Jones, R, Umerski, A. Interaction of impurities with dislocation cores in silicon, Philos. Mag. A. 2006, 37–41. Doi: 10.1080/01418619108213900.

[47] Eaksuwanchai P, Kurosaki, K, Tanusilp, S A, Ohishi, Y, Muta, H, Yamanaka, S. Thermoelectric properties of Si-NiSi2 bulk nanocomposites synthesized by a combined method of melt spinning and spark plasma sintering, J. Appl. Phys. 2017, 121, 3–8. Doi: 10.1063/1.4985283.

[48] Li Y, Buddharaju, K, Tinh, B C, Singh, N, Lee, S J. Improved vertical silicon nanowire based thermoelectric power generator with polyimide filling, IEEE Electron Device Lett 2012, 33, 715–717. Doi: 10.1109/LED.2012.2187424.

[49] Zhang T, Wu, S, Xu, J, Zheng, R, Cheng, G. High thermoelectric figure-of-merits from large-area porous silicon nanowire arrays, Nano Energy 2015, 13, 433–441. Doi: 10.1016/j.nanoen.2015.03.011.

[50] Cahill D G, Watson, S K, Pohl, R O. Lower limit to the thermal conductivity of disordered crystals, Phys. Rev. B. 1992, 46, 6131–6140. Doi: 10.1103/PhysRevB.46.6131.

[51] Nakamura Y, Isogawa, M, Ueda, T, Yamasaka, S, Matsui, H, Kikkawa, J, Ikeuchi, S, Oyake, T, Hori, T, Shiomi, J, Sakai, A. Anomalous reduction of thermal conductivity in coherent nanocrystal architecture for silicon thermoelectric material, Nano Energy 2015, 12, 845–851. Doi: 10.1016/j.nanoen.2014.11.029.

Teruyuki Ikeda

# 3 Microstructure control of thermoelectric materials

**Abstract:** Materials having structures with size scales of various orders of magnitude could cause crucial effects on their properties. In this chapter, first, fundamentals of thermodynamic theories governing phase diagrams will be given. Next, the ways to control the morphology and size scales of microstructure based on phase transformation theories will be provided.

**Keywords:** Phase diagram, Phase transformation, Solidification, Precipitation, Eutectoid reaction, Spinodal decomposition, Porous materials

## 3.1 Introduction

Materials having structures with size scales of various orders of magnitude could cause crucial effects on their properties. For example, structures in atomic scales such as crystal structures, atomic arrangements, or arrangements of point defects within crystalline lattice; those in nano- to micrometer scale such as artificial superlattice structures, structures composed of grain boundaries, or phase boundaries arranged with nanoscale length scales; and those in even larger length scales such as compositionally graded materials.

Since size scales in materials affect various material properties, a number of researches aiming at extracting superior material properties via optimized structures in materials are going on including those in the field of thermoelectric materials. In this decade, for example, nanostructuring, which leads to enhanced phonon scattering resulting in reduction of thermal conductivity [1], has been recognized well among researchers working with thermoelectric materials. Thereafter, it has been proposed that hierarchical structure is more effective in reduction of thermal conductivity [2]. More recently, detailed mechanisms on the roles of boundaries in reduction of lattice thermal conductivity have been greatly improved [3–5]. There are novel ideas to enhance power factors through structure control such as energy filtering [6] or modulation doping effect [7]. Developments of porous materials could be an interesting research field for thermoelectrics since a porous material with open pores could have quite large area of surface in contact with fluid flowing through the porous material and hence has a quite high heat exchange ability, which might be useful for improving performance of thermoelectric devices. Thus,

**Teruyuki Ikeda,** College of Engineering, Ibaraki University, Ibaraki, Japan

https://doi.org/10.1515/9783110596526-003

the control of micro/nanostructures of thermoelectric materials is an attractive research field, which is full of potential and could lead to a next stage of studies on thermoelectric materials.

In studying such useful functions of materials or practically using such materials, it is necessary to know how respective microstructures can be obtained, controlled, and stabilized as well as to understand the physical mechanism on origins of the functions. Clues for those can be found in materials science, which has its origin in metallurgy and is described using keywords such as thermodynamics, atomic diffusion, and phase transformations. This chapter focuses on microstructure control of thermoelectric materials based on simple phase transformation theories.

## 3.2 Until formation of microstructures

Microstructure in a sample reflects its thermal history (Figure 3.1). Accordingly, in order to understand what is the origin of microstructure, one needs to know the equilibrium state of each temperature where the sample has undergone, for which equilibrium phase diagrams are of considerable help.

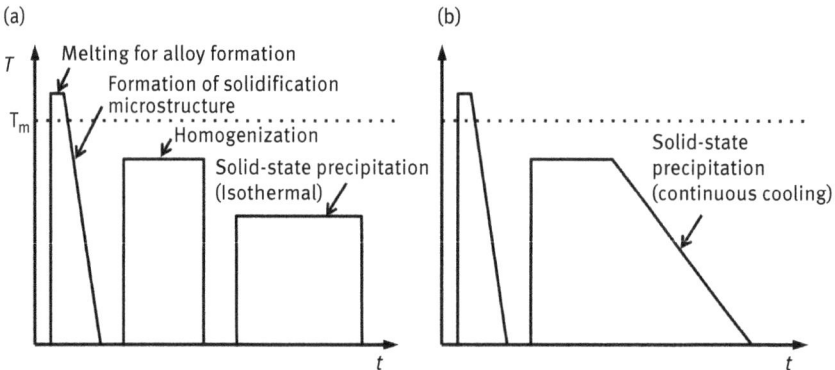

**Figure 3.1:** Examples of thermal histories until precipitates of secondary phase are formed by a solid-state precipitation reaction. An alloy is synthesized by melting, homogenized, and subject to an isothermal annealing (a) or continuous cooling (b) to obtain precipitates.

In addition to what have been the equilibrium state of the sample at each temperature in the thermal history, atomic diffusion, change in chemical free energy, elastic and chemical energies associated with the formation of boundaries, and mode of phase transformation affect the morphology and size of resulting microstructure. There are also microstructures resulting from mechanical processing: rolling texture, polycrystalline structure obtained by sintering fine powder, oriented crystals due to heat treatment under a uniaxial stress, and so on.

Resulting microstructure, which reflects its thermal history, is maintained due to frozen diffusion if the sample is kept at low temperatures, but keeps evolution if atomic diffusion cannot be neglected. Microstructures can be described using chemical compositions, characteristic length scale, and morphology as shown in Figure 3.2. Since formation of microstructure reflects the equilibrium phase diagram and processing parameters of the sample, if there is unknown information on those factors, then microstructure could be a clue.

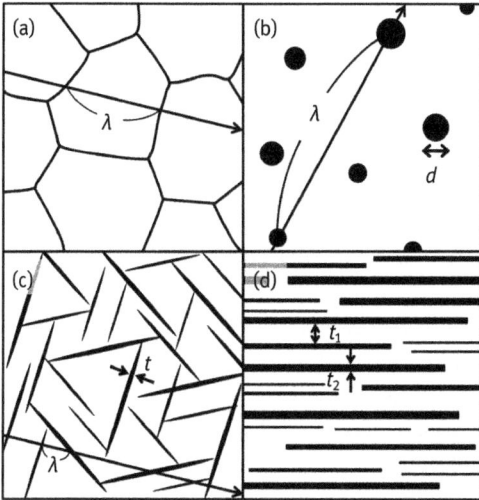

**Figure 3.2:** Morphology and characteristic length scales of various microstructures: (a) polycrystalline structure, (b) microstructure with sphere-shaped particles of secondary phase dispersed, (c) microstructure with plate-shaped particles of secondary phase dispersed, and (d) lamellar structure. $\lambda$ is a distance between two neighboring cross points of a line in a random direction and a boundary. $d$, $t$, $t_1$, $t_2$ are the diameter of the spherical particle, thickness of plates, and thickness of phases 1 and 2, respectively.

## 3.3 Free energy and phase equilibria

An equilibrium phase diagram shows what phase exists on specific conditions: temperature, chemical composition, and pressure. In this chapter, the pressure is fixed at the atmospheric pressure and hence the pressure axis is not used. The existence of phases is not just random, but there is a law that governs what state should appear in what range. A phase diagram is not only drawn based on experimental data but also explained based on a thermodynamic theory.

Equilibrium phase diagrams show phase boundaries. This section discusses how they are determined in theory. All that one need here is a principle that the free energy of a system is lowest in the equilibrium state at a given chemical composition. The

Gibbs free energy $G$ is represented using enthalpy $H$, temperature $T$, and entropy $S$ as follows:

$$G = H - TS. \tag{3.1}$$

In Figure 3.3, the free energy of the $\alpha$ phase (a phase corresponds to a crystal structure) is lower than that of the $\beta$ phase at the molar fraction of B atoms, $x_{B,1}$. Therefore, one obtains the $\alpha$ phase at the equilibrium state. Splitting the free energy at this composition to partial quantities, $\mu_{A,1}$ and $\mu_{B,1}$, for constituent elements, A and B, gives

$$G_\alpha(x_{B,1}) = x_{A,1}\mu_{A,1} + x_{B,1}\mu_{B,1}, \tag{3.2}$$

where $x_{A,1} = 1 - x_{B,1}$. From Figure 3.3(a), one finds that $\mu_{A,1}$ and $\mu_{B,1}$ are the intercepts of the $x_A = 0$ and $x_B = 0$ axes. $\mu_{A,1}$ and $\mu_{B,1}$ are called chemical potentials of elements A and B at $x_{B,1}$ being free energies per unit quantity of atoms.

In Figure 3.3(b), we consider a case where the free energy of the $\beta$ phase is lower than that of the $\alpha$ phase at high B concentrations. The composition, $x_{B,1}$, is assumed to be located between the compositions at which the lowest free energies of the $\alpha$ and $\beta$ phases take lowest values, respectively. If a sample takes the $\alpha$ state, its free energy is given by $G_\alpha$ ($x_{B,1}$), while if it takes the $\beta$ state, the free energy is given by $G_\beta$ ($x_{B,1}$). Comparing $G_\alpha$ ($x_{B,1}$) and $G_\beta$ ($x_{B,1}$), $G_\alpha$ ($x_{B,1}$) < $G_\beta$ ($x_{B,1}$) from the figure. But if the $\alpha$ and $\beta$ phases are mixed, it seems to give a lower free energy than the single phase states. If the $\alpha$ and $\beta$ phases are mixed at arbitrary fractions, $x_{B,\alpha}$ and $x_{B,\beta}$, respectively, the fractions of atoms in the $\alpha$ and $\beta$ phases need to be equal to $f_\alpha : f_\beta$ ($= x_{B,\beta} - x_{B,1} : x_{B,1} - x_{B,\alpha}$) in the figure under the overall composition of $x_{B,1}$. This is a so-called lever rule. The free energy for this state is given by

$$G(x_{B,1}) = f_\alpha G_\alpha(x_{B,\alpha}) + f_\beta G_\beta(x_{B,\beta}). \tag{3.3}$$

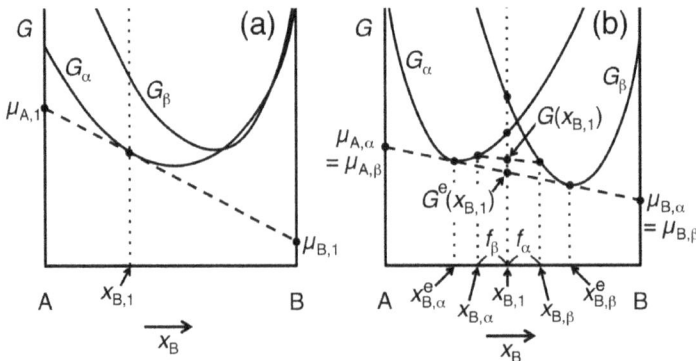

Figure 3.3: The dependence of the free energy of an A–B binary system on the concentration of B, $x_B$. At the composition $x_{B,1}$, while the alloy exhibits a single phase of $\alpha$ in the case of (a), the alloy is decomposed to two phases of $\alpha$ and $\beta$ at compositions $x^e_{B,\alpha}$ and $x^e_{B,\beta}$, respectively. For details, see text.

However, this does not yield the lowest free energy yet. The lowest free energy state, that is, the equilibrium state, is attained when the common tangent line of the curves $G_\alpha$ and $G_\beta$ is drawn and the sample is composed of the $\alpha$ and $\beta$ phases at $x^e_{B,\alpha}$ and $x^e_{B,\beta}$, which are the compositions of the tangent points, respectively. This state is stable because of the lowest free energy and hence $x^e_{B,\alpha}$ and $x^e_{B,\beta}$ give the phase boundary compositions. In addition, the chemical potentials in the $\alpha$ and $\beta$ phases are found to be equal for both A and B atoms, which means these two phases are in equilibrium:

$$\mu_{A,\alpha} = \mu_{A,\beta}, \tag{3.4a}$$
$$\mu_{B,\alpha} = \mu_{B,\beta}. \tag{3.4b}$$

Thus, for equilibrium of two phases, the lowest free energy condition is equivalent to the condition where chemical potentials of constituent elements in the two phases are equal.

The vertical axis of a typical phase diagram is of temperature. Therefore, in order to draw phase boundary curves as functions of temperature, phase boundary compositions need to be determined at various temperatures based on free energy curves. From eq. (3.1), the free energy of a high entropy phase undergoes a great variation with temperature. This is evident in melting, where a liquid phase, which has large entropy, tends to have lower free energy than the solid phase at high temperatures.

# 3.4 Basic understanding of phase transformations

## 3.4.1 Solidification

Figure 3.4 shows the relation between free energy curves and phase diagrams that have typical solidification reactions. Figure 3.4(a) shows the case where the interaction enthalpy between different kinds of atoms is lower than that between the same kind of atoms, meaning that there are attractive interactions between different kinds of atoms. Figures 3.4(b) is for a system where atomic interactions between constituent elements are neither attractive nor repulsive. (c) is for a repulsive combination. Strictly, (a), (b), and (c) correspond to negative, zero, and positive interaction parameters ($\Omega_{AB} = H_{AB} - (H_{AA} + H_{BB})/2$, where $H_{ij}$ are interaction enthalpies between nearest neighbor $i$ and $j$ atoms), respectively, in a regular solution model. In the following arguments, values of $\Omega_{AB}$ for liquid and solid states are assumed to be close to each other.

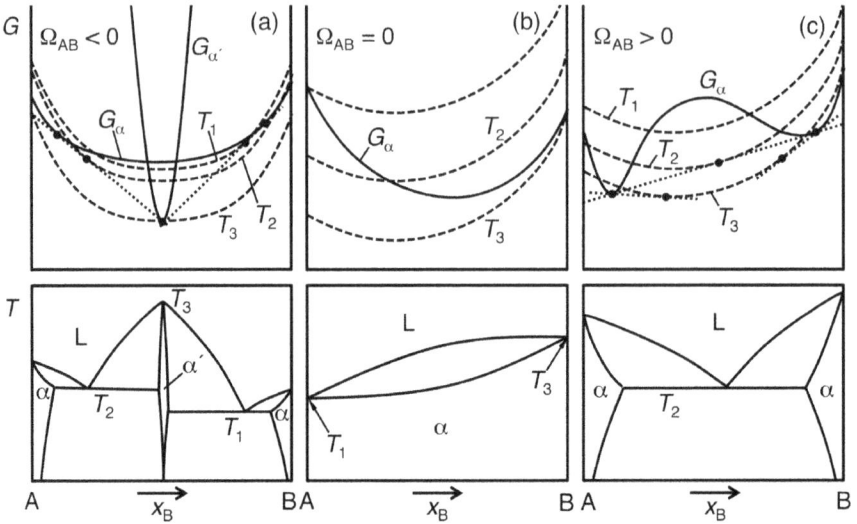

**Figure 3.4:** The relation between the interaction parameter, $\Omega_{AB}$, and the phase diagram. In cases of $\Omega_{AB} < 0$ (a) and $\Omega_{AB} \approx 0$ (b), the free energy curves are downward convex functions. For $\Omega_{AB} > 0$, it is possible that the free energy curve partially shows an upward curvature (c). For $\Omega_{AB} < 0$, interactions between different kinds of atoms are attractive and hence ordered phase $\alpha'$ could be more stable than the random alloy (a). For $\Omega_{AB} = 0$ and $\Omega_{AB} > 0$, phase diagrams of a complete solid solution (b) and with a eutectic reaction (c) are typical ones, respectively. In the diagrams of free energy versus composition, solid and broken curves show solid and liquid states, respectively, Regarding temperature, the relation $T_1 > T_2 > T_3$ holds for the system shown in the diagrams.

In binary alloys, configurational entropy takes a maximum value around the middle composition since the number of configuration exhibits the maximum. Thus, the entropy term $-TS$ in the free energy shows a downward curvature in its composition dependence regardless of combination of constituent elements. For $\Omega_{AB} > 0$ and $\Omega_{AB} < 0$, the enthalpy term is upward and downward convex functions, respectively, while it is a line between the enthalpy values at the pure element states of A and B for $\Omega_{AB} = 0$. Considering these behaviors of the enthalpy and entropy terms, the free energy curve can partially be an upward convex function only for $\Omega_{AB} > 0$ as shown in Figure 3.4(c), while it is a downward convex function for other cases.

Because, in general, entropy is smaller for a solid phase than for a liquid phase, the interaction parameter affects more prominently the composition dependence of free energy in a solid phase than in a liquid phase. Here, it is assumed that the composition dependence of free energy in a liquid phase is a downward convex function regardless of $\Omega_{AB}$.

Next, the temperature dependence of free energy is considered. Entropy of a liquid phase is larger than that of a solid phase, and free energy of a liquid phase depends on temperature more prominently than that of a solid phase. Thus, free energy decreases more significantly in a liquid phase than in a solid phase as temperature increases.

What effects can appear in phase diagrams from difference in the sign of $\Omega_{AB}$?

For $\Omega_{AB} > 0$ as shown in Figure 3.4(c), there is a region of upward curvature and hence phase separation occurs at low temperatures. By increasing the temperature, the free energy curve for liquid becomes tangent to the common tangent line of the A-rich and B-rich regions in the free energy curve of solid phase at a specific temperature. This condition gives the eutectic temperature. Similar reaction occurs for a system where A and B take different crystal structures in their elemental states while in this case the free energy of the solid phase is not expressed as one curve but two curves for respective elements (phases). Pseudobinary thermoelectric material systems such as the PbTe–SrTe, PbTe–MgTe, PbTe–Ag$_2$Te, and PbTe–Sb$_2$Te$_3$ systems are included in this type, for which there are many studies of nanostructuring utilizing phase separation. Even in cases of $\Omega_{AB} > 0$, if the absolute value of $\Omega_{AB}$ is small, a complete solid solution may form without showing phase separation. In such cases, however, phase separation may occur at low temperatures in a mode of spinodal decomposition depending on composition, which will be discussed later.

If $\Omega_{AB} = 0$ as shown in Figure 3.4(b), A and B atoms tend to arrange in a completely random manner. Such a solid solution is called an ideal solution. Free energy of the solid state is a downward convex function; hence, constituent elements are soluble with each other at the whole composition range in the solid state. Examples in thermoelectric materials include the Si–Ge system.

A negative interaction parameter ($\Omega_{AB} < 0$) also gives a free energy curve with a downward curvature. The difference from cases with $\Omega_{AB} = 0$ is that interactions between neighboring atoms are negative for those between different kinds of atoms. This gives rise to ordering in atomic arrangements and hence the free energy of the ordered phase tends to be lower than that of the solid solution phase (Figure 3.4(a)), resulting in the formation of an intermediate phase.

## 3.4.2 Solid-state precipitation

Here, we consider a system that has a positive interaction enthalpy ($\Omega_{AB} > 0$) and hence causes a phase separation. In general, a contribution of the enthalpy term to free energy is relatively larger at low temperatures while the entropy term shows a reverse trend. Therefore, the trend of enthalpy term of upward curvature is reflected on the composition dependence of free energy more prominently at low temperatures as shown in Figure 3.5. For this reason, solubility tends to be lower at lower temperatures. Accordingly, an alloy was annealed at a high temperature as solution heat treatment causes solid-state precipitation of the solute element when it is annealed at a lower temperature where the solute concentration exceeds the solubility at the temperature. This reaction gives a microstructure, where particles of the solute element are dispersed in the matrix phase as shown in Figure 3.2(b) and (c).

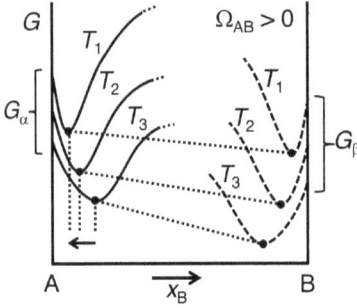

**Figure 3.5:** The dependence of the free energy of an A–B binary system on the concentration of B, $x_B$. Solid lines and broken lines are for the $\alpha$ and $\beta$ phases, respectively. If there is a region with an upward curvature due to a positive interaction parameter ($\Omega_{AB} > 0$), the composition range of such a region is widened as temperature decreases resulting in the decrease of solubility, which is shown by an arrow in the figure. In this case, it is possible to utilize a solid-state precipitation reaction.

## 3.4.3 Eutectoid reaction

In this mode of phase transformation, a single solid phase at a high temperature is decomposed to two solid phases at a low temperature. Eutectic reaction is a similar reaction, where a liquid phase at a high temperature is involved instead of a solid phase as the reactant. The formation of pearlite in steel (the Fe–C system), austenite (fcc) → ferrite (bcc) + cementite ($Fe_3C$), is well known. Examples of thermoelectric material systems include the decomposition of $Pb_2Sb_6Te_{11}$ [8], which produces PbTe and $Sb_2Te_3$, and $\alpha$-$FeSi_2$ [9], which does $\beta$-$FeSi_2$ and Si. Typically, lamellar structure (Figure 3.2(d)) is obtained due to this class of reaction.

## 3.4.4 Spinodal decomposition

This class of reaction occurs on conditions that constituent elements form a solid solution at high temperatures but decomposed at low temperatures, which requires that the interaction parameter is positive ($\Omega_{AB} > 0$) and the absolute value of it is small. While the free energy has downward curvature in the whole composition in a system at high temperatures, the free energy curve could have an upward convex at low temperatures due to larger contribution of the enthalpy term, which is an upward convex function because of positive $\Omega_{AB}$. It is assumed that an alloy that has a uniform composition around the middle composition due to annealing at a high temperature is quenched to a low temperature, where the free energy curve has an upward convex and atomic diffusion cannot be neglected. In such a system, once fluctuation in composition occurs, it lowers the free energy of the system as long as the free energy has an upward curvature $\partial^2 G/\partial x_B^2 < 0$, as shown in Figure 3.6(a). Therefore, a small compositional fluctuation can be a trigger of the phase separation and the compositional gap is amplified as time proceeds. Examples of thermoelectric materials include the PbTe–PbS [10] and PbTe–$AgSbTe_2$ [11] systems.

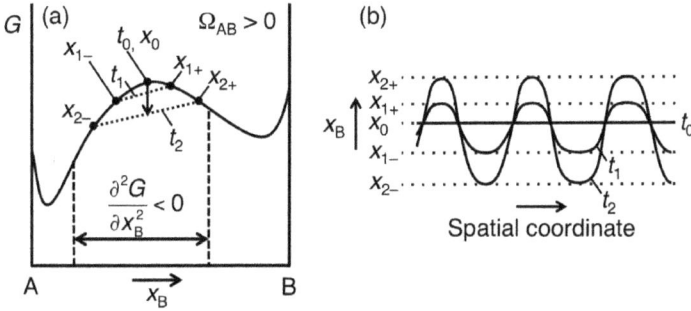

**Figure 3.6:** If elemental A and B take a same crystal structure and their interaction is repulsive ($\Omega_{AB} > 0$), then there could be a region with an upward curvature as shown in (a). In this case, a slight fluctuation in composition lowers free energy and is hence amplified leading to phase separation. This mode of phase separation is called a spinodal decomposition.

## 3.5 Controlling microstructures of thermoelectric materials via phase transformations

### 3.5.1 Relation between microstructure and thermoelectric properties

Efficiency of thermoelectric conversion is a monotonically increasing function of dimensionless thermoelectric figure of merit, $zT$:

$$zT = \frac{S^2 \sigma T}{\kappa} = \frac{S^2}{L} \frac{1}{1 + \frac{\kappa_L}{\kappa_E}}, [12] \tag{3.5}$$

where $S$, $\sigma$, $K$, $L$, and $T$ stand for the Seebeck coefficient, electrical conductivity, thermal conductivity, Lorenz number, and absolute temperature. Thermal conductivity has contributions from electrons, $\kappa_E$, and lattice vibrations, $\kappa_L$, and hence expressed as $\kappa = \kappa_E + \kappa_L$. To enhance thermoelectric conversion efficiency, materials that show high $zT$ values are desired. Thus, a number of researches are going on to enhance $zT$, which are roughly classified into two kinds: one is studies to increase the power factor $S^2\sigma$ including band engineering, and career concentration tuning, and the other is structural studies to reduce the ratio $\kappa_L/\kappa_E$. Table 3.1 summarizes the strategies to enhance $zT$ that are considered to be closely related to microstructures. As shown in the table, it is essential to control the morphology and/or size scales of the microstructure in order to enhance the effect. Regarding the size scale, all the strategies listed in the table require structure of a nanometer scale or a scale close to nanometers. In this section, we discuss the control of the morphologies and size of microstructure related to Table 3.1.

**Table 3.1:** Strategies for improving performance of thermoelectric materials via microstructural control and required condition.

| Strategy | Quantity affected | Effect | Examples of microstructure | Necessary condition to enhance the effect |
|---|---|---|---|---|
| Polycrystalline or nanocomposite structure [1] | $\dfrac{\kappa_L}{\kappa_E}$ | Enhances phonon scattering at boundaries leading to reduced phonon mean free paths resulting in reduction of $\kappa_L$ | Poly crystals Secondary phase particles dispersed | Decrease in $\lambda$ |
| Modulation doping [7] | $\dfrac{\kappa_L}{\kappa_E}$ | Reduces carrier scattering of dopants by separating carriers from dopants resulting in increase in $\kappa_E$ | Secondary phase particles dispersed | Decrease in $\lambda$ |
| Porous structure | Heat transfer coefficient | Makes the area of surface in contact with large thermal fluid | Porous material | Decrease in $d$ Increase in $p$ |

## 3.5.2 Polycrystalline structure

First, we consider a polycrystalline structure in a material that is homogeneous in terms of chemical composition, which is comprised of grain boundaries. This class of structure is obtained by solidification, crystallization of amorphous, recrystallization, sintering powder material, and so on.

In solidification with a relatively low cooling rate, nuclei of solid phase are formed with a high nucleation rate in the region where a degree of undercooling is high or interfacial energy is low. They grow roughly in the direction of heat flow resulting in a columnar structure. Controlling the velocity of the liquid/solid interface to be slow on a condition with a relatively large temperature gradient, resultant grains tend to be extremely a large single crystal. In a slow cooled sample, the crystal growth orientation exhibits a specific relation with a solidification direction (preferential growth orientation). Instead, a fast cooling can cause nucleation all over the undercooled liquid region resulting in the formation of a polycrystalline structure; hence, there is no specific orientation relations between neighboring grains.

Crystallization of amorphous materials can give nanosized polycrystalline structure in an early stage of crystallization. There are several routes for fabrication of amorphous, including techniques utilizing cooling from a liquid state, sublimation, and a solid-state reaction [13].

Plastic deformation of a crystal leads to the formation dislocations. Dislocations are accompanied with the strain of the crystal and hence their formation increases the crystal energy. When a plastically deformed sample is heat-treated, dislocations with opposite signs will annihilate each other due to a slip or climbing motion if the dislocation density is not too high. On the other hand, if the dislocation density is high, new crystals with less strain are formed instead of recovery, which is called recrystallization, and grow. Thus, if a sample is severely deformed, the nucleation rate is high resulting in a polycrystalline structure with fine grains [14].

To obtain a polycrystalline structure, the three processes utilize spontaneous crystallization from high free energy state. In these processes, the difference in the free energies between the high-energy state and the equilibrium state, that is, the degree of nonequilibrium, governs the nucleation rate and thus the grain size. On the other hand, in sintering powder material, grain size depends on the granularity of power. There are various fabrication techniques, which depend on the resulting granularity [15].

## 3.5.3 Composite structures

### 3.5.3.1 Precipitation structures

Structures where particles of a secondary phase are dispersed in a matrix phase can be obtained by solidification or solid-state reactions. In general, solid-state reactions give smaller microstructures than reactions where solidification is involved. Fine microstructures due to solid-state reactions often show features in a nanometer scale. In this section, solid-state precipitation is discussed.

Figure 3.7 shows an example of a phase diagram where a solid-state reaction can occur. For an alloy at the composition shown by the arrow, while a single phase state of the α phase is stable at high temperatures, it is not so at low temperatures, where a two-phase state composed of the α and β phases is stable. A solution heat treatment at a high temperature followed by a low-temperature heat treatment results in the nucleation and growth of the β phase. In this process, the solute concentration exceeds the solubility in the equilibrium state in the alloy right after the temperature is changed to the low temperature until the solid-state precipitation begins. This state is called a supersaturated solid solution, which is also understood as an undercooled state if one considers in the direction of the temperature axis. A solid-state precipitation reaction can occur at any compositions, $c$, up to the maximum solubility. The lever rule gives the molar ratio of the matrix α, $x_\alpha$, and the precipitate β phases, $x_\beta$, as follows:

$$\frac{x_\alpha}{x_\beta} = \frac{c_\beta - c}{c - c_\alpha}, \tag{3.6}$$

where $c_\alpha$ and $c_\beta$ stand for the solute concentrations of the α and β phases in the microstructure. Thus, there is freedom in the choice of volume fractions of the matrix and the precipitate phases in the resulting microstructures. Developments of thermoelectric materials sometimes require a complex tuning of thermoelectric properties. The high degree of freedom in choice of the volume fractions of constituent phases in microstructures might have an advantage.

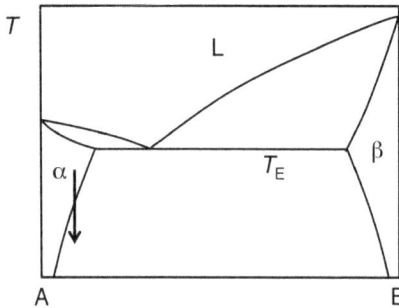

**Figure 3.7:** An equilibrium phase diagram of the binary A–B system. At the composition shown by an arrow in the figure, microstructure with secondary phase particles dispersed in a matrix phase and that with lamellae are obtained by continuous precipitation and discontinuous precipitation, respectively.

Morphology of precipitates can be classified into spheres, plates, and needles. Comparing them for a same volume fraction of the secondary phase, plate morphology gives the largest interfacial area per volume or the smallest average distance between neighboring cross points of straight lines in a random direction and interfaces of microstructure [16]. Interfacial energy between matrix and precipitate phases and strain energy due to the formation of precipitates are involved in determining morphology of precipitates.

Here, control of size scales of precipitation microstructure is discussed. If nucleation is assumed to occur homogeneously, a nucleation rate, $\dot{N}$, is expressed as follows:

$$\dot{N} \propto \exp\left(-\frac{\Delta G^*}{kT}\right) \exp\left(-\frac{\Delta E_D}{kT}\right),$$ (3.7)

where $\Delta E_D$, $k$, and $T$ are the activation energy of atomic diffusion, the Boltzmann constant, and absolute temperature, according to a classical nucleation theory [17]. The critical free energy for nucleation, $\Delta G^*$, is expressed using the interfacial energy, $y$, the chemical driving force, $\Delta G_c$, and the strain energy, $\Delta G_E$, as follows:

$$\Delta G^* = \frac{16}{3} \frac{\pi y^3}{(\Delta G_c + \Delta G_E)^2}.$$ (3.8)

Thus, the factor $\exp(-\Delta G^*/kT)$ in eq. (3.7) depends on the interfacial energy, strain energy, chemical driving force, and temperature. While the interfacial energy and strain energy do not depend much on the temperature, the chemical driving force is approximately proportional to undercooling $\Delta T$ [17], that is, $\Delta G_c \propto \Delta T$. Therefore, $\Delta G^*$ decreases with increasing $\Delta T$ resulting in the increase in the factor $\exp(-\Delta G^*/kT)$ in eq. (3.7). On the other hand, the factor $\exp(-\Delta E_D/kT)$ decreases with increasing $\Delta T$, that is, decreasing $T$. Thus, because of the opposite trends of $\exp(-\Delta G^*/kT)$ and $\exp(-\Delta E_D/kT)$ in eq. (3.7), the nucleation rate shows a maximum value at some undercooling.

The above argument describes what the size scales of precipitation microstructure are determined for a fixed alloy composition. Now we consider what should be taken into account in choice of a material system. The chemical driving force is expressed using heat of solution, $\Delta H_s$, as follows:

$$\Delta G_c = \Delta H_s \frac{\Delta T}{T_s}, [18]$$ (3.9)

where $T_s$ is a solvus temperature. For a fixed $\Delta T$, an alloy system having a larger $\Delta H_s$ gives larger $\Delta G_c$; hence, it is considered that such an alloy system with a large heat of solution tends to show fine precipitation microstructures. $\Delta H_s$ is a quantity dependent on an alloy system.

### 3.5.3.2 Lamellar structures

Lamellar structures can be obtained by eutectic, eutectoid, or discontinuous precipitation reactions. Reactions are expressed as follows:

$$\text{eutectic: } L \rightarrow \alpha + \beta \tag{3.10a}$$

$$\text{eutectoid: } \gamma \rightarrow \alpha + \beta \tag{3.10b}$$

$$\text{discontinuous precipitation: } \alpha \rightarrow \alpha' + \beta. \tag{3.10c}$$

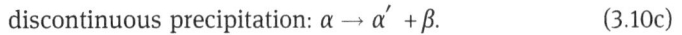

In all these reactions, a homogeneous single phase is decomposed into two different phases with chemical compositions varying from the composition of the mother phase. Figure 3.8(a) is an example phase diagram of a binary system showing a eutectic reaction.

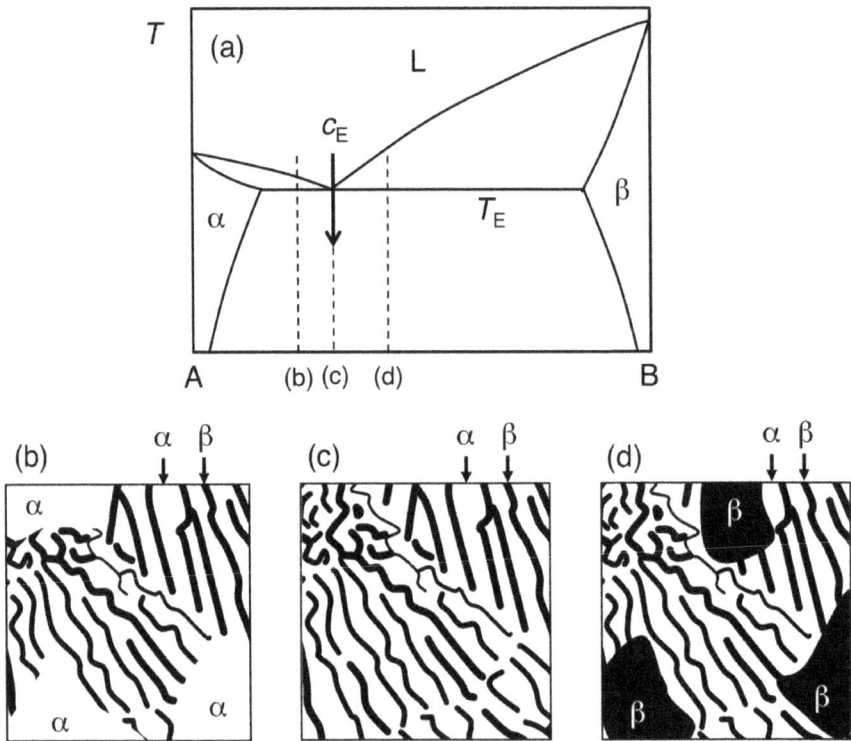

**Figure 3.8:** An example of a eutectic phase diagram (a). Microstructures as shown in (b), (c), and (d) can be obtained at compositions indicated in (a). At hypoeutectic (b) and hypereutectic (d) compositions, primary crystals of the α and β phase are observed in the microstructures (b) and (d), respectively.

Because eutectic and eutectoid reactions are invariant reactions, the composition where such a reaction occurs is indicated at a single point on a phase diagram, as shown at $c_E$ in Figure 3.8(a). At other compositions, formation of primary crystals normally occurs in advance of the eutectic reaction as shown in Figure 3.8(b) or (d), resulting in an inhomogeneous microstructure. The similar thing occurs in a eutectoid system. However, if the degree of undercooling is large due to fast temperature decrease and so on, lamellar structure could be obtained at compositions other than $c_E$ without the formation of primary crystals. Nevertheless, since compositions where homogeneous lamellar structure can be obtained are anyway limited to a narrow range close to $c_E$, the fraction of the two phases, $x_\alpha/x_\beta$, is close to

$$\frac{x_\alpha}{x_\beta} = \frac{c_\beta - c_E}{c_E - c_\alpha}, \tag{3.11}$$

where $c_\alpha$ and $c_\beta$ are concentrations of B in the $\alpha$ and $\beta$ phases, respectively, implying that it is difficult to control the $x_\alpha/x_\beta$ ratio.

Lamellar structure can also be observed in a composition, which is of neither eutectic nor eutectoid but of that shown by the arrow in Figure 3.7. Such a reaction is called a discontinuous precipitation. Normally, the secondary phase ($\beta$) nucleates at such a composition when the temperature goes down below the solvus line and grows with decreasing temperature. This reaction is accompanied by the continuous decrease in solute concentration in the matrix phase by increasing the volume fraction of precipitates. In contrast, "discontinuous" precipitation is accompanied by discontinuous change in the matrix composition as the reaction proceeds. This mode of precipitation occurs by phase separation of the $\alpha$ phase involving coupled growth of the $\alpha'$ and $\beta$ phases, which is similar to eutectoid reactions with respect to morphology while the $\alpha'$ phase, one of the reaction products, is the same phase as the $\alpha$ phase but is just different in chemical composition in contrast to eutectoid reactions. This reaction tends to occur with a high degree of undercooling [19]. Because discontinuous precipitation is not an invariant reaction, that is, the degree of freedom is not zero, the chemical compositions where this class of reaction occurs are not limited to an invariant composition.

In the growth of lamellar structure, the interface between the mother phase and the lamellae, the reaction product, moves into the mother phase as the reaction proceeds. The chemical composition is partitioned to the two compositions of the constituent phases in the lamellar. The lamellae are often parallel to the direction of the growth but are not necessarily so in cases of eutectoid reactions. If the direction of the growth of lamellae is externally controlled, typically by controlling the temperature gradient, it might be possible to make lamellae align in one direction. The alignment of lamellae is readily conducted for eutectic reactions by a unidirectional solidification method. But, it is not as easy for eutectoid reactions while there are some reports [20]. The difficulty is possibly related to the smaller difference in the free energies due to smaller difference entropies between reactant and reaction products in eutectoid reactions than eutectic reactions. A very large temperature gradient might be a key to align eutectoid lamellae.

The mechanisms of eutectic and eutectoid reactions are similar; hence, equations that describe size scales are similar to each other. For both reactions, the lamellar spacing, $\lambda$, is related to undercooling, $\Delta T$,

$$\lambda \propto \frac{\gamma T_E}{\Delta H \Delta T}, [19, 21] \tag{3.12}$$

and is related to growth rate, $v$,

$$v\lambda^2 = \text{const.}, [19, 21] \tag{3.13}$$

where $\Delta H$ is the enthalpy change of the phase transformation. According to eq. (3.12), in cases of the phase transformations due to isothermal heat treatment, the lamellar spacing decreases with decreasing heat treatment temperature or increasing the enthalpy change of the phase transformation ($=\Delta H/T_E$). According to eq. (3.13), the lamellar spacing decreases by increasing the velocity. Examples include a case of unidirectional solidification or heat treatments at a constant velocity in a fixed temperature gradient.

## 3.5.4 Porous structures

Porous materials have features such as light weight, impact energy absorption ability, sound absorption ability, and high specific surface area. Among such features, high specific surface area might be useful for improving performance of thermoelectric conversion devices; if porous material has open pores, fluid flows through the pores and the high specific surface area could directly lead to enhanced heat exchange abilities. Actually, "lotus copper," which has unidirectionally aligned elongated pores, is used as high-performance heat sinks [22]. For thermoelectric conversion system, high heat exchange ability could lead to a high power or compactness of the system. Modification of surfaces of thermoelectric devices in contact with thermal fluid by making them porous could be an attractive field of research. The fabrication technique for materials with lotus structure as shown in Figure 3.9 has developed originally for metals and alloys [22], which utilizes difference in solubilities of an element that takes a gas form in its elemental state such as hydrogen, nitrogen, and oxygen in solid and liquid states of a metal or an alloy as shown in Figure 3.10. If a material is solidified in hydrogen atmosphere, for example, hydrogen molecules that cannot be dissolved in solidification due to the decrease in solubilities in solidification form pores. This reaction is a sort of invariant reactions called gas-evolution crystallization reaction [23], which is analogous to the eutectic reaction in the sense that the growth of reaction products is coupled but is different, and in that one of the reaction products is a gas phase. Because pores are elongated and aligned in the direction of solidification, their alignment direction can be controlled by controlling the heat flow direction.

So far, various metals and alloys including copper, nickel, iron, stainless steel, and nickel-based superalloys have been made porous by this technique [22]. The materials that can be made porous are not limited to metals and alloys. Lotus alumina

**Figure 3.9:** Schematic view of a lotus metal. Elongated pores are aligned in one direction.

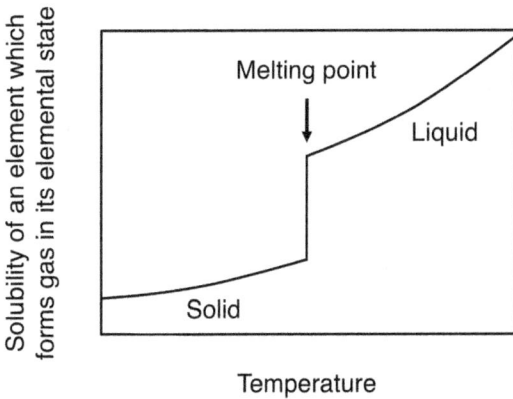

**Figure 3.10:** Typical temperature dependence of solubility across the melting temperature of an element that forms gas in its elemental state such as hydrogen and nitrogen.

[24] and silicon [25, 26] have also been fabricated. However, materials other than metallic materials have not studied much for making porous using this technique so far.

For controlling porous structure, experimental parameters could be the total pressure of gasses, the partial pressure of a dissolved element, the temperature gradient in solidification, and the solidification velocity. In the following, we consider a case where hydrogen and argon mixture is used as an example. In this case, first, we should take into account the solubilities of hydrogen in the liquid and solid states of an alloy, which follows the Sieverts' law:

$$c_{H,L} = k_{H,L} p_{H2}^{1/2},$$ (3.14a)

$$c_{H,S} = k_{H,S} p_{H2}^{1/2},$$ (3.14b)

where $c_{H,L}$ and $c_{H,S}$ are the concentrations of hydrogen in the liquid and solid alloys, $k_{H,L}$ and $k_{H,S}$ are the equilibrium constants of hydrogen for dissolution in the liquid and solid states, respectively, and $p_{H2}$ is the partial pressure of hydrogen. The hydrogen gas corresponding to the difference $c_{H,L} - c_{H,S}$ can form pores. Second, the pressure of hydrogen in pores, $p_{H2'}$, should be balanced with the total pressure, $p_{H2} + p_{Ar}$. Therefore, the volume of hydrogen pores, $v_p$, is inversely proportional to the total pressure according to the Boyle's law:

$$v_p \propto (p_{H2} + p_{Ar})^{-1}.$$ (3.15)

These equations are related to the volume fraction of pores (porosity). Furthermore, the size of pores is considered to be dependent on solidification velocity and the gas pressures. See ref [22]. for detailed analysis.

## 3.6 Concluding remarks

In this chapter, the relation among equilibrium phase diagrams, thermodynamics, and microstructures was discussed. The desired microstructure can be obtained under a careful control of thermal history of a sample. We hope that understanding backgrounds of microstructures will help control them with a high degree of precision or freedom. Regarding thermoelectric properties, understanding relations to microstructures is on the way to development in many cases. For better understanding of their impacts on thermoelectric properties, it is essential to quantify microstructures, which involves much efforts. More and more accumulation of experimental data is desired as well as theoretical considerations.

If nanostructures can be controlled as one wishes, it would contribute to improvements of thermoelectric materials due to various effects as shown in Table 3.1. While microstructure control is an attractive field in facilitating use of thermoelectric materials, there are challenges to be addressed. One of them is their structural stability at high temperatures. A study on long time stability of nanostructures of thermoelectric materials is scarce but is essential to practical use. Nanostructuring is essentially accompanied by increased interfacial density, hence, inevitably lead to a state with high interfacial energy. To address this issue, a kinetic or thermodynamic approach or both might be needed including but not limited to:

(1) A material that has a melting temperature high enough compared with the operating temperature should be chosen. In this case, heat treatments for nanostructuring should be done at a temperature high enough compared with the operating temperature.

(2) The interfaces should be stabilized. For that interfacial energy needs to be lowered by segregation of impurity elements on interfaces, utilizing coherent phase boundaries, or utilizing interfaces with a chemically low-phase boundary energy.

# References

[1]  Kanatzidis M G., Nanostructured Thermoelectrics: The New Paradigm?, Chem. Mater. 2010, 22, 648.
[2]  Biswas K, He, J, Blum, I D, Wu, C-I, Hogan, T P, Seidman, D N, Dravid, V P, Kanatzidis, M G., High-performance bulk thermoelectrics with all-scale hierarchical architectures, Nature 2012, 489, 414.
[3]  Zhang Y, Ke, X, Chen, C, Yang, J, Kent, P R C., Nanodopant-Induced Band Modulation in AgPbmSbTe$_2$þm-Type Thermoelectrics, Phys. Rev. Lett. 2011, 106, 206601.
[4]  Kim S I, Lee, K H, Mun, H A, Kim, H S, Hwang, S W, Roh, J W, Yang, D J, Shin, W H, Li, X S, Lee, Y H, Snyder, G J, Kim, S W., Dense dislocation arrays embedded in grain boundaries for high-performance bulk thermoelectrics, Science 2015, 348, 109.
[5]  Wu Y, Chen, Z, Nan, P, Xiong, F, Lin, S, Zhang, X, Chen, Y, Chen, L, Ge, B, Pei, Y., Lattice Strain Advances Thermoelectrics, Joule 2019, 3, 1276.
[6]  Zide J M O, Vashaee, D, Bian, Z X, Zeng, G, Bowers, J E, Shakouri, A, Gossard, A C., Demonstration of electron filtering to increase the Seebeck coefficient in In$_{0.53}$Ga$_{0.47}$As/In$_{0.53}$Ga$_{0.28}$Al$_{0.19}$As superlattices, Phys. Rev. 2006, 74B, 205335.
[7]  Zebarjadi M, Joshi, G, Zhu, G, Yu, B, Minnich, A, Lan, Y, Wang, X, Dresselhaus, M, Ren, Z, Chen, G., Power factor enhancement by modulation doping in bulk nanocomposites, Nano. Lett. 2011, 11, 2225.
[8]  Ikeda T, Collins, L A, Ravi, V A, Gascoin, F S, Haile, S M, Snyder, G J., Self-Assembled Nanometer Lamellae of Thermoelectric PbTe and Sb$_2$Te$_3$ with Epitaxy-like Interfaces, Chem. Mater. 2007, 19, 763.
[9]  Kimura Y, Takeno, K, Mori, A, Chai, Y-W. J., Fabrication of β-FeSi$_2$-Based Thermoelectric Composite Alloys by Oxidation and Reduction Reactions during Sintering of Eutectoid Si and Iron Oxide Powder, Jpn. Inst. Met. 2015, 79, 613.
[10] Androulakis J, Lin, C-H, Kong, H-J, Uher, C, Wu, C-I, Hogan, T, Cook, B A, Caillat, T, Paraskevopoulos, K M, Kanatzidis, M G. J., Spinodal Decomposition and Nucleation and Growth as a Means to Bulk Nanostructured Thermoelectrics: Enhanced Performance in Pb$_{1-x}$Sn$_x$Te-PbS, Am. Chem. Soc. 2007, 129, 9780.
[11] Ikeda T, Snyder, G J Unpublished work 2019.
[12] Dames C, Chen, G. Thermal conductivity of nanostructured thermoelectric materials, In: Rowe DM, editor, Thermoelectrics Handbook: Macro to Nano, CRC Press, 2005.
[13] Inoue A., Production Techniques of Amorphous Alloys, Materia Japan 1996, 35, 244.
[14] Horita Z, Furukawa, M, Langdon, T G, Nemoto, M., Equal-Channel Angular Pressing (ECAP): A Novel Method for Microstructural Control, Materia Japan 1998, 37, 767.
[15] Antony L V M, Reddy, R G. J., Processes for Production of High-Purity Metal Powders, Miner. Metals Mater. Soc. 2003, 55, 14.

[16] Ikeda T, Marolf, N J, Bergum, K, Toussaint, M B, Heinz, N A, Ravi, V A, Snyder, G J., Size control of $Sb_2Te_3$ Widmanstätten precipitates in thermoelectric PbTe, Acta. Mater. 2011, 59, 2679.

[17] Doherty R D. Diffusive phase transformations in the solid state, In: Haasen RWCaP, editor, Physical Metallurgy, Vol: Vol. II, Amsterdam, Elsevier Science, 1996, 1364.

[18] Russell K C., Nucleation in solids: the induction and steady state effects, Adv. Colloid Interface Sci. 1980, 13, 205.

[19] Enomoto M. Phase Transformations in Metals: An Introduction to Science of Materials Microstructure, Tokyo, Uchida Rokakuho, 2005.

[20] Ridley N., A review of the data on the interlamellar spacing of pearlite, Metall. Trans. 1984, 15A, 1019.

[21] Chalmers B. Principle of Solidification, New York, Wiley, 1964.

[22] Nakajima H., Fabrication, properties and application of porous metals with directional pores, Prog. Mater. Sci. 2007, 52, 1091.

[23] Nakajima H., Invariant Reaction of Liquid → Solid + Gas —Gas-Evolution Crystallization Reaction—, Mater. Trans. 2001, 42, 1827.

[24] Ueno S, Lin L M, Nakajima H., Formation Mechanism of Porous Alumina With Oriented Cylindrical Pores Fabricated by Unidirectional Solidification, Am. Ceram. Soc. 2008, 91, 223.

[25] Nakahata T, Nakajima, H., Fabrication of lotus-type porous silicon by unidirectional solidification in hydrogen, Mater. Sci. Eng.: A 2004, 384, 373.

[26] Ikeda T, Nagano, T, Ide, T, Nakajima, H Porous thermoelectric materials and their applications. 37th International and 16th European Conference on Thermoelectrics. Caen, France, 2018.

Nader Farahi, Dao Y Nhi Truong, Christian Stiewe, Holger Kleinke, and Eckhard Müller

# 4 Half-Heusler thermoelectric materials for waste heat recovery

**Abstract:** Heusler materials have attracted substantial attention over decades because of their impressive magnetic and electronic properties. Half-Heusler compounds, attributed to their semiconducting properties, are promising candidates for thermoelectric devices in the medium to high temperature range. In this regard, the chapter provides a thorough summary on the properties of high efficiency half-Heusler thermoelectrics. In the first section, the crystal structure and molecular orbital diagrams of full-Heusler and half-Heusler materials are analyzed for a better understanding of their electronic properties. Following, phase diagram, electronic structure, and thermoelectric properties of n-type XNiSn-based and p-type XCoSb half-Heusler compounds (X = Ti, Zr, Hf) are discussed in detail. Subsequently, recent progress in high performance NbFeSb-based and ZrCoBi-based materials are addressed, and some suggestions are provided for further investigations. Lastly, the development of thermoelectric modules based on half-Heusler materials is summarized to demonstrate the advancements towards industrial applications.

**Keywords:** Half-Heusler, TiNiSn, TiCoSb, NbFeSb, Thermoelectric

## 4.1 Introduction to half-Heusler compounds

The material class of Heusler compounds has attracted great research interests in the last few decades after the first compound of $MnCu_2Sn$ was discovered by Fritz Heusler in 1903 [1]. This vast family is currently consisting of more than 1,500 members from the large combination of almost all elements in the periodic table [2]. Since Heusler compounds possess many desirable physical properties, ranging from nonmagnetic semiconductors [3], to half-metallic ferromagnets [4], compensated ferrimagnets [5], superconductors [6], and topological insulators [7], several

**Nader Farahi, Dao Y Nhi Truong, Christian Stiewe,** Institute of Materials Research, Linder Höhe, German Aerospace Center (DLR), Köln, Germany
**Holger Kleinke,** Department of Chemistry and Waterloo Institute for Nanotechnology, University of Waterloo, Waterloo, ON, Canada
**Eckhard Müller,** Institute of Materials Research, Linder Höhe, German Aerospace Center (DLR), Köln, Germany; Institute of Inorganic and Analytical Chemistry, Justus-Liebig University Giessen, Giessen, Germany

https://doi.org/10.1515/9783110596526-004

applications for these materials have been explored over the years, including thermoelectrics [8], solar cells [9], spintronic devices [10], shape memory alloys [11], magnetic tunnel junctions [12], magneto-optical gadgets [13], and multifunctional equipment [14, 15], that is, a mixture of magneto-optical, magneto-electronic, magneto-structural, and magneto-caloric properties. A detailed discussion on the physical properties and potential applications of Heusler materials can be found in the review of Graf and co-authors [16]. This chapter mainly focuses on the compounds that are appealing for thermoelectric applications.

The nomenclature for Heusler compounds broadly varies depending on the convenience, for example, alphabetically, electronegativity ordering, or randomly. Here, the elements will be ordered by their electronegativity according to the International Union of Pure and Applied Chemistry nomenclature system.

The full Heusler, or Heusler, compounds are the ones with the general formula of $XY_2Z$, where $X$ and $Y$ are transition metals or rare earth elements, and $Z$ is a main group element. Moreover, $X$ can also be an alkaline earth element, such as Be or Mg, and is the most electropositive element, while $Z$ is the most electronegative one. The Heusler structure can be described as an interpenetration of four face-centered cubic sublattices, which corresponds to a $MnCu_2Al$ prototype structure, and a space group of $Fm\bar{3}m$. The $X$, $Y$, and $Z$ atoms take up the Wyckoff positions of 4a (1/2, 1/2, 1/2), 4b (1/4, 1/4, 1/4), and 4c (3/4, 3/4, 3/4), as well as 4d (0, 0, 0), respectively (Figure 4.1(a)). For bonding descriptions, two approaches are commonly applied. In the first approach, the $X$ and $Z$ atoms are considered to form a rock salt (NaCl) structure, where the $Y$ atoms fill all the tetrahedral holes. In the other approach, the $Y$ and $Z$ atoms are described to create an antifluorite (CaF₂) structure, where the $X$ atoms occupy all the octahedral holes. Here, the later will

(a)                                          (b)

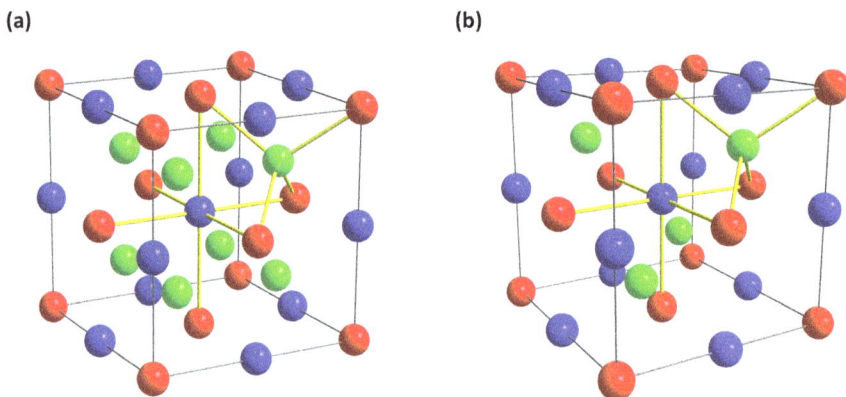

**Figure 4.1:** Crystal structure of (a) full-Heusler $XY_2Z$, and (b) half-Heusler $XYZ$ compounds, where $X$, $Y$, and $Z$ atoms are in blue, green, and red colors, respectively. Yellow lines represent the octahedral and tetrahedral coordinates for $X$ and $Y$ atoms, respectively.

be used to explain the bonding characters in full Heusler compounds using molecular orbital (MO) theory.

A derivative from the Heusler structure is the inverse Heusler structure [17], which resembles the $HgCu_2Ti$ prototype structure, where $Y$ is the most electropositive element, and $Z$ is the most electronegative one. Consequently, $Y$ and $Z$ atoms form a rock salt structure, while $X$ and $Y$ alternatively take up all the tetrahedral holes.

Figure 4.2 shows the illustration of MO diagrams for full-Heusler compounds. For one formula unit of $XY_2Z$, one set of four bonding and one set of four antibonding orbitals are generated from the hybridization of the $s$ and $p$ states of the $Y$ and $Z$ atoms (Figure 4.2(a)). Similarly, the interactions between the $d$ states from the two $Y$ atoms result in two sets of bonding and antibonding orbitals with five MOs in each set (Figure 4.2(b)). Finally, the $d$ atomic orbitals of $X$ atom form the nonbonding orbitals, usually located between the bonding and antibonding $d$ orbitals of the two $Y$ atoms, for example, $VFe_2Al$ [16]. Semiconducting properties are usually a consequence of the small energy gap between the split $d$ states of $X$ atom due to the octahedral field from its $Z$ neighbors (Figure 4.2(c)). It should be noted that the $d$ states of the $Y$ atoms also show a splitting between the $dx^2-y^2$, $dz^2$ and the $dxy$, $dxz$, $dyz$ orbitals due to the tetrahedral field created by the $Z$ atoms, which has been simplified in Figure 4.2(b) and (c). Consequently, both the highest occupied molecular orbital

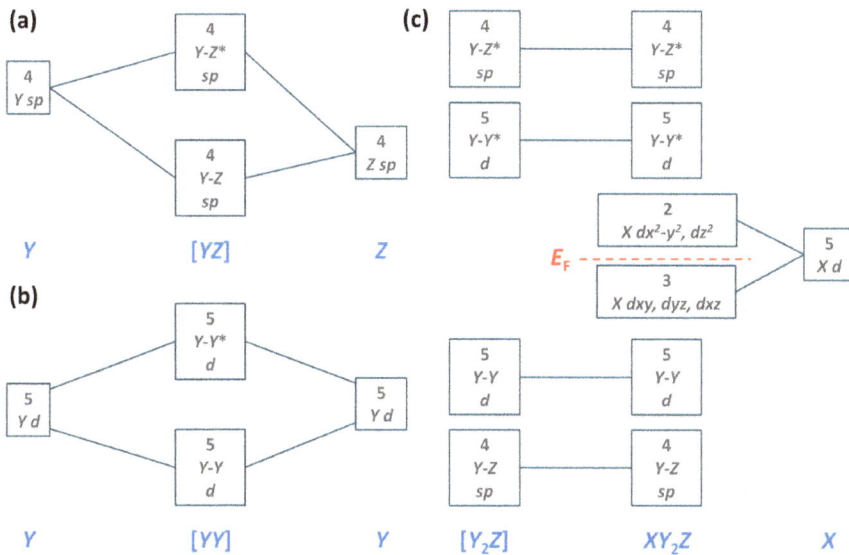

**Figure 4.2:** Schematic illustration of MO diagrams for one formula unit of $XY_2Z$ full-Heusler compound: (a) interactions between $s$ and $p$ states in $Y$ and $Z$ atoms; (b) interactions between $d$ states of two $Y$ atoms; and (c) full MO diagram including contributions of $d$ states from the $X$ atom. $E_F$ denotes the Fermi level.

(HOMO) and the lowest unoccupied molecular orbital (LUMO) have predominantly the character of the $X$ $d$ orbitals [16]. Since there are 12 MOs to be filled below the Fermi level, electronically stable full-Heusler compounds with 24 valence electrons are commonly found in nature, for example, VFe₂Al, or TiFe₂Sn, which have been demonstrated via theoretical calculations [18, 19]. However, the other two nonbonding $d$ orbitals from $X$ atoms can also be filled without significantly affecting the bond strengths, electronically less stable full-Heusler compounds with valence electron counts (VEC) up to 28 exist, for example, Mn₂RhAl (VEC = 26), Mn₂RhGe (VEC = 27), Mn₂RhSb (VEC = 28) [20], and TiNi₂Sn (VEC = 28) [21, 22]. As a consequence of the splitting of the $X$ $d$ states due to the weak octahedral field, full-Heusler compounds possess considerably small band gap. Therefore, these materials are less attractive for thermoelectric applications, but have been widely known to show all kinds of magnetic behavior, for example, ferromagnetism [23], ferrimagnetism [24], or half-metallic ferromagnetism [25]. Among the full-Heusler compounds, mainly VFe₂Al material has been thoroughly studied for thermoelectric purpose [26, 27].

The half-Heusler (or semi-Heusler) compounds are formulated as $XYZ$, where $X$ and $Y$ are transition metals or rare earth elements, and $Z$ is the main group element, for example, Si, Ge, Sn, As, Sb, or Bi. It should be noted that in some cases, $X$ can also be the main group element, such as Li. As mentioned previously, the elements are ordered according to their electronegativity, where $X$ is the most electropositive element, and $Z$ is the most electronegative one, for example, NbFeSb or TiNiSn. These compounds crystallize in the same crystal structure as the full-Heusler compounds, where one of the tetrahedral sites is empty, commonly referred to as MgAgAs-type structure. Consequently, the half-Heusler structure is a noncentrosymmetric cubic structure with the space group $F\bar{4}3m$. The $X$, $Y$, and $Z$ atoms are located at the Wyckoff positions of 4a (1/2, 1/2, 1/2), 4b (1/4, 1/4, 1/4), and 4c (0, 0, 0), respectively (Figure 4.1(b)). The structure can be described either as a rock salt (NaCl) structure formed by the $X$ and $Z$ atoms, where the $Y$ atoms fill half of the tetrahedral holes, or as a zinc blende (ZnS) structure made of $Y$ and $Z$ atoms, where the $X$ atoms occupy all the octahedral holes. The second approach will be used in this section to reveal the nature of the band gap in half-Heusler materials using MO diagrams.

The total VEC of the constituent elements strongly affects the physical properties of half-Heusler materials, which can vary from 8 to 22 [28]. The most electronically stable compounds possess either 8 or 18 valence electrons. There also exist half-Heusler compounds with other VEC, such as 17-electron compounds like TiFeSb [29], or 19-electron compounds like NbCoSb [30].

For one formula unit $XYZ$ of the eight-electron compounds (known as Nowotny–Juza phases), where both $X$ and $Z$ are the main group elements, and $X$ is usually Li, a set of four bonding and four antibonding orbitals is formed from the $s$ and $p$ hybridized atomic orbitals of $Y$ and $Z$ atoms, which are separated by a distinct energy gap (Figure 4.3(a)). The $s$ orbital of the $X$ atom is located above the antibonding states of the [$YZ$] substructure (Figure 4.3(b)). Both the HOMO and the LUMO have the

**(a)**

**(b)**

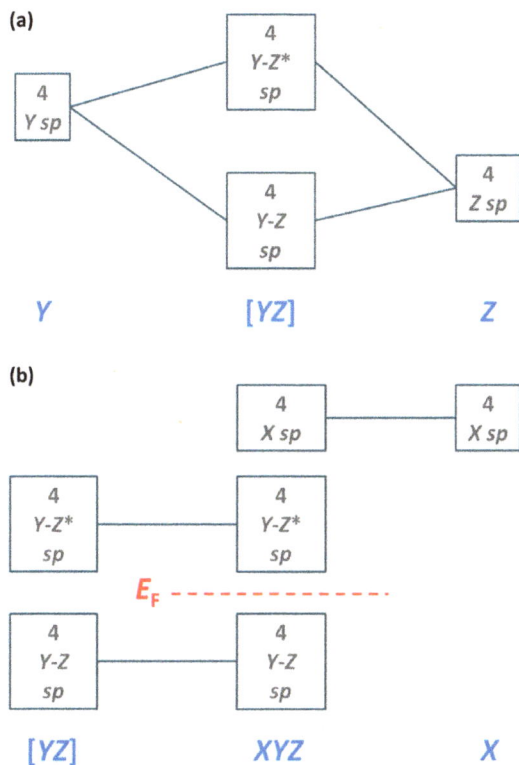

**Figure 4.3:** Schematic illustration of MO diagrams for one formula unit of *XYZ* half-Heusler compound with eight valence electrons: (a) interactions between *s* and *p* states in *Y* and *Z* atoms; (b) full MO diagram including contributions of *s* and *p* states from the *X* atom. $E_F$ denotes the Fermi level.

characters of [*YZ*] *s* and *p* orbitals. A total VEC of 8, therefore, is needed to fill all the bonding orbitals. Because of the strong valence interactions between *Y* and *Z* atoms, the band gap in these compounds is usually large, and the electrons are highly local-ized [28]. As a result, the half-Heusler compounds with eight valence electrons, such as LiAlSi and LiAlGe, are barely explored for thermoelectric materials, even though they possess a high Seebeck coefficient and a low thermal conductivity due to the rattling of Li atom [31].

The most common VEC of 18 electrons is acquired for thermoelectric half-Heusler materials. All the 18-electron compounds are semiconductors with a narrow band gap. The nature of the band gap in one formula unit of *XYZ* can be explained using MO theory [32, 33]. According to the linear combination of atomic orbitals (LCAO) method, the *s* and *p* states of *Z* atom hybridize with the *s* and *p* states of *Y* atom to form a set of four low-energy bonding orbitals, as well as a set of four high-energy antibonding orbitals (Figure 4.4(a)). The *d* orbitals of *Y* atom interact with those of

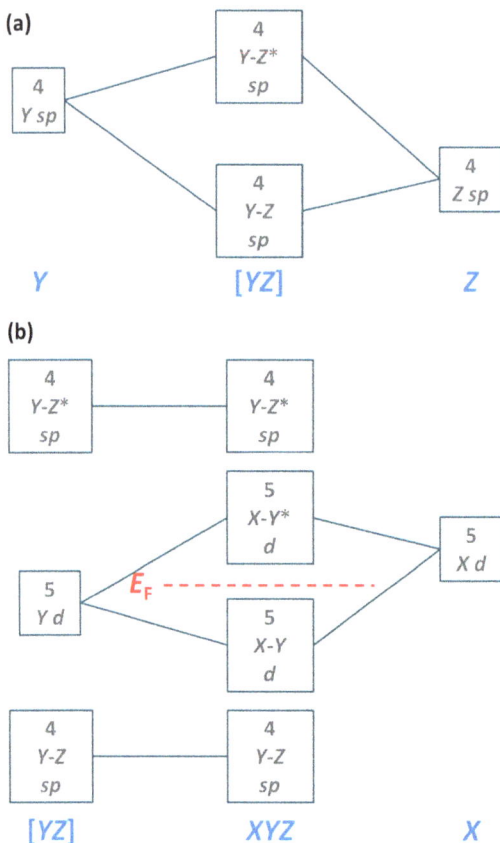

**Figure 4.4:** Schematic illustration of MO diagrams for one formula unit of *XYZ* half-Heusler compound with 18 valence electrons: (a) interactions between *s* and *p* states in *Y* and *Z* atoms; (b) full MO diagram including interactions between *d* states in *X* and *Y* atoms. $E_F$ denotes the Fermi level.

*X* atom to generate one set of five bonding orbitals and one set of five antibonding ones (Figure 4.4(b)). The *d* orbitals of *Y* atom exhibit a tetrahedral splitting, while those of *X* atom display an octahedral splitting, both are due to the *Z* surroundings, which has been neglected for simplification in Figure 4.4. Thus, there are a total of nine bonding MOs to be filled with 18 electrons. As a result, the 18-electron compounds are electronically stable since all the bonding states are occupied, while all the antibonding ones are empty. Consequently, the HOMO has mainly the character of the *Y d* orbitals, and the LUMO has that of the *X d* orbitals. Therefore, the size of the band gap depends on the electronegativity difference between *X* and *Y* elements under the influences of the *Z* element, which has been verified by linear muffin-tin orbital calculations [28]. Theoretical calculations based on nonspin-polarized electronic band structure have confirmed the MO theory for half-Heusler compounds

[34]. There exists strong covalent bonding between $Y$ and $Z$ atoms and ionic interactions between $X$ and $[YZ]$ substructure, which has been supported by the charge density calculations [8].

Half-Heusler compounds are potential candidates for thermoelectric applications in the medium to high temperature ranges. They are not only environmentally friendly and have sufficient thermoelectric performance [35], but also possess high mechanical strength and thermal stability, which make the TE module manufacturing feasible. Theoretically, the shorter the thermoelectric legs are, the higher the power density is, but the thermal stress acting on them also increases with decreasing leg length, resulting in a shorter device lifetime. Therefore, half-Heusler materials with their exceptional strength [36] (compared to $Bi_2Te_3$ or skutterudites) are very attractive for developing thermoelectric modules with both high power density and long lifetime. Moreover, for thermoelectric applications at high temperature, these compounds possess a performance comparable to SiGe and PbTe, but are less expensive due to the absence of germanium and tellurium [37].

Approximately a decade ago, Yang et al. [38] have thoroughly studied the electronic structure and electrical transport properties of 36 half-Heusler compounds, from which the potential candidates for thermoelectric materials have been suggested for experimental investigations. Later, Carrete and coworkers [39] have performed a high-throughput calculation of over 79,057 half-Heusler compounds, from which 75 of mechanically and thermodynamically stable materials have been selected for the evaluation of lattice thermal conductivity. Subsequently, the electrical transport properties of these compounds with nanograined structure have been investigated using an ab initio approach [40]. The results have shown that about 15% of those are predicted to possess $zT > 2$ at high temperature.

Half-Heusler compounds offer many opportunities for tuning electrical and thermal properties through substitutions at the three lattice sites separately or simultaneously. Usually, the substitutions at the $X$ and $Y$ sites induce mass fluctuation and strain field effects due to atomic size differences leading to a reduction in lattice thermal conductivity, while those at the $Z$ site alter the carrier concentration. Changing chemical compositions in half-Heusler compounds can tune the band gap from 0 eV to approximately 4 eV [16]. Also, a small amount of disorder due to the atomic distribution on different lattice sites significantly affects the electronic structure, which consequently modifies the transport properties [41, 42]. Electronic structure calculations in half-Heusler compounds performed by Tobola et al. [43] predicted metal-like behavior to semiconductor-like behavior and then again to metal-like behavior transitions with changing compositions due to alternating the total VECs. The theory was then confirmed from electrical property measurements for several materials.

Generally, half-Heusler compounds possess relatively high electrical conductivity and Seebeck coefficient. However, their performance has been at a moderate level for a long time due to the high thermal conductivity [8]. Intensive investigations to improve the performance of half-Heusler materials have been carried out over the years. Recently, Yang et al. [44] have thoroughly reviewed the strategies for tuning transport properties in thermoelectric materials. The authors have particularly mentioned the successful usage of energy barrier scattering in half-Heusler compounds with metallic full-Heusler inclusions to enhance the thermopower [21], and that of efficient nanostructuring in half-Heusler alloys, thanks to their long phonon mean free path. Moreover, Huang et al. [8] have summarized the approaches to enhance thermoelectric performance of half-Heusler alloys, leading to a prediction that preservation of the nanopowders' original grain sizes during the hot pressing step or introduction of alloy scattering will further reduce thermal conductivity, and a decrease in the usage of hafnium without degrading thermoelectric performance will enhance the cost-effectiveness.

Currently, the highest $zT$ value for $p$-type half-Heusler materials has been reported to be around 1.6 at 1,200 K for the compositions of $(Nb_{1-x}Ta_x)_{0.8}Ti_{0.2}FeSb$ ($x = 0.36$ and $0.40$) [45], and for $n$-type ones to reach 1.5 at 825 K for the materials of $Ti_{0.5}Zr_{0.25}Hf_{0.25}NiSn + 1$ wt.% densification aid [46], and 1.5 at 700 K for the $(Zr_{0.5}Hf_{0.5})_{0.5}Ti_{0.5}NiSn_{0.998}Sb_{0.002}$ composition [47]. The $X$NiSn and $X$CoSb systems, where $X$ is Ti, Zr, or Hf, have been thoroughly investigated for thermoelectric materials, and will be discussed in detail later in Section 4.2 and 4.3, respectively. Other materials with recently reported high $zT$ values such as NbFeSb and ZrCoBi are presented in Section 4.4.

# 4.2 $X$NiSn-based half-Heusler compounds ($X$ = Ti, Zr, Hf)

As previously mentioned, $X$NiSn ($X$ = Ti, Zr, Hf) is formulated according to the electronegativity of the elements, where $X$ ($X$ = Ti, Zr, Hf), Ni, and Sn occupy the Wyckoff positions of 4a (1/2, 1/2, 1/2), 4b (1/4, 1/4, 1/4), and 4c (0, 0, 0), respectively (Figure 4.1(b)). Considering the $s$ and $d$ orbitals of $X$ ($X$ = Ti, Zr, Hf) (four valence electrons) and Ni (10 valence electrons), as well as $s$ and $p$ orbitals of Sn (four valence electrons) per formula unit, all these compounds possess a total of 18 valence electrons, which makes them thermodynamically stable with a semiconducting character [16].

The experimental investigation combined with the CALPHAD modeling on the Ti–Ni–Sn phase diagram performed by Gürth et al. [48] revealed the formation of four ternary compounds. Among those four, the $TiNi_{1+x}Sn$ half-Heusler phase exhibited a limited solubility range with respect to Ni ($0.02 \le x \le 0.06$ at 950 °C), whereas the full-Heusler $TiNi_2Sn$ compound shows a large homogeneity range with respect to the Ti, Ni, and Sn variation. On the contrary to the theoretical results obtained by Page et al. [49] which showed low solubility of excess Ni in $X$NiSn ($X$ = Ti, Zr, Hf) half-Heusler alloys, an experimental study by Tang et al. [50] confirmed the approximate excess amount of Ni ($0.00 \leq x \leq 0.06$ at 950 °C) in $TiNi_{1+x}Sn$. The electron probe microanalysis (EPMA) results corroborated that the excess value even extends to $0.00 \leq x \leq 0.25$ at 1,450 °C for $XNi_{1+x}Sn$ ($X = Zr$, Hf) [51–53]. Since the discovery of $n$-type semiconductor properties of TiNiSn in the mid-1980s [54], numerous investigations were performed on the thermoelectric properties of this compound. Environmentally friendly and cost-effective nature of the constituent elements are among the main reasons that have attracted the attention of the researchers for thermoelectric waste heat recovery applications [55]. The TiNiSn-based material's stability and performance make them suitable for applications with hot side temperatures between 700 and 900 K.

So far, to obtain TiNiSn-based compounds with high purity, solid state reaction synthesis [56–59], mechanical alloying [60, 61], combustion synthesis [62, 63], microwave heating [64–66], and various melting processes [67–74] have been investigated. Since the mechanical alloying process cannot provide enough energy to successfully produce pure TiNiSn-based compounds because of the high melting points (above 1,700 K) of the involved elements such as Ti, Zr, Hf, and Ni, engagement of a high temperature process seems inevitable. Hitherto, among the high temperature processes, arc melting of the pure elements under Ar [47, 75–92] is the most extensively utilized method. It is worth mentioning that using the stoichiometric amounts of elements is not the right approach due to the high vapor pressure of some constituent elements such as Sn that can evaporate conspicuously during the melting. Therefore, excess amounts of high vapor pressure elements need to be added to recompense the loss. Although the arc melting process is usually applied more than once on the ingot to improve the homogeneity, the final product may still contain congruently melting side phases, such as full-Heusler $TiNi_2Sn$, $Ti_6Sn_5$, and unreacted elemental Sn [93–95] due to the incongruent melting character of TiNiSn-based compounds. Therefore, the melting process is commonly followed by a high temperature (above 1,000 K) long annealing step to achieve a high phase purity [89, 93–95].

The synthesis route seems to become more crucial in controlling microstructure formation while synthesizing $(Ti_{1-x-y}Zr_xHf_y)NiSn$ solid solutions. The melting step creates Ti-rich and Ti-poor regions during solidification that are kinetically trapped by the low diffusion rate of Ti, Zr, and Hf and are stable upon heat treatment [96]. Based on the calculated pseudo-ternary phase diagram, the solubility of Ti in $(Zr_xHf_y)NiSn$ is governed by temperature and the Zr to Hf ratio. Due to the larger lattice mismatch between the Zr-rich and Ti-rich half-Heusler phases compared to the Ti-rich and Hf-rich alloys, the highest solubility of Ti in Zr-rich phase is obtained at higher temperature (above 850 K) in comparison to 500 K for the Hf-rich phase. The high temperature stability of the microstructures with varying Ti, Zr, and Hf content can arise from the combination of low mobilities of Ti, Zr, and Hf together with the small Ti vacancy concentration and a large hopping barrier (1.9–3 eV), which make vacancy-mediated hopping mechanism a highly unlikely scenario [96]. A recent experimental study by Schwall et al. [97] on the $(Ti_{1-x-y}Zr_xHf_y)NiSn$ system revealed that it is not a full solid solution system. The synthesis method used in the study was a combination of multiple arc melting of stoichiometric ratios of the elements and 1 week annealing at 1,223 K in an evacuated quartz ampule. It was found out that in total, stable half-Heusler phases could be summarized into 15 distinct compositions, and their formation is governed by the covalent contribution of the constituent elements to the bonding dominated by their chemical character rather than the atomic sizes [97]. The list of the detected stable (Ti,Zr,Hf)NiSn-based half-Heusler phases is as follows: TiNiSn, ZrNiSn, HfNiSn, $Zr_{0.78}Hf_{0.22}$NiSn, $Zr_{0.55}Hf_{0.45}$NiSn, $Zr_{0.36}Hf_{0.64}$NiSn, $Ti_{0.81}Zr_{0.19}$NiSn, $Ti_{0.66}Zr_{0.34}$NiSn, $Ti_{0.29}Zr_{0.71}$NiSn, $Ti_{0.83}Hf_{0.17}$NiSn, $Ti_{0.64}Hf_{0.36}$NiSn, $Ti_{0.24}Hf_{0.76}$NiSn, $Ti_{0.21}Zr_{0.40}Hf_{0.39}$NiSn, $Ti_{0.43}Zr_{0.28}Hf_{0.29}$NiSn, and $Ti_{0.68}Zr_{0.17}Hf_{0.15}$NiSn (composition error range ~3%) [97].

Figure 4.5 illustrates the band structure of TiNiSn-based half-Heusler materials. The origin of the formation of the band gap in these compounds comes from the interaction between the $d$ atomic orbitals of Ni and the $d$ atomic orbitals of Ti, Zr, or Hf, respectively, which results in the splitting of the $d$ orbitals into bonding and antibonding MOs, separated from each other by an energy gap [28, 33]. The top of the valence band is dominated by the Ni $d$ states, and the major contributions to the bottom of the conduction band stem from the $d$ orbitals of Ti, Zr, or Hf.

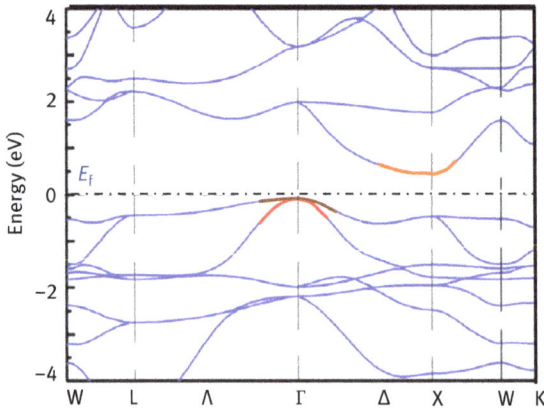

**Figure 4.5:** Band structure of TiNiSn-based half-Heusler compound (adapted with permission from [101]).

The hybridization of $d$ orbitals between Ti, Zr, or Hf and Ni determines the size of the band gap. Since the electronegativity of Ti, Zr, and Hf are quite comparable, all three compounds exhibit a similar calculated band gap of around 0.5 eV [98]. However, by moving from Ti to Hf, the experimental band gap values vary from ~0.12 to ~0.22 eV, respectively [83]. The smaller observed empirical band gap can be attributed to the structural imperfections and defects such as Ni interstitials, which can introduce in-gap states [42, 99, 100].

With respect to the band structure, all the undoped $X$NiSn-based half-Heusler compounds ($X$ = Ti, Zr, Hf) demonstrate a semiconducting behavior. The electrical conductivity of the samples at room temperature is between $100\,\Omega^{-1}\cdot\text{cm}^{-1}$ for TiNiSn and $77\,\Omega^{-1}\cdot\text{cm}^{-1}$ for HfNiSn [67, 102]. The electrical conductivity of all the samples increases with increasing temperature which is a typical behavior of nondegenerate semiconductors. The Seebeck coefficient of $X$NiSn-based half-Heusler compounds ($X$ = Ti, Zr, Hf) increases with increasing temperature, showing an intrinsic $n$-type character with negative room temperature values ranging between $-124\,\mu\text{V}\cdot\text{K}^{-1}$ for HfNiSn and $-176\,\mu\text{V}\cdot\text{K}^{-1}$ for ZrNiSn [82]. While the thermal conductivity of all the samples decreases with increasing temperature, all of the undoped ternary compounds exhibit a relatively high thermal conductivity at room temperature typically above $5\,\text{W}\cdot\text{m}^{-1}\cdot\text{K}^{-1}$ [67, 81, 82]. The maximum reported thermoelectric figure of merit $zT_{max}$ for the undoped ternary samples is roughly from 0.3 to 0.4 at 750 K depending on the synthesis technique [64, 103].

Comparing the above-mentioned thermoelectric properties to the current state-of-the-art materials, undoped TiNiSn-based ternary half-Heusler compounds manifest a low electrical conductivity and a high thermal conductivity, which leads to a low figure of merit, $zT$, even at high temperatures (above 700 K). Enhancing the electrical conductivity can be accomplished through increasing the charge carrier density via doping. The doping can be applied on all three $X$, Ni, and Sn atomic positions. Thus far, for the $n$-type dopant on the $X$ site ($X$ = Ti, Zr, Hf), Sc [104, 105], Y [106], La [107], Ce [108], Lu [109], Mn [110, 111], V [46, 82, 112], Nb [46, 82, 113, 114], and Ta [115, 116]; on the Ni site, Cr and Fe [82, 117], Co [118–120], Pd [92], Pt [121], and Cu [118]; and also on the Sn site, In [81, 82], Ge [122], Sb [47, 57, 66, 106, 123–130], and Bi [131, 132] were investigated. Among all the dopants, Sb appeared to be the most efficient and prominent one. Due to the intrinsic $n$-type character of TiNiSn-based half-Heusler compounds, $p$-type doping attempts such as Lu or Y on the Ti site [82, 109], Ga or In on the Sn site [82] were not as successful in achieving high efficiency thermoelectric materials. Although doping ZrNiSn with Co and Ir was successful in making $p$-type material, the thermoelectric properties are not promising considering the obtained $zT_{max}$ of 0.3 at 1,000 K.

In addition to doping, carrier density can be increased by the in situ formation of the full-Heusler metallic phases [57, 125], which can act as carrier donors to the half-Heusler matrix. Due to the compensating effect on the Seebeck coefficient, tuning electrical conductivity through carrier concentration should be utilized with caution. According to the previous studies, the optimized $n$-type doping concentration for Sb on $X$NiSn-based half-Heusler compounds is usually between 0.5 and 2 at.% [46, 133, 134]. The optimized doping content can increase the carrier concentration of undoped sample from $\sim10^{19}$ to $\sim10^{21}\,\text{cm}^{-3}$ [103]. The room temperature electrical conductivity and Seebeck values of some $n$-type materials are depicted in Table 4.1. As mentioned, doping with Sb can significantly increase the electrical conductivity (by a factor of 10).

**Table 4.1:** Electrical conductivity and Seebeck coefficient of some *n*-type TiNiSn-based half-Heusler materials at room temperature.

| Reference | Composition | Electrical conductivity ($\Omega^{-1}\cdot cm^{-1}$) | Seebeck coefficient ($\mu V\cdot K^{-1}$) |
|---|---|---|---|
| [64] | TiNiSn | 150 | −140 |
| [50] | $TiNiSn_{0.99}$ | 200 | −140 |
| [50] | $TiNi_{1.06}Sn_{0.98}$ | 160 | −198 |
| [126] | $Ti_{0.5}Zr_{0.5}NiSn$ | 227 | −280 |
| [126] | $Ti_{0.5}Hf_{0.5}NiSn$ | 250 | −220 |
| [101] | $Zr_{0.5}Hf_{0.5}NiSn$ | 500 | −120 |
| [135] | $Ti_{0.5}Zr_{0.25}Hf_{0.25}NiSn$ | 196 | −306 |
| [50] | $TiNiSn_{0.979}Sb_{0.011}$ | 1,440 | −113 |
| [126] | $Ti_{0.5}Zr_{0.5}NiSn_{0.98}Sb_{0.02}$ | 2,941 | −81 |
| [103] | $Ti_{0.8}Hf_{0.2}NiSn_{0.99}Sb_{0.01}$ | 2,200 | −105 |
| [81] | $Zr_{0.5}Hf_{0.5}NiSn_{0.99}Sb_{0.01}$ | 1,250 | −196 |
| [47] | $Ti_{0.5}Zr_{0.25}Hf_{0.25}NiSn_{0.99}Sb_{0.01}$ | 1,200 | −150 |

The undoped *X*NiSn-based half-Heusler compounds generally have high Seebeck coefficient, which is assigned to the sharp states near the Fermi energy $E_F$ in the density of states (DOS) originating from the hybridized *d* orbitals [33, 38, 136]. Nanosized crystals in (Hf, Zr)NiSn appeared to have the same effect near the $E_F$ in the DOS due to an increase in cell volume that decreases the orbital overlap between the atoms [85] and antisite defects in ZrNiSn [87]. In contrast to Ta and Nb, 1 at.% of V doping in (Hf,Zr)NiSn [137] enhances the effective mass $m^\star$ (from $2.16\cdot m_e$ to $2.6\cdot m_e$) and thus the Seebeck coefficient possibly due to introducing a resonant state near the Fermi energy.

Although high thermal conductivity might look like the Achilles heel of *X*NiSn-based half-Heusler compounds once compared with other conventional classes of thermoelectric materials such as tellurides, clathrates, and chalcogenides, there are strategies that can quite significantly reduce the thermal conductivity. With respect to the high symmetry crystal structure that facilitates the phonon transport, the main contribution to the thermal conductivity in half-Heusler materials originates from the lattice portion. Traditionally, to increase the phonon scattering, isoelectronic substitution of heavy elements is considered to be an effective approach due to introducing disorder in the form of mass or strain field fluctuation in the lattice. For instance, substituting Hf and Zr on Ti site or Pt and Pd on Ni site can lead to a significant reduction (up to 40%) of thermal conductivity [92, 121]. From a practical point of view, the Ti site substitution is more favorable and cost-effective for the large-scale industrial application. Another approach to increase the phonon scattering is through increasing boundary scattering via increasing the grain boundary density. It should be taken into account that, depending on the grain sizes, introducing more boundaries can also affect the mobility of charge carriers and consequently the electrical conductivity. In Sb-doped TiNiSn samples, grain boundary scattering seemed to have a major effect on reducing the room temperature lattice thermal conductivity (~3.7 W·m$^{-1}$·K$^{-1}$) once the grain sizes are below 20 μm; on the other hand, for samples with larger grain sizes, the lattice thermal conductivity (~9.4 W·m$^{-1}$·K$^{-1}$) appeared to be primarily affected by alloy and impurity scatterings [130]. The same effect on thermal conductivity is also observed on ball-milled $Zr_{0.25}Hf_{0.75}NiSn_{0.99}Sb_{0.01}$ sample with nanometer size grains (~200 nm) [79]. Besides nanostructuring [138], additional grain boundaries can be incorporated via nanoparticle inclusion. For example, addition of high temperature stable oxide such as $ZrO_2$ from 3 to 9 vol.% to $Zr_{0.5}Hf_{0.5}Ni_{0.8}Pd_{0.2}Sn_{0.99}Sb_{0.01}$ [139] consistently reduced the thermal conductivity from ~4.5 to ~3.4 W·m$^{-1}$·K$^{-1}$ at room temperature. Other nanoinclusions such as y-$Al_2O_3$ [140] and metallic full-Heusler phases [57, 125] were also shown to act as phonon scattering centers and reduce the thermal conductivity. Figure 4.6 illustrates the thermal conductivity of various XNiSn-based half-Heusler compounds.

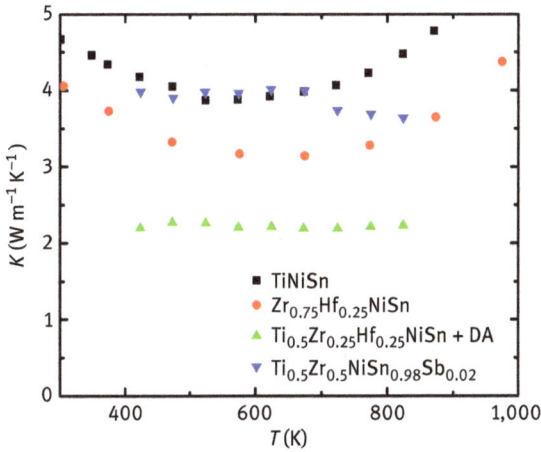

Figure 4.6: Thermal conductivity of various XNiSn (X = Ti, Zr, Hf)-based half-Heusler compounds [46, 64, 114]; DA denotes densification aid.

Excess amount of Ni in TiNiSn-based half-Heusler materials can lead to either in situ full-Heusler formation [141, 142] or randomly filled interstitial positions [50], which in both cases can enhance phonon scattering, thus reducing the thermal conductivity by ~30% and ~55%, respectively. In situ full-Heusler phases [57] also showed to have a constructive effect on the electronic transport properties, especially at high temperatures due to their metallic character and coherent boundaries, which sustain the carrier mobility.

The combination of all properties manifests itself in the thermoelectric figure of merit, zT. Figure 4.7 exhibits some of the hitherto reported high-performance n-type

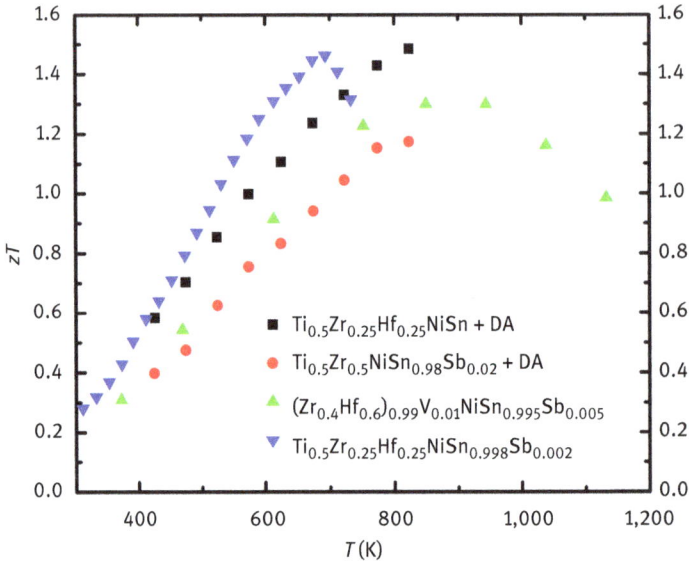

**Figure 4.7:** Figure of merit, $zT$, of some state-of-the-art $X$NiSn ($X$ = Ti, Zr, Hf)-based half-Heusler compounds [46, 47, 137]; DA denotes densification aid.

$X$NiSn-based half-Heusler compounds. While the $zT$ of 1.5 at 700 K for the Sb-doped $Ti_{0.5}Zr_{0.25}Hf_{0.25}$NiSn composition [47] has not been achieved at the same temperature, a similar value was recently reported by Rogl et al. at 825 K [46].

# 4.3 $X$CoSb-based half-Heusler compounds ($X$ = Ti, Zr, Hf)

Among the most extensively studied half-Heusler compounds are $X$CoSb-based materials ($X$ = Ti, Zr, Hf). The number of valence electrons of $X$ ($X$ = Ti, Zr, Hf), Co, and Sb are 4, 9, and 5, respectively, which sums up to 18 electrons, making the compound a stable nondegenerate semiconductor. The edges of the valence band and conduction band are mainly comprised of the $d$ orbitals of Co and $X$, respectively [16]. Reported band gap sizes for TiCoSb range from 0.82 eV [143], 0.95 eV [119], to 1.06 eV [144, 145], and the value can be affected by the electronegativity difference between the transition metals, that is, $X$, Co, and their substitutes.

The combined isothermal sections of Ti–Co–Sb system at 870 and 1,070 K investigated by Stadnyk et al. [146] are shown in Figure 4.8. As demonstrated, TiCoSb and $Ti_5Co_{0.46}Sb_{2.54}$ are the two stable phases at those temperatures. Since no stable $TiCo_2Sb$ full-Heusler phase is observed and reported for this system, excess Co only results in the in situ formation of Ti–Co and Co–Sb binary phases. It should be

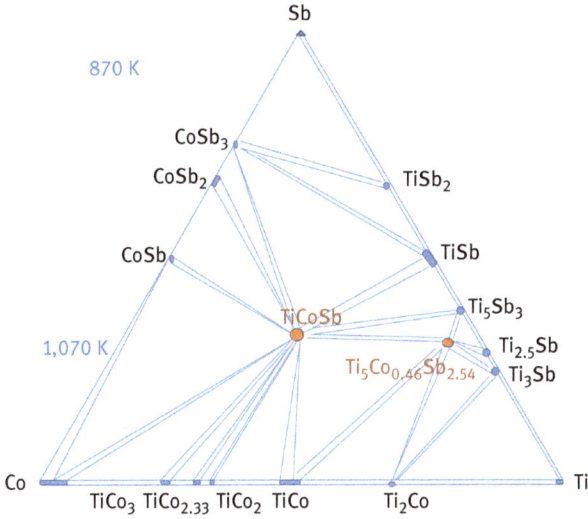

**Figure 4.8:** The isothermal sections of Ti–Co–Sb system at 870 and 1,070K (reproduced with permission from [146]).

noted that the information regarding the binary system of Ti–Sb was taken from a previously reported work of Massalski [147], but the recent phase diagram constructed by Tavassoli et al. [148] shows the existence of $Ti_2Sb$ and $Ti_{11}Sb_8$ phases while the $Ti_{2.5}Sb$ phase was not observed. Due to the high melting points of Ti (1,943 K) and Co (1,768 K), most of the synthesis methods include a melting process, such as arc melting [149–152] or induction melting [153], proceeded by an annealing step [154]. In addition to the melting methods, pure TiCoSb phase seems to form after 12 h of mechanical alloying [155], or even promptly after around 1 min of microwave synthesis [64]. Unfortunately, the metallic CoSb phase is detected on pressed samples prepared by the previously mentioned unconventional techniques, which led to inferior thermoelectric properties [64, 155, 156].

Although moving from Ti to Hf in XCoSb-based half-Heusler materials enhances the room temperature electrical conductivity from ~20 to ~200 $\Omega^{-1} \cdot cm^{-1}$ [157], the values are still not satisfactory for thermoelectric applications. Taking into account the low carrier concentration in the order of $10^{18}$ $cm^{-3}$ [158], the electrical conductivity can be improved via tuning the charge carrier concentration. To increase the carrier concentration, doping with Sc (theoretical evaluation) [159], V [151, 160, 161], Nb [162, 163], and Ta [158] on the Ti site; Fe [164, 165], Ni [149, 152, 166], Pd [167], and Pt [157] on the Co site; and Ge [153] and Sn [133, 168–175] on the Sb site have already been investigated. Even though TiCoSb-based half-Heusler materials exhibit intrinsic room temperature n-type conductivity [176, 177], they can successfully be changed into a p-type conductor by partial substitution with

Fe, Ge, or Sn. In fact, Sn-doped samples are the most explored $p$-type TiCoSb-based half-Heusler materials. Besides carrier concentration, boosting carrier mobility by in situ nanodomains inclusion such as InSb [164, 178], full-Heusler phase [179, 180], and ex situ modulation doping such as incorporating $Cu_2Te$-based chalcogenides [181] lead to the enhancement of electrical conductivity. For instance, 25 wt.% addition of $Cu_{1.96}Ni_{0.04}Se_{0.03}Te_{0.97}$ to $Hf_{0.7}Ti_{0.3}CoSn_{0.2}Sb_{0.8}$ sample resulted in simultaneous increase in room temperature electrical conductivity (from ~220 to ~475 $\Omega^{-1}\cdot cm^{-1}$) and the Hall mobility (from 18.6 to 22.4 $cm^2\cdot V^{-1}\cdot s^{-1}$). Table 4.2 presents some of the highest electrical transport properties reported so far for $XCoSb$-based half-Heusler materials ($X$ = Ti, Zr, Hf). $ZrCoSn_{0.15}Sb_{0.85}$ [175] demonstrates one of the highest values for electrical conductivity with ~1,700 $\Omega^{-1}\cdot cm^{-1}$ in comparison to the values of ~1,400 $\Omega^{-1}\cdot cm^{-1}$ for $Hf_{0.5}Zr_{0.5}CoSn_{0.3}Sb_{0.7}$ [168] and ~1,200 $\Omega^{-1}\cdot cm^{-1}$ for $HfCoSn_{0.15}Sb_{0.85}$ [133], which could be attributed to its high carrier concentration. Interestingly, the Sn-doped $Zr_{0.5}Hf_{0.5}CoSb$ solid solution shows a higher mobility than that of the ones only containing Zr or Hf.

Table 4.2: Electrical conductivity, carrier concentration and mobility of some $XCoSb$-based half-Heusler materials ($X$ = Ti, Zr, Hf) at room temperature [133, 168, 175].

| Nominal composition | Carrier concentration $(cm^{-3})$ | Mobility $(cm^2\ V^{-1}\ s^{-1})$ | Electrical conductivity $(\Omega^{-1}\ cm^{-1})$ |
|---|---|---|---|
| $ZrCoSn_{0.15}Sb_{0.85}$ | $2.37 \times 10^{21}$ | 4.4 | ~1,700 |
| $HfCoSn_{0.15}Sb_{0.85}$ | $1.48 \times 10^{21}$ | 5.0 | ~1,200 |
| $Zr_{0.5}Hf_{0.5}CoSn_{0.3}Sb_{0.7}$ | $6.70 \times 10^{20}$ | 13.0 | ~1,400 |

Unlike the $XNiSn$-based half-Heusler compounds, $XCoSb$-based compounds ($X$ = Ti, Zr, Hf) show different tendency for the Seebeck coefficient. Although all the $XCoSb$-based materials exhibit a negative Seebeck coefficient value at room temperature, the absolute Seebeck coefficient is drastically decreasing from ~ −300 $\mu V\cdot K^{-1}$ for TiCoSb to ~−70 $\mu V\cdot K^{-1}$ and ~−5 $\mu V\cdot K^{-1}$ for ZrCoSb and HfCoSb, respectively [64, 176, 182]. Alloying seems to improve the Seebeck coefficient value of (Zr,Hf)CoSb alloys. One way to improve the Seebeck coefficient while increasing the carrier concentration is through DOS modification near the Fermi energy $E_F$. Since the DOS of the valence band near the Fermi level mainly consists of the $d$ character of Co, a small Ni substitution (≤1 at.%) appeared to initiate an impurity state [183] near the $E_F$, hence, improving the Seebeck coefficient. For instance, substituting Ni in $Ti_{0.5}Zr_{0.25}Hf_{0.25}CoSb$ solid solution can simultaneously increase the carrier concentration and room temperature Seebeck coefficient (Table 4.3) [166].

Scattering the low energy carriers while allowing the high energy ones to contribute to the conduction process, known as "energy filtering," seemed to be an

**Table 4.3:** Carrier concentration and Seebeck coefficient of Ni-doped $Ti_{0.5}Zr_{0.25}Hf_{0.25}CoSb$ compounds at room temperature [149].

| Nominal composition | Carrier concentration $(cm^{-3})$ | Seebeck coefficient $(\mu V \cdot K^{-1})$ |
|---|---|---|
| $Ti_{0.5}Zr_{0.25}Hf_{0.25}CoSb$ | $1.1 \times 10^{19}$ | ~−40 |
| $Ti_{0.5}Zr_{0.25}Hf_{0.25}Co_{0.99}Ni_{0.01}Sb$ | $9.9 \times 10^{19}$ | ~−75 |
| $Ti_{0.5}Zr_{0.25}Hf_{0.25}Co_{0.97}Ni_{0.03}Sb$ | $4.1 \times 10^{20}$ | ~−203 |

effective method to enhance the Seebeck coefficient without deteriorating the electrical conductivity. Previous investigations showed that introducing nanophases such as in situ full-Heusler compound [184] or InSb [178] can lead to such improvement in the Seebeck coefficient. As an example, the room temperature electrical conductivity and Seebeck coefficient values of $Ti_{0.5}Zr_{0.25}Hf_{0.25}Co_{0.95}Ni_{0.05}Sb$ alloys with various InSb contents (1, 3, and 7 at.%) are shown in Table 4.4.

**Table 4.4:** Electrical conductivity and Seebeck coefficient of $Ti_{0.5}Zr_{0.25}Hf_{0.25}Co_{0.95}Ni_{0.05}Sb$ nanocomposites at room temperature [178].

| InSb content (at.%) | Electrical conductivity $(\Omega^{-1} \cdot cm^{-1})$ | Seebeck coefficient $(\mu V \cdot K^{-1})$ |
|---|---|---|
| 0% | 240 | −146 |
| 1% | 210 | −192 |
| 3% | 230 | −180 |
| 7% | 250 | −161 |

The thermal conductivity of undoped TiCoSb at 323 K is around 25 $W \cdot m^{-1} \cdot K^{-1}$, which decreases to around 21 and ~15 $W \cdot m^{-1} \cdot K^{-1}$ by moving to ZrCoSb and HfCoSb, respectively [176]. This could be due to consistent reduction in both the longitudinal $v_L$ and shear $v_S$ sound velocities once moving toward the heavier cation [176]. Indisputably, lower thermal conductivity values are required for TiCoSb-based half-Heusler materials to be employed for thermoelectric applications. Since doping and alloying with heavier and larger elements increases the phonon scattering via mass fluctuation and strain disorder, making Ti, Zr, and Hf mixed solid solution is one of the most effective methods to reduce the thermal conductivity. For instance, at 323 K, the thermal conductivity of $Ti_{0.25}Zr_{0.375}Hf_{0.375}CoSb$ sample is ~6 $W \cdot m^{-1} \cdot K^{-1}$, which is 60–75% lower than that of the nonalloyed ones [176]. In addition, this is not confined to the elements substituting Ti as previous studies [184, 185] revealed that replacing Co by Rh in TiCoSb-based half-Heusler materials can significantly reduce the room temperature lattice thermal conductivity (from ~18 to ~7 $W \cdot m^{-1} \cdot K^{-1}$). Introducing nonisoelectronic doping such as Ta on Ti site [158], Ni on Co site [166, 186], Ge [153] and Sn [174] on Sb site also appears to systematically reduce the thermal conductivity of the TiCoSb-based compounds. In addition to alloying, nanostructuring via ball milling increased the grain boundary scattering, which led to more than 30% reduction (from ~3.7 to ~2.2 $W \cdot m^{-1} \cdot K^{-1}$) in the lattice thermal conductivity of $Zr_{0.5}Hf_{0.5}CoSn_{0.2}Sb_{0.8}$ at 700 K [187]. Grain boundary scattering can also be augmented by adding nanoparticles such as $HfO_2$ [188] and $ZrO_2$ [189], which can act as scattering centers to the bulk half-Heusler matrix. Up until now, one of the lowest reported values of thermal conductivity in this system belongs to $Ti_{0.5}Hf_{0.5}CoSn_{0.15}Sb_{0.85}$, ranging from ~3 $W \cdot m^{-1} \cdot K^{-1}$ at 400 K to ~2 $W \cdot m^{-1} \cdot K^{-1}$ at 900 K [169]. The rationale behind such a relatively low thermal conductivity is due to its peculiar microstructure forming via phase segregation of this composition to $Ti_{0.42}Hf_{0.57}Co_{1.02}Sn_{0.05}Sb_{0.94}$ and $Ti_{0.82}Hf_{0.21}Co_{1.05}Sn_{0.57}Sb_{0.35}$ half-Heusler phases [169].

To summarize this section, the thermoelectric figure-of-merit values, $zT$, of some high efficiency XCoSb-based half-Heusler materials (X = Ti, Zr, Hf) are depicted in Figure 4.9. All the compounds in Figure 4.9 show a maximum $zT_{max}$ larger than unity, with the Sn-doped $Ti_{0.5}Hf_{0.5}CoSb$ demonstrating the highest value of ~1.3 at 900 K. Both nanostructured Sn-doped $Ti_{0.2}Hf_{0.8}CoSb$ and $Ti_{0.12}Zr_{0.44}Hf_{0.44}CoSb$ samples have their maximum $zT_{max}$ of ~1 at 1,073 K.

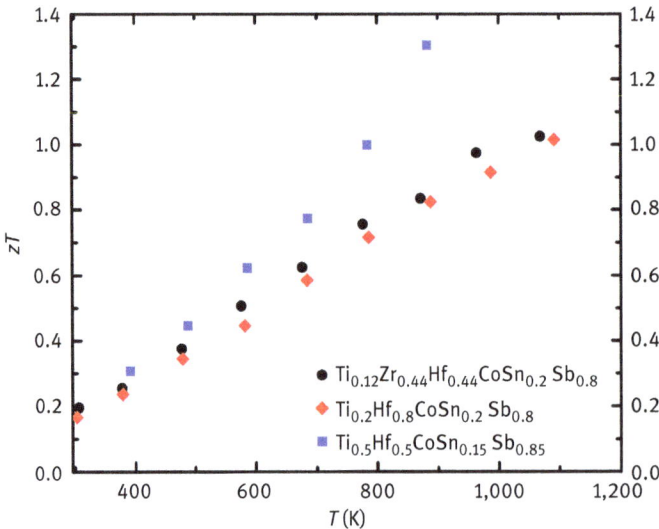

**Figure 4.9:** Thermoelectric figure of merit, $zT$, of some high efficiency XCoSb-based half-Heusler materials (X = Ti, Zr, Hf) [169, 173, 190].

## 4.4 Recent high-performance half-Heusler compounds

This section is dedicated to the quite recent high-performance half-Heusler thermo-electric materials. Among them, NbFeSb-based compounds have been the focus of attention in recent years. Following the half-Heusler structure previously mentioned, Nb, Fe, and Sb occupy the 4a, 4b, and 4c Wyckoff positions, respectively. Considering the contribution from Nb (five valence electrons) and Fe (eight valence electrons), together with the $s$ and $p$ orbitals of Sb (five valence electrons), the compound complies with the 18-electron rule, which exhibits nondegenerate semiconducting behavior in an undoped state [32].

The NbFeSb compound exhibits a high-temperature stable single-phase composition (lattice parameter of ~5.95 Å [191–193]) with a low tolerance toward off-stoichiometric variation. An early study by Lomnitska suggested the high homogeneity of $Nb_{0.3}Fe_{0.4}Sb_{0.3}$ phase with the random occupation of Fe in Nb and Sb positions [194]. Later, an EPMA investigation by Tavassoli et al. revealed that $Nb_{0.3}Fe_{0.4}Sb_{0.3}$ composition does not exist as a homogeneous phase but is a combination of $Nb_{33.4}Fe_{33.3}Sb_{33.3}$ half-Heusler phase and an Fe-rich-based solid solution $Nb_{0.1}Fe_{96.8}Sb_{3.1}$ [195]. In case of excess Nb, $NbFe_2$ and $Nb_3Sb$ were observed, while $NbSb_2$ was detected with an excess amount of Sb after annealing at 873 K [195]. It is worth mentioning that one of the main differences between NbFeSb and TiNiSn system is the nonexistence of the $NbFe_2Sb$ full-Heusler phase compared to $TiNi_2Sn$ [194, 196].

Thus far, the arc melting [197] and levitation melting [45] followed by a milling or annealing step were utilized for the synthesis of NbFeSb-based half-Heusler compounds. Due to the high vapor pressure of Sb at elevated temperatures, an excess amount of Sb needs to be used to compensate for the loss [37]. In an undoped state, NbFeSb exhibits mainly an intrinsic $p$-type character [198], with a relatively narrow indirect band gap of ~0.54 eV [199, 200]. Figure 4.10 illustrates the band structure and DOS of NbFeSb constructed from the first principle calculation [193]. As shown in the DOS, the valence band maximum and conduction band minimum are dominated by the $d$ states of Fe and Nb, respectively, with the overall higher DOS near the valence band maximum than that of the conduction band minimum [193]. Studies showed that introducing Ta or V on the Nb site can affect the band structure, hence, the size of the band gap. For instance, increasing the V substitution in an undoped NbFeSb sample can reduce the $E_g$ to 0.34 eV [199]. Indeed, due to the character of the valence band maximum, this phenomenon is not limited to the Nb site, and calculation for substituting Ru on the Fe site also showed a reduction in the band gap to ~0.36 eV [201]. Consequently, it is worth examining the effect of doping in addition to cosubstitution on the Nb and Fe sites on the band structure and thermoelectric properties of NbFeSb-based materials.

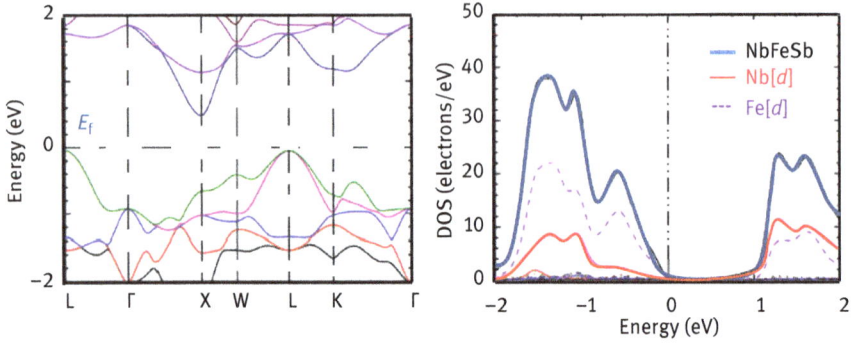

**Figure 4.10:** Band structure and density of states (DOS) of NbFeSb (adapted with permission from [193]).

NbFeSb exhibits an electrical conductivity lower than 20 $\Omega^{-1}\cdot cm^{-1}$ at room temperature, which is not comparable to the high-efficiency thermoelectric materials. In order to enhance the electrical conductivity, doping on various sites can be applied to increase the charge carrier concentration. So far, Sc [202], Ce [203], Ti [198, 204–207], Zr [208], Hf [209, 210], and Mo[197, 211], on the Nb site; Mn [212], Co [197, 213], and Ir [214] on the Fe site; and Sn [215, 216] on the Sb site were investigated. Except for Co and Ir, all other dopants act as a hole donor, hence leading to *p*-type conductors. Among all the dopants, Hf demonstrated to be the most auspicious one with the best performance achieved by $Nb_{0.88}Hf_{0.12}FeSb$. The reason lies in the fact that Hf seems to have higher doping efficiency than Zr and Ti, so less Hf is required to gain the same hole concentration. This could result in less alloy scattering of the charge carriers, thus higher carrier mobility at the same carrier concentration. For instance, at the same carrier concentration of ~2 × $10^{21}$ $cm^{-3}$, the room temperature electrical conductivity of $Nb_{0.88}Hf_{0.12}FeSb$ is around 6,000 $\Omega^{-1}\cdot cm^{-1}$ compared to the 4,900 $\Omega^{-1}\cdot cm^{-1}$ value of $Nb_{0.86}Zr_{0.14}FeSb$ [217]. Although depending on the synthesis procedure, there have been reports of *n*-type character (due to antiphase boundary) for undoped NbFeSb [197, 218], *n*-type doping appears not to be successful in this system. This could be attributed to the low band degeneracy ($N_V$) of only three at the conduction band minimum, which makes it less likely to obtain a large effective mass ($m^\star$) without depreciating the mobility ($\mu$) based on the following formulas:

$$m^* = N_V^{2/3} m_b^*, \qquad \mu = 1/\left(m_b^*\right)^{5/2}$$

where $m_b^*$ represents the band effective mass [193, 219].

The Seebeck coefficient of doped NbFeSb seems to be in accord with the obtained value from a single parabolic band (SPB) model with a steady effective mass of $m^\star$ = ~6.9 $m_e$. Considering the simplified Mott relation for heavily doped half-Heusler compounds, increasing the carrier concentration leads to the reduction in

Seebeck coefficient at 300 K, from ~115 to ~85 $\mu$V·K$^{-1}$ and from ~125 to ~95 $\mu$V·K$^{-1}$ for $Nb_{1-x}Hf_xFeSb$ and $Zr_xNb_{1-x}FeSb$ samples, respectively [217]. Since the valence band mainly consists of $d$ states of Fe, the dopant type and co-doping on the Nb site [210] seem to have an insignificant effect on the shape of the valence band edge; thus, the values from the SPB model coincide well with the empirical Seebeck coefficient data [203].

The thermal conductivity of undoped $X$FeSb-based half-Heusler materials ($X$ = Nb, Ta) ranges from ~9 to ~16 W·m$^{-1}$·K$^{-1}$ at room temperature, which reduces at elevated temperatures [198, 217, 220]. Undoubtedly, the aforementioned thermal conductivity values are far larger than the known state-of-the-art thermoelectric materials such as tellurides and chalcogenides [221, 222]. Thermal conductivity can be reduced via increasing the point defect scattering through introducing mass and strain field fluctuation. In this regard, Hf appears to be very beneficial both as a dopant and as a point defect scattering center due to its mass difference and large radius [217]. The thermal conductivity of $Nb_{0.86}Hf_{0.14}FeSb$ sample is around ~7.5 W·m$^{-1}$·K$^{-1}$ at 300 K, which is less than half of the value of the undoped sample [217]. It is also possible to introduce disorder through isoelectronic substitution, for example, using V [45, 223] and Ta [45] on the Nb site, and Ru (theoretically) [38, 201] on the Fe site. Although substitution of V into undoped NbFeSb resulted in roughly 60% reduction of lattice thermal conductivity, its effect is more trivial (10% reduction) in the presence of heavy dopants such as Hf [217]. The reason for that could be the majority of phonon scattering coming from the Hf-induced point defects; thus, less influence can be assigned to the alloying effect from V. On the contrary, Ta seems to be a potent substitution to increase phonon scattering and reduce lattice thermal conductivity [45]. Increasing Ta content in Ti-doped NbFeSb samples substantially reduces the lattice thermal conductivity to the extent that the $Nb_{0.48}Ta_{0.32}Ti_{0.2}FeSb$ composition exhibits the lowest lattice thermal conductivity of around 1.4 W·m$^{-1}$·K$^{-1}$, which is very close to the minimum theoretical value, $\kappa_{min}$, ~0.9 W·m$^{-1}$·K$^{-1}$ [45]. Although formation of a solid solution is beneficial in reducing the thermal conductivity, its effects on dopant solubility, dopant efficiency, band gap, and band structure modification need to be taken into consideration for optimized thermoelectric performance of the materials.

To the best of our knowledge, hitherto there is no experimental report on the thermal conductivity of isoelectronic substitution on the Fe and Sb sites. First principle calculations predicted that NbRuSb has lower thermal conductivity and higher thermoelectric power factor in comparison with the Fe counterpart, which makes it a compelling candidate to explore empirically [38, 201, 219]. In order to further reduce the lattice thermal conductivity, nanostructuring [215], grain refinement [224, 225], and hierarchical scattering [206] are effective to increase phonon scattering albeit the overall performance of the material may not always enhance due to the reduced carrier mobility. A hierarchical approach in $Nb_{0.8}Ti_{0.2}FeSb$ through creating Ti-rich nanoprecipitates inside the grains effectively enhanced phonon scattering, which led to 17% reduction in room temperature thermal conductivity without notably affecting the carrier mobility (~14 $cm^2 \cdot V^{-1} \cdot s^{-1}$ at 300 K) [206].

Figure 4.11 demonstrates the thermoelectric properties for some of the high-efficiency $X$FeSb-based materials ($X$ = Nb, Ta). The lowest total thermal conductivity of ~2.9 $W \cdot m^{-1} \cdot K^{-1}$ at 975 K belongs to the $Ti_{0.16}V_{0.1}Ta_{0.74}FeSb$ sample. At 1,000 K, the highest electrical conductivity of ~1,100 $\Omega^{-1} \cdot cm^{-1}$ and Seebeck coefficients of ~230 $\mu V \cdot K^{-1}$ are associated with $Ti_{0.2}Nb_{0.48}Ta_{0.32}FeSb$ and $Ti_{0.16}V_{0.1}Ta_{0.74}FeSb$ compositions, respectively. The combination of all the properties yields the figure of merit, $zT$, with $Nb_{0.88}Hf_{0.12}FeSb$, $Ti_{0.2}Nb_{0.48}Ta_{0.32}FeSb$, and $Ti_{0.16}V_{0.1}Ta_{0.74}FeSb$ demonstrating record $zT_{max}$ values of ~1.5, 1.6, and 1.52, correspondingly [45, 217, 220].

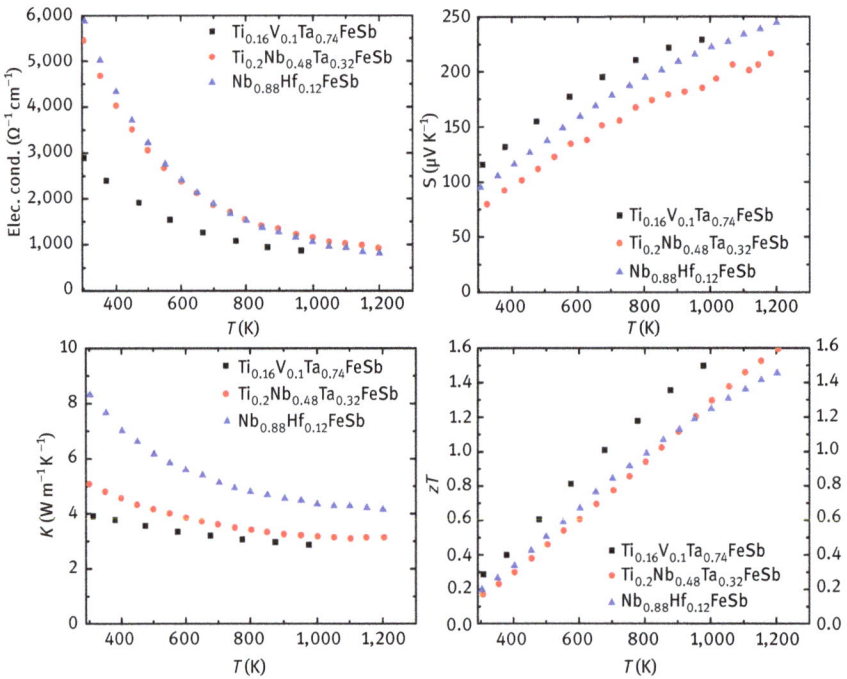

**Figure 4.11:** Thermoelectric properties for some of the high efficiency $X$FeSb-based materials ($X$ = Nb, Ta) [45, 214, 217].

It is worth mentioning that recently a $ZrCoSn_{0.2}Sb_{0.15}Bi_{0.65}$ half-Heusler compound was discovered, demonstrating an impressive $zT_{max}$ of ~1.4 at 973 K among the $p$-type half-Heusler materials [226]. This outstanding performance is the combination of large power factor originating from its high band degeneracy of $N_V$ = 10 at the valence band maximum and low thermal conductivity, which is the amalgam of both its low sound velocity (weak bonding of heavy Bi atoms), and mass fluctuation on the Bi site (with

Sn and Sb substitutions) as well as the grain boundary scattering (~250 nm average grain size). This discovery unequivocally opens new avenues for further investigation on this material system. Here, we would like to suggest the following:

- To further enhance the phonon scattering, substitution of Ti or Hf on the Zr site and Rh or Ir on the Co site can be explored.
- The possibility of *p*-type co-doping with Sc, Fe, and Ge and *n*-type doping with V, Nb, and Ta can be investigated.
- Ternary phase diagram study for the Zr–Co–Bi system to identify the possible stable phases for full-Heusler/half-Heusler composites can be considered.

## 4.5 Conclusions

An overview about half-Heusler materials for thermoelectrics at high temperature has been presented in this chapter, focusing on the systems thoroughly investigated in the past as well as current state-of-the-art compounds. A general introduction about half-Heusler alloys, including their crystal structure and electronic system, has been given at first to explain the attractiveness of this material class for thermoelectric researchers. Subsequently, the progress to enhance thermoelectric performance of TiNiSn, TiCoSb, and recent materials with high $zT$ is summarized in the next three sections, respectively.

Aside from improving thermoelectric performance of half-Heusler materials, it is worth mentioning that strong efforts on half-Heusler-based module development have been carried out. Bartholomé et al. [227] have successfully upscaled the synthesis process to produce kilogram batches of half-Heusler materials. They also obtained consistent electrical power generation and module's internal resistance from two different modules during several operational cycles under harsh conditions, for example, up to 130 times of heat treatments cycling, or a fast heating rate of 100 K min$^{-1}$, which indicated the long-term stability and reproducible module assembly process. Thermoelectric devices with various configurations ranging from unileg module [228], to unicouples [189, 229], fully assembled modules [227, 230], and thermoelectric generator (TEG) systems [231] have been developed and characterized, especially the 1 kW TEG made from nanostructured bulk half-Heusler materials has been implemented to an automotive diesel engine for recovering the exhaust waste heat [231], and eight TEGs consisting of 60 half-Heusler-based unicouples have been integrated into the combustion region of a residential boiler to illustrate a combined heat and power microsystem of ~95 W electrical power output [230]. Intensive works have also been conducted during the last 2 years for the development and integration of half-Heusler modules into luxury German vehicles [232]. The search for a good electrode material and an effective joining method for the production of half-Heusler-based thermoelectric

Table 4.5: A summary of half-Heusler-based thermoelectric modules.

| Reference (Year) | Materials | | Electrode | Number of legs | Efficiency @ $\Delta T$ |
|---|---|---|---|---|---|
| | n-Type | p-Type | | | |
| [189] (2011) | Nanostructured $Zr_{0.4}Hf_{0.6}NiSn_{0.995}Sb_{0.005}$ | Nanostructured $Zr_{0.7}Hf_{0.3}CoSn_{0.3}Sb_{0.7}$ + nano-$ZrO_2$ | – | 2 | 8.7% @ 657 K |
| [227] (2014) | $Zr_{0.4}Hf_{0.6}NiSn_{0.98}Sb_{0.02}$ | $Zr_{0.5}Hf_{0.5}CoSn_{0.2}Sb_{0.8}$ | Cu-based alloys | 14 | 5% @ 500 K |
| [231] (2015) | Nanostructured $Ti_xZr_{0.25}Hf_{0.75-x}NiSn_{0.99}Sb_{0.01}$ | Nanostructured $Ti_xHf_{1-x}CoSb_{0.8}Sn_{0.2}$ | Cu | 56 (per module) | 2.1% @ 339 K |
| [229] (2016) | Nanostructured $Zr_{0.25}Hf_{0.75}NiSn_{0.99}Sb_{0.01}$ | Nanostructured $Zr_{0.5}Hf_{0.5}CoSn_{0.2}Sb_{0.8}$ | Ti | 2 | 8.9% @ 678 K |
| [230] (2016) | Nanostructured $Ti_xZr_{0.25}Hf_{0.75-x}NiSn_{0.99}Sb_{0.01}$ | Nanostructured $Ti_xHf_{1-x}CoSb_{0.8}Sn_{0.2}$ | Cu | 120 | 5.3% @ 500 K |

devices have gained considerable achievements [229, 233]. Attempts have been made to segment half-Heusler alloys with other materials to improve the efficiency on a broadened operational temperature range, for example, $Bi_2Te_3$ [234, 235], and misfit-layered cobaltite $Ca_3Co_4O_{9+\delta}$ [236, 237]. Two-stage and three-stage cascaded TEGs employing half-Heusler materials for the high temperature stage have been theoretically considered [238]. Moreover, a theoretical model to optimize heat exchanger parameters and overall configuration of half-Heusler-based TEGs for automotive waste heat applications in order to maximize the fuel efficiency has been reported [239]. Table 4.5 presents a summary on the efficiency of the thermoelectric modules built entirely from half-Heusler materials over the past few years. With this promising progress, optimization of half-Heusler materials and development of half-Heusler-based modules for thermoelectric applications will continue to be active in the coming years.

**Acknowledgments:** The authors kindly acknowledge the support from the endorsement of the DLR Executive Board Member for Space Research and Technology and the support from the Young Research Group Leader Program.

# References

[1]   Heusler F. Über Manganbronze und über die Synthese magnetisierbarer Legierungen aus unmagnetischen Metallen, Angew. Chemie, 17(9), 260–264, Feb. 1904.

[2]   Graf T, Parkin, S S P, Felser, C. Heusler compounds – A material class with exceptional properties, IEEE Trans. Magn. 2011, 47(2), 367–373.

[3]   Feng Y, Cui, Z, Wei, M, Wu, B, Azam, S. Spin gapless semiconductor–nonmagnetic semiconductor transitions in Fe-doped $Ti_2CoSi$: First-principle calculations, Appl. Sci. 2018, 8(11), 2200.

[4]   Bombor D. et al. Half-metallic ferromagnetism with unexpectedly small spin splitting in the Heusler compound $Co_2FeSi$, Phys. Rev. Lett. 2013, 110(6), 66601.

[5]   Nayak A K. et al. Design of compensated ferrimagnetic Heusler alloys for giant tunable exchange bias, Nat. Mater. 2015, 14, 679–684.

[6]   Klimczuk T. et al. Superconductivity in the Heusler family of intermetallics, Phys. Rev. B 2012, 85(17), 174505.

[7]   Yan B, de Visser, A. Half-Heusler topological insulators, MRS Bull. 2014, 39(10), 859–866.

[8]   Huang L, Zhang, Q, Yuan, B, Lai, X, Yan, X, Ren, Z. Recent progress in half-Heusler thermoelectric materials, Mater. Res. Bull. 2016, 76, 107–112.

[9]   Kieven D. et al. Preparation and properties of radio-frequency-sputtered half-Heusler films for use in solar cells, Thin Solid Films 2011, 519(6), 1866–1871.

[10]  Hirohata A, Sagar, J, Lari, L, Fleet, L R, Lazarov, V K. Heusler-alloy films for spintronic devices, Appl. Phys. A 2013, 111(2), 423–430.

[11]  Ahamed Khan R, Ghomashchi, R, Xie, Z, Chen, L. Ferromagnetic shape memory Heusler materials: Synthesis, microstructure characterization and magnetostructural properties, Materials (Basel). 2018, 11(6), 988.

[12]  Kämmerer S, Thomas, A, Hütten, A, Reiss, G. $Co_2MnSi$ Heusler alloy as magnetic electrodes in magnetic tunnel junctions, Appl. Phys. Lett. 2004, 85(1), 79–81.

[13]  Ricci F. et al. Optical and magneto-optical properties of ferromagnetic full-Heusler films: Experiments and first-principles calculations, Phys. Rev. B 2007, 76(1), 14425.

[14]  Dubenko I. et al. Multifunctional properties related to magnetostructural transitions in ternary and quaternary Heusler alloys, J. Magn. Magn. Mater. 2015, 383, 186–189.

[15]  Vivas R J C. et al. Experimental evidences of enhanced magnetocaloric properties at room temperature and half-metallicity on $Fe_2MnSi$-based Heusler alloys, Mater. Chem. Phys. 2016, 174, 23–27.

[16]  Graf T, Felser, C, Parkin, S S P. Simple rules for the understanding of Heusler compounds, Prog. solid state Chem. 2011, 39(1), 1–50.

[17]  Ma J. et al. Computational investigation of inverse Heusler compounds for spintronics applications, Phys. Rev. B 2018, 98(9), 94410.

[18]  Fujii S, Ienaga, Y, Ishida, S, Asano, S. Roles of excess atoms in electronic and magnetic properties of $Fe_{2+x}V_{1-x}Z$ (Z= Al, Ga), J. Phys. Soc. Japan 2003, 72(3), 698–704.

[19]  Buffon M L C. et al. Thermoelectric performance and the role of anti-site disorder in the 24-electron Heusler $TiFe_2Sn$, J. Phys. Condens. Matter 2017, 29(40), 405702.

[20]  Ren Z, Liu, Y, Li, S, Zhang, X, Liu, H. Site preference and electronic structure of $Mn_2RhZ$ (Z= Al, Ga, In, Si, Ge, Sn, Sb): a theoretical study, Mater. Sci. 2016, 34(2), 251–259.

[21]  Douglas J E. et al. Enhanced thermoelectric properties of bulk TiNiSn via formation of a $TiNi_2Sn$ second phase, Appl. Phys. Lett. 2012, 101(18), 183902.

[22]  Wambach M. et al. Unraveling self-doping effects in thermoelectric TiNiSn half-Heusler compounds by combined theory and high-throughput experiments, Adv. Electron. Mater. 2016, 2(2), 1–9.

[23]  Luo H. et al. Ferromagnetism in the Mn-based Heusler alloy $Mn_2NiSb$, J. Appl. Phys. 2009, 105(10), 103903.

[24]  Meinert M, Schmalhorst, J-M, Reiss, G, Arenholz, E. Ferrimagnetism and disorder of epitaxial $Mn_{2-x}Co_xVAl$ Heusler compound thin films, J. Phys. D. Appl. Phys. 2011, 44(21), 215003.

[25]  Han J, Wang, Z, Xu, W, Wang, C, Liu, X. Investigation of half-metallic ferromagnetism in Heusler compounds $Co_2VZ$ (Z= Ga, Ge, As, Se), J. Magn. Magn. Mater. 2017, 442, 80–86.

[26]  Mikami M, Matsumoto, A, Kobayashi, K. Synthesis and thermoelectric properties of microstructural Heusler $Fe_2VAl$ alloy, J. Alloys Compd. 2008, 461(1–2), 423–426.

[27]  Mikami M, Kobayashi, K, Kawada, T, Kubo, K, Uchiyama, N. Development and evaluation of high-strength $Fe_2VAl$ thermoelectric module, Jpn. J. Appl. Phys. 2008, 47(3), R, 1512.

[28]  Kandpal H C, Felser, C, Seshadri, R. Covalent bonding and the nature of band gaps in some half-Heusler compounds, J. Phys. D. Appl. Phys. 2006, 39(5), 776.

[29]  Zhonghong L, Jian, M, Jingchuan, Z, Rongda, Z. Fabrication and thermoelectric properties of TiFeSb system half-Heusler compounds, Rare Met. Mater. Eng. 2013, 42(8), 1742–1745.

[30]  Huang L. et al. A new n-type half-Heusler thermoelectric material NbCoSb, Mater. Res. Bull. 2015, 70, 773–778.

[31]  Barth J. et al. Investigation of the thermoelectric properties of LiAlSi and LiAlGe, J. Electron. Mater. 2010, 39(9), 1856–1860.

[32]  Zhou J. et al. Large thermoelectric power factor from crystal symmetry-protected non-bonding orbital in half-Heuslers, Nat. Commun. 2018, 9(1), 1–9.

[33]  Zeier W G. et al. Engineering half-Heusler thermoelectric materials using Zintl chemistry, Nat. Rev. Mater. 2016, 1(6).

[34]  Jung D, Koo, H J, Whangbo, M H. Study of the 18-electron band gap and ferromagnetism in semi-Heusler compounds by non-spin-polarized electronic band structure calculations, J. Mol. Struct. THEOCHEM 2000, 527(1–3), 113–119.

[35]   Casper F, Graf, T, Chadov, S, Balke, B, Felser, C. Half-Heusler compounds: novel materials for energy and spintronic applications, Semicond. Sci. Technol. 2012, 27(6), 63001.
[36]   Rogl G. et al. Mechanical properties of half-Heusler alloys, Acta Mater. 2016, 107, 178–195.
[37]   Chen S, Ren, Z . Recent progress of half-Heusler for moderate temperature thermoelectric applications, Mater. Today, 16(10), 387–395, Oct. 2013.
[38]   Yang J, Li, H, Wu, T, Zhang, W, Chen, L, Yang, J. Evaluation of half-Heusler compounds as thermoelectric materials based on the calculated electrical transport properties, Adv. Funct. Mater. 2008, 18(19), 2880–2888.
[39]   Carrete J, Li, W, Mingo, N, Wang, S, Curtarolo, S. Finding unprecedentedly low-thermal-conductivity half-Heusler semiconductors via high-throughput materials modeling, Phys. Rev. X 2014, 4(1), 11019.
[40]   Carrete J, Mingo, N, Wang, S, Curtarolo, S. Nanograined half-Heusler semiconductors as advanced thermoelectrics: An Ab initio high-throughput statistical study, Adv. Funct. Mater. 2014, 24(47), 7427–7432.
[41]   Xie -H-H. et al. Interrelation between atomic switching disorder and thermoelectric properties of ZrNiSn half-Heusler compounds, CrystEngComm 2012, 14(13), 4467–4471.
[42]   Xie H. et al. The intrinsic disorder related alloy scattering in ZrNiSn half-Heusler thermoelectric materials, Sci. Rep. 2014, 4, 6888.
[43]   Tobola J, Jodin, L, Pecheur, P, Venturini, G. Unusual electron structure and electron transport properties of some disordered half-Heusler phases, J. Alloys Compd. 2004, 383(1–2), 328–333.
[44]   Yang J. et al. On the tuning of electrical and thermal transport in thermoelectrics: an integrated theory–experiment perspective, NPJ Comput. Mater. 2016, 2(1), 15015.
[45]   Yu J. et al. Unique role of refractory Ta alloying in enhancing the figure of merit of NbFeSb thermoelectric materials, Adv. Energy Mater. 2018, 8(1), 1701313.
[46]   Rogl G. et al. (V, Nb)-doped half Heusler alloys based on {Ti, Zr, Hf} NiSn with high ZT, Acta Mater. 2017, 131, 336–348.
[47]   Sakurada S, Shutoh, N. Effect of Ti substitution on the thermoelectric properties of (Zr,Hf) NiSn half-Heusler compounds, Appl. Phys. Lett. 2005, 86(8), 082105.
[48]   Gürth M. et al. On the constitution and thermodynamic modelling of the system Ti-Ni-Sn, RSC Adv. 2015, 5(112), 92270–92291.
[49]   Page A, Uher, C, Poudeu, P F, Van der Ven, A . Phase separation of full-Heusler nanostructures in half-Heusler thermoelectrics and vibrational properties from first-principles calculations, Phys. Rev. B, 92(17), 174102, Nov. 2015.
[50]   Tang Y. et al. Impact of Ni content on the thermoelectric properties of half-Heusler TiNiSn, Energy Environ. Sci. 2018, 11(2), 311–320.
[51]   Romaka V V. et al. Peculiarities of structural disorder in Zr- and Hf-containing Heusler and half-Heusler stannides, Intermetallics 2013, 35, 45–52.
[52]   Sauerschnig P. et al. On the constitution and thermodynamic modelling of the system Zr-Ni-Sn, J. Alloys Compd., 742, 1058–1082, Apr. 2018.
[53]   Romaka V A. et al. Effect of the accumulation of excess Ni atoms in the crystal structure of the intermetallic semiconductor n-ZrNiSn, Semiconductors, 47(7), 892–898, Jul. 2013.
[54]   Skolozdra E E, Stadnyk, R V, Starodynova, Y V. Crystal structure and magnetic properties of M'M;Sn compounds (M' = titanium, zirconium, hafnium, niobium; M = cobalt, nickel), Ukr. Fiz. Zhurnal (Russian Ed.) 1986, 31(8), 1258–1261.
[55]   Yang J, Potential applications of thermoelectric waste heat recovery in the automotive industry, in ICT 2005. 24th International Conference on Thermoelectrics, 2005., 2005, pp. 170–174.
[56]   Makongo J P A. et al. Thermal and electronic charge transport in bulk nanostructured $Zr_{0.25}Hf_{0.75}NiSn$ composites with full-Heusler inclusions, J. Solid State Chem. 2011, 184(11), 2948–2960.

[57] Makongo J P A. et al. Simultaneous large enhancements in thermopower and electrical conductivity of bulk nanostructured half-Heusler alloys, J. Am. Chem. Soc. 2011, 133(46), 18843–18852.

[58] Huang X Y, Xu, Z, Chen, L D. The thermoelectric performance of ZrNiSn/ZrO$_2$ composites, Solid State Commun. 2004, 130(3–4), 181–185.

[59] Shen Q. et al. Effects of partial substitution of Ni by Pd on the thermoelectric properties of ZrNiSn-based half-Heusler compounds, Appl. Phys. Lett. 2001, 79(25), 4165–4167.

[60] Tillard M, Berche, A, Jund, P, Tillard, M, Berche, A, Jund, P. Synthesis of Pure NiTiSn by Mechanical Alloying: An Investigation of the Optimal Experimental Conditions Supported by First Principles Calculations, Metals (Basel). 2018, 8(10), 835.

[61] Germond J, Structural Characterization and Thermoelectric Performance of ZrNiSn Half-Heusler Compound Synthesized by Mechanical Alloying, University of New Orleans, 2010.

[62] Hu T. et al. Interpreting the Combustion Process for High-Performance ZrNiSn Thermoelectric Materials, ACS Appl. Mater. Interfaces 2018, 10(1), 864–872.

[63] Xing Y. et al. Self-propagation high-temperature synthesis of half-Heusler thermoelectric materials: reaction mechanism and applicability, J. Mater. Chem. A 2018, 6(40), 19470–19478.

[64] Birkel C S. et al. Rapid microwave preparation of thermoelectric TiNiSn and TiCoSb half-Heusler compounds, Chem. Mater. 2012, 24(13), 2558–2565.

[65] Birkel C S. et al. Improving the thermoelectric properties of half-Heusler TiNiSn through inclusion of a second full-Heusler phase: Microwave preparation and spark plasma sintering of TiNi$_{1+x}$Sn, Phys. Chem. Chem. Phys. 2013, 15(18), 6990–6997.

[66] Lei Y, Cheng, C, Li, Y, Wan, R, Wang, M . Microwave synthesis and enhancement of thermoelectric figure of merit in half-Heusler TiNiSb$_x$Sn$_{1-x}$, Ceram. Int., 43(12), 9343–9347, Aug. 2017.

[67] Hohl H. et al. A new class of materials with promising thermoelectric properties: MNiSn (M = Ti, Zr, Hf), Mater. Res. Soc. Proc. 1997, 478, 109–114.

[68] Xiao K, Zhu, T-J, Yu, C, Yang, S-H, Zhao, X-B. P-type doping of Hf$_{0.6}$Zr$_{0.4}$NiSn half-Heusler thermoelectric materials prepared by levitation melting and spark plasma sintering, J. Mater. Res. 2011, 26(15), 1913–1918.

[69] Yu C, Zhu, T J, Shi, R Z, Zhang, Y, Zhao, X B, He, J. High-performance half-Heusler thermoelectric materials Hf$_{1-x}$Zr$_x$NiSn$_{1-y}$Sb$_y$ prepared by levitation melting and spark plasma sintering, Acta Mater. 2009, 57(9), 2757–2764.

[70] Xie H H, Yu, C, Zhu, T J, Fu, C G, Jeffrey Snyder, G, Zhao, X B. Increased electrical conductivity in fine-grained (Zr,Hf)NiSn based thermoelectric materials with nanoscale precipitates, Appl. Phys. Lett. 2012, 100(25), 0–4.

[71] Yu C, Zhu, T, Xiao, K, Shen, J, Zhao, X. Microstructure and thermoelectric properties of (Zr,Hf) NiSn based half-Heusler alloys by melt spinning and spark plasma sintering, Funct. Mater. Lett. 2010, 03(04), 227–231.

[72] Kimura Y, Ueno, H, Mishima, Y. Thermoelectric properties of directionally solidified half-Heusler (M$^a_{0.5}$,M$^b_{0.5}$)NiSn (M$^a$, M$^b$= Hf, Zr, Ti) alloys, J. Electron. Mater. 2009, 38(7), 934–939.

[73] Kimura Y, Tanoguchi, T, Kita, T . Vacancy site occupation by Co and Ir in half-Heusler ZrNiSn and conversion of the thermoelectric properties from n-type to p-type, Acta Mater., 58(13), 4354–4361, Aug. 2010.

[74] Ouardi S. et al. Transport and thermal properties of single- and polycrystalline NiZr$_{0.5}$Hf$_{0.5}$Sn, Appl. Phys. Lett. 2011, 99(15), 152112.

[75] Populoh S, Aguirre, M H, Brunko, O C, Galazka, K, Lu, Y, Weidenkaff, A. High figure of merit in (Ti,Zr,Hf)NiSn half-Heusler alloys, Scr. Mater. 2012, 66(12), 1073–1076.

[76] Bhattacharya S. et al. Effect of boundary scattering on the thermal conductivity of TiNiSn-based half-Heusler alloys, Phys. Rev. B 2008, 77(18), 184203.

[77] Chen S, Lukas, K C, Liu, W, Opeil, C P, Chen, G, Ren, Z. Effect of Hf concentration on thermoelectric properties of nanostructured N-type half-Heusler materials $Hf_xZr_{1-x}NiSn_{0.99}Sb_{0.01}$, Adv. Energy Mater. 2013, 3(9), 1210–1214.

[78] Joshi G. et al. Enhancement of thermoelectric figure-of-merit at low temperatures by titanium substitution for hafnium in n-type half-Heuslers $Hf_{0.75-x}Ti_xZr_{0.25}NiSn_{0.99}Sb_{0.01}$, Nano Energy 2013, 2(1), 82–87.

[79] Joshi G, Yan, X, Wang, H, Liu, W, Chen, G, Ren, Z. Enhancement in thermoelectric figure-of-merit of an N-type half-Heusler compound by the nanocomposite approach, Adv. Energy Mater. 2011, 1(4), 643–647.

[80] Bhattacharya S. et al. Effect of Sb doping on the thermoelectric properties of Ti-based half-Heusler compounds, $TiNiSn_{1-x}Sb_x$, Appl. Phys. Lett. 2000, 77(16), 2476–2478.

[81] Uher C, Yang, J, Hu, S, Morelli, D T, Meisner, G P. Transport properties of pure and doped MNiSn (M =Zr, Hf), Phys. Rev. B 1999, 59(13), 8615–8621.

[82] Hohl H, Ramirez, A P, Goldmann, C, Ernst, G, Wölfing, B, Bucher, E. Efficient dopants for ZrNiSn-based thermoelectric materials, J. Phys. Condens. Matter 1999, 11(7), 1697–1709.

[83] Aliev F G, Brandt, N B, Moshchalkov, V V, Kozyrkov, V V, Skolozdra, R V, Belogorokhov, A I. Gap at the Fermi level in the intermetallic vacancy system RBiSn(R=Ti,Zr,Hf), Zeitschrift fuer Phys. B Condens. Matter 1989, 75(2), 167–171.

[84] Katsuyama S, Matsuo, R, Ito, M. Thermoelectric properties of half-Heusler alloys $Zr_{1-x}Y_xNiSn_{1-y}Sb_y$, J. Alloys Compd. 2007, 428(1–2), 262–267.

[85] Bhardwaj A, Misra, D K, Pulikkotil, J J, Auluck, S, Dhar, A, Budhani, R C. Implications of nanostructuring on the thermoelectric properties in half-Heusler alloys, Appl. Phys. Lett. 2012, 101(13).

[86] Simonson J W, Wu, D, Xie, W J, Tritt, T M, Poon, S J. Introduction of resonant states and enhancement of thermoelectric properties in half-Heusler alloys, Phys. Rev. B 2011, 83(23), 235211.

[87] Qiu P, Yang, J, Huang, X, Chen, X, Chen, L. Effect of antisite defects on band structure and thermoelectric performance of ZrNiSn half-Heusler alloys, Appl. Phys. Lett. 2010, 96(15), 6–9.

[88] Lee P-J, Tseng, S C, Chao, L-S. High-temperature thermoelectric properties of Tix(ZrHf) 0.99–xV0.01Ni0.9Pd0.1Sn0.99Sb0.01 half-Heusler alloys, J. Alloys Compd. 2010, 496(1–2), 620–623.

[89] Lee P-J, Chao, L-S. High-temperature thermoelectric properties of $Ti_{0.5}(ZrHf)_{0.5-x}Nb_xNi_{0.9}Pd_{0.1}Sn_{0.98}Sb_{0.02}$ half-Heusler alloys, J. Alloys Compd. 2010, 504(1), 192–196.

[90] Katsuyama S, Kobayashi, T. Effect of mechanical milling on thermoelectric properties of half-Heusler $ZrNiSn_{0.98}Sb_{0.02}$ intermetallic compound, Mater. Sci. Eng. B Solid-State Mater. Adv. Technol. 2010, 166(1), 99–103.

[91] Tang M B, Zhao, J T. Low temperature transport and thermal properties of half-Heusler alloy $Zr_{0.25}Hf_{0.25}Ti_{0.5}NiSn$, J. Alloys Compd. 2009, 475(1–2), 5–8.

[92] Culp S R, Poon, S J, Hickman, N, Tritt, T M, Blumm, J. Effect of substitutions on the thermoelectric figure of merit of half-Heusler phases at 800°C, Appl. Phys. Lett. 2006, 88(4), 1–3.

[93] Gelbstein Y. et al. Thermoelectric properties of spark plasma sintered composites based on TiNiSn half-Heusler alloys, J. Mater. Res. 2011, 26(15), 1919–1924.

[94] Appel O, Schwall, M, Mogilyansky, D, Köhne, M, Balke, B, Gelbstein, Y. Effects of microstructural evolution on the thermoelectric properties of spark-plasma-sintered $Ti_{0.3}Zr_{0.35}Hf_{0.35}NiSn$ half-Heusler compound, J. Electron. Mater. 2013, 42(7), 1340–1345.

[95] Katayama T, Kim, S W, Kimura, Y, Mishima, Y. The effects of quaternary additions on thermoelectric properties of TiNiSn-based half-Heusler alloys, J. Electron. Mater. 2003, 32 (11), 1160–1165.

[96]   Page A, Van Der Ven, A, Poudeu, P F P, Uher, C. Origins of phase separation in thermoelectric (Ti, Zr, Hf)NiSn half-Heusler alloys from first principles, J. Mater. Chem. A 2016, 4(36), 13949–13956.

[97]   Schwall M, Balke, B. On the phase separation in n-type thermoelectric half-Heusler materials., Mater. (Basel, Switzerland) 2018, 11(4).

[98]   Öğüt S, Rabe, K M. Band gap and stability in the ternary intermetallic compounds NiSn M (M =Ti,Zr,Hf): A first-principles study, Phys. Rev. B 1995, 51(16), 10443–10453.

[99]   Do D T, Mahanti, S D, Pulikkoti, J J. Electronic structure of Zr–Ni–Sn systems: role of clustering and nanostructures in half-Heusler and Heusler limits, J. Phys. Condens. Matter 2014, 26(27), 275501.

[100]  Hazama H, Asahi, R, Matsubara, M, Takeuchi, T. Study of electronic structure and defect formation in $Ti_{1-x}Ni_{1+x}Sn$ half-Heusler alloys, J. Electron. Mater. 2010, 39(9), 1549–1553.

[101]  Zou D F, Xie, S H, Liu, Y Y, Lin, J G, Li, J Y. Electronic structure and thermoelectric properties of half-Heusler $Zr_{0.5}Hf_{0.5}NiSn$ by first-principles calculations, J. Appl. Phys. 2013, 113(19), 0–7.

[102]  Douglas J E. et al. Phase stability and property evolution of biphasic Ti-Ni-Sn alloys for use in thermoelectric applications, J. Appl. Phys. 2014, 115(4), 0–11.

[103]  Kim S W, Kimura, Y, Mishima, Y. High temperature thermoelectric properties of TiNiSn-based half-Heusler compounds, Intermetallics 2007, 15(3), 349–356.

[104]  Schmitt J, Gibbs, Z M, Snyder, J, Felser, C. Resolving the true band gap of ZrNiSn half-Heusler thermoelectric materials, Mater. Horiz. 2014.

[105]  Krez J. et al. Long-term stability of phase-separated half-Heusler compounds, Phys. Chem. Chem. Phys. 2015, 17(44), 29854–29858.

[106]  Zhu T J. et al. Effects of yttrium doping on the thermoelectric properties of $Hf_{0.6}Zr_{0.4}NiSn_{0.98}Sb_{0.02}$ half-Heusler alloys, J. Appl. Phys. 2010, 108(4), 0–5.

[107]  Akram R. et al. Enhanced thermoelectric properties of La-Doped ZrNiSn half-Heusler compound, J. Electron. Mater. 2015, 44(10), 3563–3570.

[108]  Li Y F, Huo, X G, He, D X, Zhao, C J, Lü, S C. Effect of rare-earth doping on the thermoelectric properties of the tin-based half-Heusler alloys, J. Inorg. Mater. 2010, 25(6), 573–576.

[109]  Romaka V A. et al. Features of the band structure and conduction mechanisms of n-HfNiSn semiconductor heavily Lu-doped, Semiconductors 2015, 49(3), 290–297.

[110]  Sanyal B, Eriksson, O, Suresh, K G, Dasgupta, I, Nigam, A K, Nordblad, P. Ferromagnetism in Mn doped half-Heusler NiTiSn: Theory and experiment, Appl. Phys. Lett. 2006, 89(21), 6–9.

[111]  Lkhagvasuren E. et al. Optimized thermoelectric performance of the n -type half-Heusler material TiNiSn by substitution and addition of Mn, AIP Adv. 2017, 7(4), 045010.

[112]  Chauhan N S. et al. Vanadium-doping-induced resonant energy levels for the enhancement of thermoelectric performance in Hf-free ZrNiSn half-Heusler alloys, ACS Appl. Energy Mater. 2018, 1(2), 757–764.

[113]  Muta H, Kanemitsu, T, Kurosaki, K, Yamanaka, S. High-temperature thermoelectric properties of Nb-doped MNiSn (M = Ti, Zr) half-Heusler compound, J. Alloys Compd. 2009, 469(1–2), 50–55.

[114]  Zhang H. et al. Thermoelectric properties of n-type half-Heusler compounds (Hf0.25Zr0.75)1-xNbxNiSn, Acta Mater. 2016, 113, 41–47.

[115]  Zhao D, Zuo, M, Wang, Z, Teng, X, Geng, H. Synthesis and thermoelectric properties of tantalum-doped ZrNiSn half-Heusler alloys, Funct. Mater. Lett. 2014, 07(03), 1450032.

[116]  Gałązka K. et al. Improved thermoelectric performance of $(Zr_{0.3}Hf_{0.7})NiSn$ half-Heusler compounds by Ta substitution, J. Appl. Phys. 2014, 115(18), 0–8.

[117]  Wang F, Fukuhara, T, Maezawa, K. Preparation of Fe-doped semiconductor NiZrSn, J. Alloys Compd. 2010, 500(2), 211–214.

[118]  Zou T, Synthesis and thermoelectric properties of chalcogenide and half-Heusler phases, University of Stuttgart, 2018.

[119] Tobola J, Pierre, J, Kaprzyk, S, Skolozdra, R V, Kouacou, M A. Crossover from semiconductor to magnetic metal in semi-Heusler phases as a function of valence electron concentration, J. Phys. Condens. Matter 1998, 10(5), 1013–1032.

[120] Romaka V A. et al. The mechanism of generation of the donor- and acceptor-type defects in the n-TiNiSn semiconductor heavily doped with Co impurity, Semiconductors 2009, 43(9), 1124–1130.

[121] Xie H. et al. Beneficial contribution of alloy disorder to electron and phonon transport in half-Heusler thermoelectric materials, Adv. Funct. Mater. 2013, 23(41), 5123–5130.

[122] Liu Y, Poudeu, P F P. Thermoelectric properties of Ge doped n-type $Ti_xZr_{1-x}NiSn_{0.975}Ge_{0.025}$ half-Heusler alloys, J. Mater. Chem. A 2015, 3(23), 12507–12514.

[123] Berry T. et al. Improving thermoelectric performance of TiNiSn by mixing MnNiSb in the half-Heusler structure, Phys. Chem. Chem. Phys. 2017, 19(2), 1543–1550.

[124] Bhattacharya S. et al. Thermoelectric properties of Sb-doping in the $TiNiSn_{1-x}Sb_x$ half-Heusler system, Proc. Intern. Conf. Thermoelectr. 1999, 18th, 336–339.

[125] Liu Y. et al. Distribution of impurity states and charge transport in $Zr_{0.25}Hf_{0.75}Ni_{1+x}Sn_{1-y}Sb_y$ nanocomposites, J. Solid State Chem. 2016, 234, 72–86.

[126] Gürth M, Rogl, G, Romaka, V V, Grytsiv, A, Bauer, E, Rogl, P. Thermoelectric high ZT half-Heusler alloys $Ti_{1-x-y}Zr_xHf_yNiSn$ ($0 \leq x \leq 1; 0 \leq y \leq 1$), Acta Mater. 2016, 104, 210–222.

[127] Chen L, Gao, S, Zeng, X, Mehdizadeh Dehkordi, A, Tritt, T M, Poon, S J. Uncovering high thermoelectric figure of merit in (Hf,Zr)NiSn half-Heusler alloys, Appl. Phys. Lett. 2015, 107(4), 4–9.

[128] Hasaka M, Morimura, T, Sato, H, Nakashima, H. Thermoelectric properties of $Ti_x(Hf_yZr_{1-y})_{1-x}NiSn_{0.998}Sb_{0.002}$ half-Heusler ribbons, J. Electron. Mater. 2009, 38(7), 1320–1325.

[129] Downie R A, Maclaren, D A, Bos, J W G. Thermoelectric performance of multiphase XNiSn (X = Ti, Zr, Hf) half-Heusler alloys, J. Mater. Chem. A 2014, 2(17), 6107–6114.

[130] Bhattacharya S, Tritt, T M, Xia, Y, Ponnambalam, V, Poon, S J, Thadhani, N. Grain structure effects on the lattice thermal conductivity of Ti-based half-Heusler alloys, Appl Phys Lett 2002, 81(1), 43–45.

[131] Appel O, Gelbstein, Y. A comparison between the effects of Sb and Bi doping on the thermoelectric properties of the $Ti_{0.3}Zr_{0.35}Hf_{0.35}NiSn$ half-Heusler alloy, J. Electron. Mater. 2014, 43(6), 1976–1982.

[132] Uher C, Yang, J, Meisner, G P, Thermoelectric properties of Bi-doped half-Heusler alloys, Proceedings of the International Conference on Thermoelectrics, vol. 18th. pp. 56–59, 1999.

[133] Rausch E, Balke, B, Stahlhofen, J M, Ouardi, S, Burkhardt, U, Felser, C. Fine tuning of thermoelectric performance in phase-separated half-Heusler compounds, J. Mater. Chem. C 2015, 3(40), 10409–10414.

[134] Chen L, Gao, S, Zeng, X, Mehdizadeh Dehkordi, A, Tritt, T M, Poon, S J . Uncovering high thermoelectric figure of merit in (Hf,Zr)NiSn half-Heusler alloys, Appl. Phys. Lett., 107(4), 041902, Jul. 2015.

[135] Shutoh N, Sakurada, S. Thermoelectric properties of the $Ti_x(Zr_{0.5}Hf_{0.5})_{1-x}NiSn$ half-Heusler compounds, Int. Conf. Thermoelectr. ICT, Proc. 2003, 312–315.

[136] Larson P, Mahanti, S D, Kanatzidis, M G. Structural stability of Ni-containing half-Heusler compounds, Phys. Rev. B – Condens. Matter Mater. Phys. 2000, 62(19), 12754–12762.

[137] Chen L, Liu, Y, He, J, Tritt, T M, Poon, S J. High thermoelectric figure of merit by resonant dopant in half-Heusler alloys, AIP Adv. 2017, 7(6).

[138] Sturm C, Jafarzadeh, P, Kleinke, H. Thermoelectric Nanomaterials, Comprehensive Nanoscience and Nanotechnology, 2nd, Vol. 1, Academic Press, 2019, 349–358.

[139] Chen L D, Huang, X Y, Zhou, M, Shi, X, Zhang, W B. The high temperature thermoelectric performances of $Zr_{0.5}Hf_{0.5}Ni_{0.8}Pd_{0.2}Sn_{0.99}Sb_{0.01}$ alloy with nanophase inclusions, J. Appl. Phys. 2006, 99(6), 64305.

[140] Huang X Y, Xu, Z, Chen, L D, Tang, X F. Effect of γ-Al2O3 Content on the Thermoelectric Performance of ZrNiSn/γ-Al2O3 Composites, Key Eng. Mater. 2003, 249, 79–82.

[141] Wang Chai Y, Kimura, Y. Nanosized precipitates in half-Heusler TiNiSn alloy, Appl. Phys. Lett. 2012, 100(3).

[142] Chai Y W, Kimura, Y. Microstructure evolution of nanoprecipitates in half-Heusler TiNiSn alloys, Acta Mater. 2013, 61(18), 6684–6697.

[143] Nanda B R K, Dasgupta, I. Electronic structure and magnetism in half-Heusler compounds, J. Phys. Condens. Matter 2003, 15(43), 7307–7323.

[144] Ouardi S. et al. Electronic structure and optical, mechanical, and transport properties of the pure, electron-doped, and hole-doped Heusler compound CoTiSb, Phys. Rev. B 2012, 86(4), 045116.

[145] Wang L L. et al. Thermoelectric performance of half-Heusler compounds TiNiSn and TiCoSb, J. Appl. Phys. 2009, 105(1).

[146] Stadnyk Y, Romaka, L, Horyn, A, Tkachuk, A, Gorelenko, Y, Rogl, P. Isothermal sections of the Ti-Co-Sn and Ti-Co-Sb systems, J. Alloys Compd. 2005, 387(1–2), 251–255.

[147] Massalski T B. Binary alloy phase diagrams, Metals Park, OH (USA), American Society for Metals, 1986.

[148] Tavassoli A. et al. Constitution of the binary M-Sb systems (M = Ti, Zr, Hf) and physical properties of $MSb_2$, Intermetallics 2018, 94, 119–132.

[149] Barth J, Schoop, M, Gloskovskii, A, Shkabko, A, Weidenkaff, A, Felser, C. Investigation of the Thermoelectric Properties of the Series $TiCo_{1-x}Ni_xSn_xSb_{1-x}$, Zeitschrift für Anorg. und Allg. Chemie 2010, 636(1), 132–136.

[150] Skovsen I. et al. Multi-temperature synchrotron PXRD and physical properties study of half-Heusler TiCoSb, Dalt. Trans. 2010, 39(42), 10154–10159.

[151] Asaad M, Buckman, J, Bos, J-W. Substitution versus full-Heusler segregation in TiCoSb, Metals (Basel). 2018, 8(11), 935.

[152] Qiu P, Huang, X, Chen, X, Chen, L. Enhanced thermoelectric performance by the combination of alloying and doping in TiCoSb-based half-Heusler compounds, J. Appl. Phys. 2009, 106(10), 1–6.

[153] Wu T, Jiang, W, Li, X, Bai, S, Liufu, S, Chen, L. Effects of Ge doping on the thermoelectric properties of TiCoSb-based p-type half-Heusler compounds, J. Alloys Compd. 2009, 467(1–2), 590–594.

[154] Sekimoto T, Kurosaki, K, Muta, H, Yamanaka, S. Annealing effect on thermoelectric properties of TiCoSb half-Heusler compound, J. Alloys Compd. 2005, 394(1–2), 122–125.

[155] Kim I H, Lee, Y G, Choi, M K, Ur, S C. Thermoelectric properties of half-Heusler TiCoSb1-xSnx synthesized by mechanical alloying process, Adv. Mater. Res. 2013, 660, 61–65.

[156] Amornpitoksuk P, Suwanboon, S. Correlation of milling time on formation of TiCoSb phase by mechanical alloying, J. Alloys Compd. 2008, 462(1–2), 267–270.

[157] Xia Y, Bhattacharya, S, Ponnambalam, V, Pope, A L, Poon, S J, Tritt, T M. Thermoelectric properties of semimetallic (Zr, Hf)CoSb half-Heusler phases, J. Appl. Phys. 2000, 88(4), 1952–1955.

[158] Zhou M, Chen, L, Feng, C, Wang, D, Li, J F. Moderate-temperature thermoelectric properties of TiCoSb-based half-Heusler compounds $Ti_{1-x}Ta_xCoSb$, J. Appl. Phys. 2007, 101(11).

[159] Stern R, Dongre, B, Madsen, G K H. Extrinsic doping of the half-Heusler compounds, Nanotechnology 2016, 27(33), 1–6.

[160] Romaka V A. et al. Mechanism of acceptor impurity initiation in the p-TiCoSb intermetallic semiconductor heavily doped with a V donor impurity. 2. Electrokinetic studies, Ukr. J. Phys. 2008, 53(9), 889–893.

[161] Asaad M, Buckman, J, Smith, R I, Bos, J W G. Thermoelectric properties and high-temperature stability of the $Ti_{1-x}V_xCoSb_{1-x}Sn_x$ half-Heusler alloys, RSC Adv. 2016, 6(61), 56511–56517.

[162] Qiu Q. et al. Grain Boundary Scattering of Charge Transport in n-Type (Hf,Zr)CoSb Half-Heusler Thermoelectric Materials, Adv. Energy Mater. 2019, 1803447, 1803447.

[163] Kawaharada Y, Uneda, H, Muta, H, Kurosaki, K, Yamanaka, S. High temperature thermoelectric properties of CoTiSb half-Heusler compounds, J. Alloys Compd. 2004, 364 (1–2), 59–63.

[164] Xie W J. et al. Significant ZT enhancement in p-type Ti(Co,Fe)Sb–InSb nanocomposites via a synergistic high-mobility electron injection, energy-filtering and boundary-scattering approach, Acta Mater. 2013, 61(6), 2087–2094.

[165] Wu T, Jiang, W, Li, X, Zhou, Y, Chen, L. Thermoelectric properties of p -type Fe-doped TiCoSb half-Heusler compounds, J. Appl. Phys. 2007, 102(10).

[166] Xie W, Jin, Q, Tang, X. The preparation and thermoelectric properties of $Ti_{0.5}Zr_{0.25}Hf_{0.25}Co_{1-x}Ni_xSb$ half-Heusler compounds, J. Appl. Phys. 2008, 103(4), 1–6.

[167] Xie W. et al. Synthesis and thermoelectric properties of (Ti,Zr,Hf)(Co,Pd)Sb half-Heusler compounds, J. Phys. D. Appl. Phys. 2009, 42(23).

[168] Ponnambalam V, Alboni, P N, Edwards, J, Tritt, T M, Culp, S R, Poon, S J. Thermoelectric properties of p -type half-Heusler alloys $Zr_{1-x}Ti_xCoSn_ySb_{1-y}$ (0.0<x<0.5; Y=0.15 and 0.3), J. Appl. Phys. 2008, 103(6), 0–5.

[169] Rausch E, Castegnaro, M V, Bernardi, F, Martins Alves, M C, Morais, J, Balke, B. Short and long range order of Half-Heusler phases in (Ti,Zr,Hf)CoSb thermoelectric compounds, Acta Mater. 2016, 115, 308–313.

[170] Yan X. et al. Enhanced thermoelectric figure of merit of p-type half-Heuslers, Nano Lett. 2011, 11, 556–560.

[171] Rausch E, Balke, B, Ouardi, S, Felser, C. Enhanced thermoelectric performance in the p-type half-Heusler $(Ti/Zr/Hf)CoSb_{0.8}Sn_{0.2}$ system via phase separation, Phys. Chem. Chem. Phys. 2014, 16(46), 25258–25262.

[172] He R, Kim, H S, Lan, Y, Wang, D, Chen, S, Ren, Z. Investigating the thermoelectric properties of p-type half-Heusler $Hf_x(ZrTi)_{1-x}CoSb_{0.8}Sn_{0.2}$ by reducing Hf concentration for power generation, RSC Adv. 2015, 4(110), 64711–64716.

[173] Yan X. et al. Stronger phonon scattering by larger differences in atomic mass and size in p-type half-Heuslers $Hf_{1-x}Ti_xCoSb_{0.8}Sn_{0.2}$, Energy Environ. Sci. 2012, 5(6), 7543–7548.

[174] Sekimoto T, Kurosaki, K, Muta, H, Yamanaka, S. Thermoelectric properties of Sn-doped TiCoSb half-Heusler compounds, J. Alloys Compd. 2006, 407(1–2), 326–329.

[175] Sekimoto T, Kurosaki, K, Muta, H, Yamanaka, S. High-thermoelectric figure of merit realized in p-type half-Heusler compounds: $ZrCoSn_xSb_{1-x}$, Japanese J. Appl. Physics, Part 2 Lett. 2007, 46(25–28).

[176] Sekimoto T, Kurosaki, K, Muta, H, Yamanaka, S. Thermoelectric and thermophysical properties of TiCoSb-ZrCoSb-HfCoSb pseudo ternary system prepared by spark plasma sintering, Mater. Trans. 2006, 47(6), 1445–1448.

[177] Sekimoto T, Kurosaki, K, Muta, H, Yamanaka, S. Thermoelectric and thermophysical properties of TiCoSb, ZrCoSb, HfCoSb prepared by SPS, Int. Conf. Thermoelectr. ICT, Proc. 2005, 2005, 335–338.

[178] Xie W J. et al. Simultaneously optimizing the independent thermoelectric properties in (Ti,Zr,Hf)(Co,Ni)Sb alloy by in situ forming InSb nanoinclusions, Acta Mater. 2010, 58(14), 4705–4713.

[179] Sahoo P. et al. Enhancing thermopower and hole mobility in bulk p-type half-Heuslers using full-Heusler nanostructures, Nanoscale 2013, 5(19), 9419–9427.

[180] Chauhan N S, Bhardwaj, A, Senguttuvan, T D, Pant, R P, Mallik, R C, Misra, D K. A synergistic combination of atomic scale structural engineering and panoscopic approach in p-type ZrCoSb-based half-Heusler thermoelectric materials for achieving high ZT, J. Mater. Chem. C 2016, 4(24), 5766–5778.

[181] Mallick M M, Vitta, S. Enhancing the thermoelectric performance of a p-type half-Heusler alloy, HfCoSb by incorporation of a band-matched chalcogenide, Cu2Te, J. Mater. Chem. A 2018, 6(30), 14709–14716.

[182] Zhao D. et al. Synthesis and thermoelectric properties of Ni-Doped ZrCoSb half-Heusler compounds, Metals (Basel). 2018, 8(1), 61.

[183] Stadnyk Y. et al. Impurity band effect on $TiCo_{1-x}Ni_xSb$ conduction: Donor impurities, J. Alloys Compd. 2006, 421(1–2), 19–23.

[184] Maji P, Takas, N J, Misra, D K, Gabrisch, H, Stokes, K, Poudeu, P F P. Effects of Rh on the thermoelectric performance of the p-type $Zr_{0.5}Hf_{0.5}Co_{1-x}Rh_xSb_{0.99}Sn_{0.01}$half-Heusler alloys, J. Solid State Chem. 2010, 183(5), 1120–1126.

[185] Zhou M, Chen, L, Zhang, W, Feng, C. Disorder scattering effect on the high-temperature lattice thermal conductivity of TiCoSb-based half-Heusler compounds, J. Appl. Phys. 2005, 98(1), 1–5.

[186] Balke B, Barth, J, Schwall, M, Fecher, G H, Felser, C. An alternative approach to improve the thermoelectric properties of half-Heusler compounds, J. Electron. Mater. 2011, 40(5), 702–706.

[187] Yan X. et al. Enhanced thermoelectric figure of merit of p-type half-Heuslers, Nano Lett. 2010, 11(2), 556–560.

[188] Hsu C C, Liu, Y N, Ma, H K. Effect of the $Zr_{0.5}Hf_{0.5}CoSb_{1-x}Sn_x/HfO_2$ half-Heusler nanocomposites on the ZT value, J. Alloys Compd. 2014, 597, 217–222.

[189] Poon S J. et al. Half-Heusler phases and nanocomposites as emerging high-ZT thermoelectric materials, J. Mater. Res. 2011, 26(22), 2795–2802.

[190] Yan X. et al. Thermoelectric property study of nanostructured p-type half-Heuslers (Hf, Zr, Ti) $CoSb_{0.8}Sn_{0.2}$, Adv. Energy Mater. 2013, 3(9), 1195–1200.

[191] Silpawilawan W, Kurosaki, K, Ohishi, Y, Muta, H, Yamanaka, S. FeNbSb p-type half-Heusler compound: Beneficial thermomechanical properties and high-temperature stability for thermoelectrics, J. Mater. Chem. C 2017, 5(27), 6677–6681.

[192] Fu C, Zhu, T, Liu, Y, Xie, H, Zhao, X. Band engineering of high performance p-type FeNbSb based half-Heusler thermoelectric materials for figure of merit zT > 1, Energy Environ. Sci. 2015.

[193] Fang T, Zheng, S, Chen, H, Cheng, H, Wang, L, Zhang, P. Electronic structure and thermoelectric properties of p-type half-Heusler compound NbFeSb: A first-principles study, RSC Adv. 2016, 6(13), 10507–10512.

[194] Lomnitska Y F. Interaction of niobium and iron with stibium, Powder Metall. Met. Ceram. 2008, 47(7–8), 447–450.

[195] Tavassoli A. et al. On the Half-Heusler compounds $Nb_{1-x}\{Ti,Zr,Hf\}_xFeSb$: Phase relations, thermoelectric properties at low and high temperature, and mechanical properties, Acta Mater. 2017, 135, 263–276.

[196] Melnyk G, Leithe-Jasper, A, Rogl, P, Skolozdra, R. The Antimony-iron-niobium (Sb-Fe-Nb) system, J. Phase Equilibria 1999, 20(2), 113–118.

[197] Young D P, Khalifah, P, Cava, R J, Ramirez, A P. Thermoelectric properties of pure and doped FeMSb (M=V,Nb), J. Appl. Phys. 2000, 87(1), 317–321.

[198] He R. et al. Achieving high power factor and output power density in p-type half-Heuslers $Nb_{1-x}Ti_xFeSb$, Proc. Natl. Acad. Sci. 2016, 113(48), 13576–13581.

[199] Page A, Poudeu, P F P, Uher, C. A first-principles approach to half-Heusler thermoelectrics: Accelerated prediction and understanding of material properties, J. Mater. 2016, 2(2), 104–113.
[200] Bhattacharya S, Madsen, G K H. A novel p-type half-Heusler from high-throughput transport and defect calculations, J. Mater. Chem. C 2016, 4(47), 11261–11268.
[201] Fang T, Zheng, S, Zhou, T, Yan, L, Zhang, P . Computational prediction of high thermoelectric performance in p-type half-Heusler compounds with low band effective mass, Phys. Chem. Chem. Phys., 19(6), 4411–4417, Feb. 2017.
[202] Stadnyk Y, Romaka, L, Gorelenko, Y, Tkachuk, A, Pierre, J, Crystal structure and electrokinetic properties of VFeSb compound, in Proceedings ICT2001. 20 International Conference on Thermoelectrics (Cat. No.01TH8589), 2011, pp. 251–253.
[203] Fang T, Zheng, S, Zhou, T, Chen, H, Zhang, P. Validity of rigid-band approximation in the study of thermoelectric properties of p-type FeNbSb-based half-Heusler compounds, J. Electron. Mater. 2017, 46(5), 3030–3035.
[204] Fu C, Zhu, T, Liu, Y, Xie, H, Zhao, X. Band engineering of high performance p-type FeNbSb based half-Heusler thermoelectric materials for figure of merit zT > 1, Energy Environ. Sci. 2015, 8(1), 216–220.
[205] Pang H J, Fu, C G, Yu, H, Chen, L C, Zhu, T J, Chen, X J. Origin of efficient thermoelectric performance in half-Heusler FeNb$_{0.8}$Ti$_{0.2}$Sb, J. Appl. Phys. 2018, 123(23), 1–8.
[206] Fu C, Wu, H, Liu, Y, He, J, Zhao, X, Zhu, T. Enhancing the figure of merit of heavy-band thermoelectric materials through hierarchical phonon scattering, Adv. Sci. 2016, 3(8), 1–7.
[207] Fu C. et al. High band degeneracy contributes to high thermoelectric performance in p-type Half-Heusler compounds, Adv. Energy Mater. 2014, 4(18).
[208] Jodin L, Tobola, J, Pecheur, P, Scherrer, H, Kaprzyk, S. Effect of substitutions and defects in half-Heusler FeVSb studied by electron transport measurements and KKR-CPA electronic structure calculations, Phys. Rev. B – Condens. Matter Mater. Phys. 2004, 70(18), 1–11.
[209] Ren W. et al. Ultrahigh power factor in thermoelectric system Nb$_{0.95}$M$_{0.05}$FeSb (M = Hf, Zr, and Ti), Adv. Sci. 2018, 5(7), 1800278.
[210] Shen J, Fu, C, Liu, Y, Zhao, X, Zhu, T. Enhancing thermoelectric performance of FeNbSb half-Heusler compound by Hf-Ti dual-doping, Energy Storage Mater. 2018, 10, 69–74.
[211] Yan J. et al. Suppression of the lattice thermal conductivity in NbFeSb-based half-Heusler thermoelectric materials through high entropy effects, Scr. Mater. 2018, 157, 129–134.
[212] Hasan R, Ur, S-C. Thermoelectric Properties of Mn-Doped FeVSb Half-Heusler System Synthesized via Mechanical Alloying, Trans. Electr. Electron. Mater. 2018, 19(4), 279–284.
[213] Fu C. et al. Electron and phonon transport in Co-doped FeV$_{0.6}$Nb$_{0.4}$Sb half-Heusler thermoelectric materials, J. Appl. Phys. 2013, 114(13), 0–7.
[214] Shen J. et al. Enhanced thermoelectric performance in n-type NbFeSb half-Heusler compound with heavy element Ir doping, Mater. Today Phys. 2019.
[215] Joshi G. et al. NbFeSb-based p-type half-Heuslers for power generation applications, Energy Environ. Sci. 2014, 7(12), 4070–4076.
[216] Zhang X. et al. Origin of high thermoelectric performance of FeNb$_{1-x}$Zr/Hf$_x$Sb$_{1-y}$Sn$_y$ alloys: A first-principles study, Sci. Rep. 2016, 6, no. July, 1–13.
[217] Fu C. et al. Realizing high figure of merit in heavy-band p-type half-Heusler thermoelectric materials, Nat. Commun. 2015, 6, 1–7.
[218] Hobbis D. et al. Structural, chemical, electrical, and thermal properties of n -type NbFeSb, Inorg. Chem. 2019, 58(3), 1826–1833.
[219] Fang T, Zhao, X, Zhu, T. Band structures and transport properties of high-performance half-Heusler thermoelectric materials by first principles, Materials (Basel). 2018, 11(5).
[220] Zhu H. et al. Discovery of TaFeSb-based half-Heuslers with high thermoelectric performance, Nat. Commun. 2019, accepted.

[221] LaLonde A D, Pei, Y, Wang, H, Jeffrey Snyder, G. Lead telluride alloy thermoelectrics, Mater. Today 2011, 14(11), 526–532.

[222] Shi Y, Sturm, C, Kleinke, H. Chalcogenides as thermoelectric materials, J. Solid State Chem. 2019, 270, 273–279.

[223] Fu C, Liu, Y, Zhao, X, Zhu, T. Are solid solutions better in FeNbSb-based thermoelectrics?, Adv. Electron. Mater. 2016, 2(12).

[224] Zou M, Li, J F, Kita, T. Thermoelectric properties of fine-grained FeVSb half-Heusler alloys tuned to p-type by substituting vanadium with titanium, J. Solid State Chem. 2013, 198, 125–130.

[225] Zou M, Li, J F, Guo, P, Kita, T. Synthesis and thermoelectric properties of fine-grained FeVSb system half-Heusler compound polycrystals with high phase purity, J. Phys. D. Appl. Phys. 2010, 43(41).

[226] Zhu H. et al. Discovery of ZrCoBi based half Heuslers with high thermoelectric conversion efficiency, Nat. Commun. 2018, 9(1), 1–9.

[227] Bartholomé K. et al. Thermoelectric modules based on half-Heusler materials produced in large quantities, J. Electron. Mater. 2014, 43(6), 1775–1781.

[228] Populoh S, Brunko, O C, Gałązka, K, Xie, W, Weidenkaff, A. Half-Heusler (TiZrHf)NiSn unileg module with high powder density, Materials (Basel). 2013, 6(4), 1326–1332.

[229] Joshi G, Poudel, B. Efficient and robust thermoelectric power generation device using hot-pressed metal contacts on nanostructured half-Heusler alloys, J. Electron. Mater. 2016, 45(12), 6047–6051.

[230] Zhang Y, Wang, X, Cleary, M, Schoensee, L, Kempf, N, Richardson, J. High-performance nanostructured thermoelectric generators for micro combined heat and power systems, Appl. Therm. Eng. 2016, 96, 83–87.

[231] Zhang Y. et al. High-temperature and high-power-density nanostructured thermoelectric generator for automotive waste heat recovery, Energy Convers. Manag. 2015, 105, 946–950.

[232] Rosenberger M, Dellner, M, Kluge, M, Tarantik, K R. Vehicle integration of a thermoelectric generator, MTZ Worldw. 2016, 77(4), 36–43.

[233] Ngan P H. et al. "On the challenges of reducing contact resistances in thermoelectric generators based on half-Heusler alloys,", J. Electron. Mater. 2016, 45(1), 594–601.

[234] Crane D T, LaGrandeur, J W. Progress report on BSST-led US Department of Energy automotive waste heat recovery program, J. Electron. Mater. 2010, 39(9), 2142–2148.

[235] Bjørk R. The universal influence of contact resistance on the efficiency of a thermoelectric generator, J. Electron. Mater. 2015, 44(8), 2869–2876.

[236] Hung L T. et al. High performance p-type segmented leg of misfit-layered cobaltite and half-Heusler alloy, Energy Convers. Manag. 2015, 99, 20–27.

[237] Hung L T. et al. Segmented thermoelectric oxide-based module for high-temperature waste heat harvesting, Energy Technol. 2015, 3(11), 1143–1151.

[238] Kanimba E, Pearson, M, Sharp, J, Stokes, D, Priya, S, Tian, Z. A modeling comparison between a two-stage and three-stage cascaded thermoelectric generator, J. Power Sources 2017, 365, 266–272.

[239] Kempf N, Zhang, Y. Design and optimization of automotive thermoelectric generators for maximum fuel efficiency improvement, Energy Convers. Manag. 2016, 121, 224–231.

Kei Hayashi, Yuzuru Miyazaki, Wataru Saito, Masataka Kubouchi,
Yohei Ogawa, Shogo Suzuki, Yuta Hayashibara, and Ikumi Ando
# 5 Thermoelectric properties of Mg$_2$Si and its derivatives: effects of lattice defects and secondary phases

**Abstract:** Since the discovery of Mg$_2$Si as a promising thermoelectric (TE) material, extensive studies have been devoted to improve the TE performance of Mg$_2$Si and its derivatives. The chapter review recent progress in the understanding of lattice defects and secondary phases in these materials, which greatly influence the conduction type and electrical transport properties.

**Keywords:** Magnesium silicide, Thermoelectric properties, Lattice defects, Secondary phase

Magnesium silicide (Mg$_2$Si) and its derivatives have attracted much attention as potential thermoelectric materials [1–46]. Mg$_2$Si belongs to the antifluorite structure with the space group of $Fm\bar{3}m$ as shown in Figure 5.1. Its lattice constant is 6.338 Å [47]. Mg and Si atoms are located at the 8$c$ (1/4 1/4 1/4) and 4$a$ (0 0 0) sites, respectively. Mg$_2$Si is an

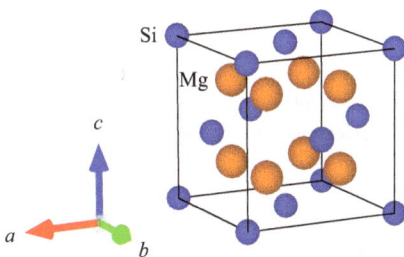

**Figure 5.1:** Crystal structure of Mg$_2$Si.

n-type semiconductor with an indirect band gap of 0.77 eV [2]. Since its electrical conductivity is as low as 11 S cm$^{-1}$ at 540 K [48], many attempts have been made to improve its thermoelectric performance by optimizing its carrier density through partial substitution or addition of other elements. Additionally, it has been subjected to lower thermal conductivity in many studies by partially substituting heavy elements for Si.

Figure 5.2 shows the plot of reported temperature dependence of the dimensionless figure of merit, $zT$, as follows: (a) n-type Mg$_2$Si, Mg$_2$Ge, or Mg$_2$Sn samples,

**Kei Hayashi, Yuzuru Miyazaki, Wataru Saito, Masataka Kubouchi, Yohei Ogawa, Shogo Suzuki, Yuta Hayashibara, Ikumi Ando,** Department of Applied Physics, Graduate School of Engineering, Tohoku University, Miyagi, Japan

https://doi.org/10.1515/9783110596526-005

(a)

(b)

(c)

(d)

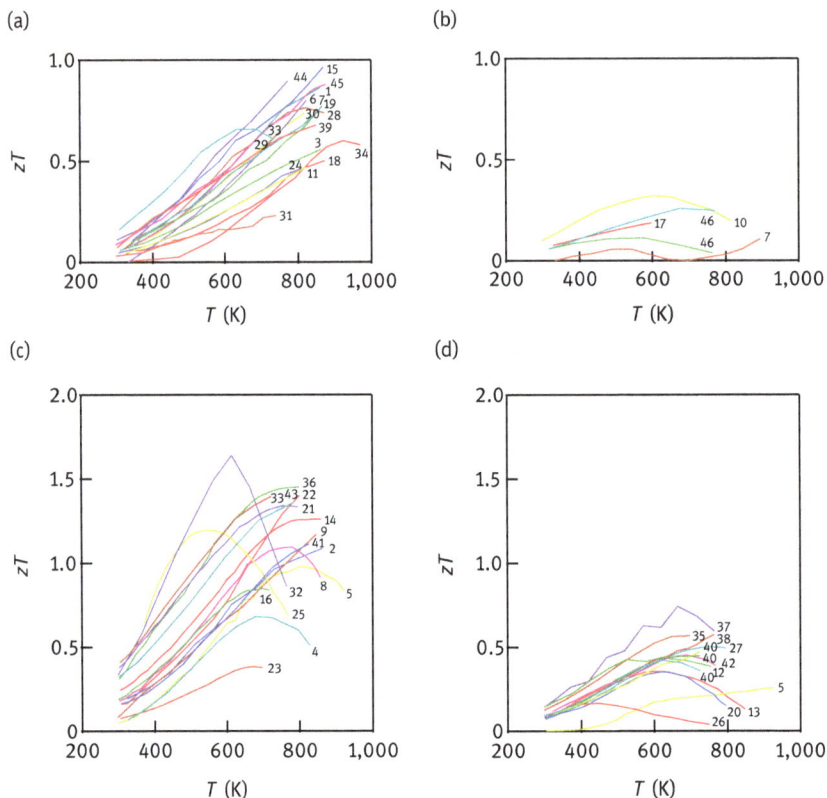

**Figure 5.2:** Dimensionless figure of merit, $zT$, of (a) n-type $Mg_2Si$, $Mg_2Sn$, or $Mg_2Ge$ samples; (b) p-type $Mg_2Si$ or $Mg_2Ge$ samples; (c) n-type $Mg_2(Si,Ge,Sn)$ samples; and (d) p-type $Mg_2(Si,Ge, Sn)$ samples. The number next to each data indicates the corresponding reference.

(b) p-type $Mg_2Si$ or $Mg_2Sn$ samples, (c) n-type $Mg_2(Si,Ge,Sn)$ samples, and (d) p-type $Mg_2(Si,Ge,Sn)$ samples. Higher $zT$ values are obtained for $Mg_2(Si,Ge,Sn)$ samples (Figure 5.2(c) and (d)) compared to $Mg_2Si$, $Mg_2Sn$, or $Mg_2Ge$ samples (Figure 5.2(a) and (b)). This is mainly due to a decrease in the thermal conductivity of $Mg_2Si$ caused by the Sn and/or Ge partial substitution for Si. In addition, n-type $Mg_2Si$, $Mg_2Sn$, or $Mg_2Ge$ samples (Figure 5.2(a)) tend to have higher $zT$ values than the p-type samples (Figure 5.2(b)). The same holds for the $zT$ values of n-type $Mg_2(Si,Ge,Sn)$ samples (Figure 5.2(c)) against those of p-type $Mg_2(Si,Ge,Sn)$ samples (Figure 5.2(d)). Among the n-type samples, the highest $zT$ value of 1.63 at 615 K is recorded for a porous sample of Sb-doped $Mg_2Si_{0.5}Sn_{0.5}$ [32]. Regarding the p-type samples, an $Mg_2Li_{0.025}Si_{0.4}Sn_{0.6}$ sample showed the highest $zT$ value of ~0.7 at 675 K [37].

The presence of lattice defects is one possible reason for the higher $zT$ values for n-type $Mg_2Si$, $Mg_2Ge$, and $Mg_2(Si,Ge,Sn)$ samples than p-type ones. Relations between thermoelectric properties and lattice defects have been reported for other materials.

For instance, Mg$_3$Sb$_2$ is inherently of p-type at the stoichiometric composition due to the existence of Mg vacancy (V$_{Mg}$). Thus, it is difficult to make an n-type Mg$_3$Sb$_2$ sample in trying to dope electron carriers by partial substitution. To overcome this difficulty, Tamaki et al. [49] attempted to increase the formation energy of V$_{Mg}$ by adding excess Mg when preparing Mg$_3$Sb$_2$ samples, and succeeded in achieving a high $zT$ value of 1.51 at 716 K for an n-type Mg$_{3.2}$Sb$_{1.5}$Bi$_{0.49}$Te$_{0.01}$ sample. Regarding conventional bismuth-telluride-based materials, the amounts of antisite defects and vacancies were controlled by changing their chemical compositions. Consequently, high thermoelectric performance was realized for both n-type and p-type bismuth-telluride-based materials [50, 51]. In a case of ZnSb, it is reported that Zn deficiency, which acts as an acceptor, enhances the p-type thermoelectric performance [52]. Note that the lattice defects have a role of scattering electrical transport carriers and phonons. Such effects have been discussed for Mg$_3$Sb$_2$ [53], bismuth-tellurides [50, 51], ZnSb [52], SnTe [54], SnSe [55], and so on. However, on the following pages, we discuss the effects of lattice defects mainly on the conduction type and electrical transport properties of Mg$_2$Si and its derivatives.

Theoretically, Kato et al. [56] investigated the formation energy of possible lattice defects in Mg$_2$Si as a function of the Fermi energy under different processing conditions (Si-rich and Mg-rich conditions; Figures. 5.3(a)–(f)). Regarding the lattice defects, interstitial Mg and Si defects at the 4$b$ (1/2 1/2 1/2) site (Mg$_i$ and Si$_i$, respectively), V$_{Mg}$ at the 8$c$ (1/4 1/4 1/4) site, Si vacancy at the 4$a$ (0 0 0) site (V$_{Si}$), and antisite defects (Mg$_{Si}$ and Si$_{Mg}$) were considered. Figure 5.3(a) and (b) pointed out that the formation energy of Mg$_i$ was lower than that of Si$_i$. Regarding vacancy defects (Figure 5.3(c) and (d)), V$_{Si}$ showed a lower formation energy than V$_{Mg}$ in the low Fermi energy region, while in the high Fermi energy region the formation energy of V$_{Mg}$ was lower than that of V$_{Si}$. Figure 5.3(e) and (f) shows the formation energy of the antisite defects. The Si$_{Mg}$ antisite defect exhibited lower formation energy than the Mg$_{Si}$ antisite defect under the Si-rich condition. In contrast, under the Mg-rich condition, the formation energy of Mg$_{Si}$ was lower than that of Si$_{Mg}$ in the low Fermi energy region. In the high Fermi energy region, the formation energy of Si$_{Mg}$ was lower than that of Mg$_{Si}$. The Fermi energy of Mg$_2$Si was calculated at ~0.28 eV, where Mg$_i$ showed the lowest formation energy among the above lattice defects in Mg$_2$Si. Jund et al. [57] also reported that Mg$_i$ is the most stable lattice defect in Mg$_2$Si, except for the MgSi divacancy.

The charge states of the lattice defect are depicted in Figure 5.3(a)–(f). At the Fermi energy of ~0.28 eV, the charge state of Mg$_i$ is +2. This means that Mg$_i$ acts as a donor of electron carriers in Mg$_2$Si. Liu et al. [58] calculated the formation energy of lattice defects in Mg$_2$Si, Mg$_2$Ge, and Mg$_2$Sn, and evaluated the defect concentration and electron/hole carrier density at 800 K (Figure 5.4(a)–(c)). In the figures, the lowest and highest chemical potentials correspond to the Si-rich and Mg-rich conditions, respectively. Moreover, electrons are dominant conduction carriers in Mg$_2$Si regardless of the processing conditions due to the highest Mg$_i$ concentration. Thus, Mg$_2$Si is inherently an n-type semiconductor. On the other hand, Mg$_2$Ge and Mg$_2$Sn become

(a)

Interstitial defect (Si rich)

(b)

Interstitial defect (Mg rich)

(c)

Vacancy defect (Si rich)

(d)

Vacancy defect (Mg rich)

(e)

Antisite defect (Si rich)

(f)

Antisite defect (Mg rich)

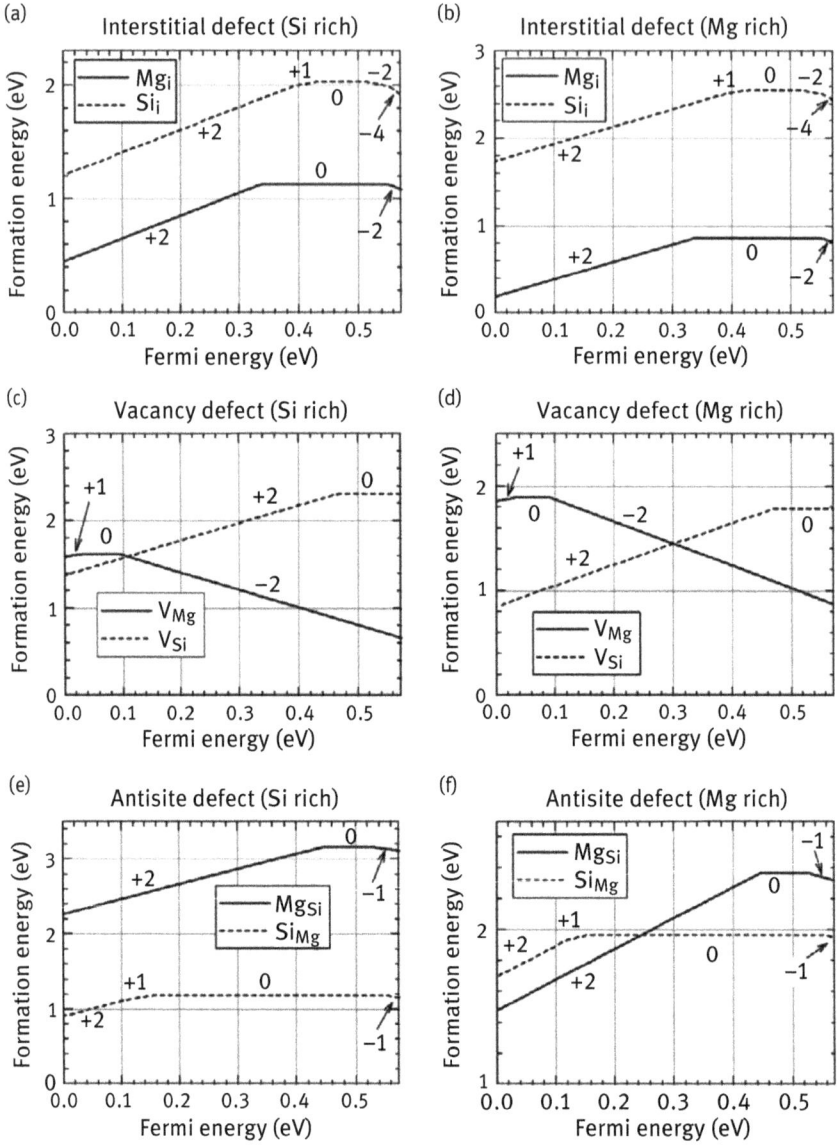

**Figure 5.3:** Formation energy of interstitial defects at (a) Si-rich and (b) Mg-rich conditions, vacancy defects at (c) Si-rich and (d) Mg-rich conditions, and antisite defects at (e) Si-rich and (f) Mg-rich conditions. The charge states of the lattice defects are depicted. (Reprinted from Kato et al. [56], Copyright 2009, with permission from IOP publishing.).

both n-type and p-type semiconductors depending on the balance of electron and hole carriers generated by $Mg_i$ and $V_{Mg}$, respectively. Based on the above calculation

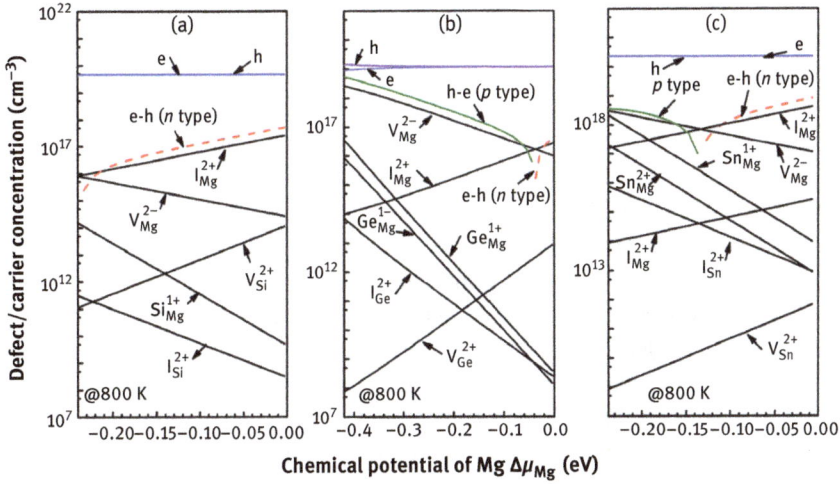

**Figure 5.4:** Concentration of lattice defects and carriers of (a) Mg$_2$Si, (b) Mg$_2$Ge, and (c) Mg$_2$Sn. Conduction carriers are denoted as "e" and "h" for electrons and holes, respectively. The difference between electron and hole concentrations is expressed as "e–h." (Reprinted from Liu et al. [58], Copyright 2015, with permission from WILEY-VCH Verlag GmbH & Co. KGaA, Weiheim.).

results, hole carrier doping in Mg$_2$Si is difficult because the hole carriers are partly canceled by the electron carriers originating from the ionized Mg$_i$.

The effect of Mg$_i$ on the electronic structure of Mg$_2$Si can be calculated using first principles theory, for example, with the full-potential code WIEN2k [59] by adopting the local exchange correlation potential in the generalized gradient approximation. For this calculation, a structure model containing Mg$_i$ is needed. Figure 5.5 illustrates

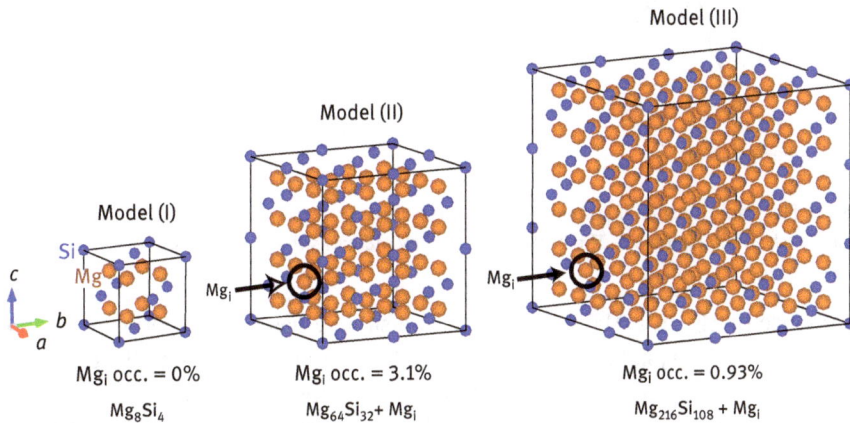

**Figure 5.5:** Three crystal structure models for the calculation of electronic density of states of Mg$_2$Si with and without Mg$_i$.

three crystal structure models: model (I) is "$Mg_8Si_4$," that is, $Mg_2Si$ without $Mg_i$, whereas models (II) and (III) are supercell structures expressed as "$Mg_{64}Si_{32} + Mg_i$" and "$Mg_{216}Si_{108} + Mg_i$," corresponding to the $Mg_i$ occupancy of 3.1% and 0.94%, respectively. We calculated the electronic density of states (DOS) for these models to examine their conduction type. The number of $k$-points in the Brillouin zone for models (I), (II), and (III) are 1059, 120, and 10, respectively.

Figure 5.6 shows the DOS for the three crystal structure models drawn in Figure 5.5. A broken line at 0 eV indicates the Fermi level. In the case of model (I), that is, $Mg_2Si$ without $Mg_i$, the Fermi level is located inside the band gap, consistent with a previous study [60]. This result indicates that $Mg_2Si$ is an intrinsic semiconductor. On the other hand, in models (II) and (III), that is, $Mg_2Si$ with $Mg_i$, the Fermi level shifts toward the conduction band. The enlarged DOS around the Fermi level shown in Figure 5.7 verifies this shift with increasing $Mg_i$ occupancy. Moreover, an additional DOS peak appears at the bottom of the conduction band in models (II) and (III). Since the additional DOS peak increases with increasing $Mg_i$ occupancy, we can conclude that it corresponds to an $Mg_i$ band. The Fermi level is situated between the $Mg_i$ band and the conduction band. Thus, $Mg_2Si$ with $Mg_i$ is actually an n-type semiconductor and $Mg_i$ acts as a donor to introduce electron carriers into $Mg_2Si$, which was also suggested by Hirayama et al. [61] from the DOS of model (II) calculated with the pseudopotential code QUANTUM ESPRESSO [62, 63] and by Imai et al. [64] from the DOS of model (III) calculated using the pseudopotential code CASTEP [65].

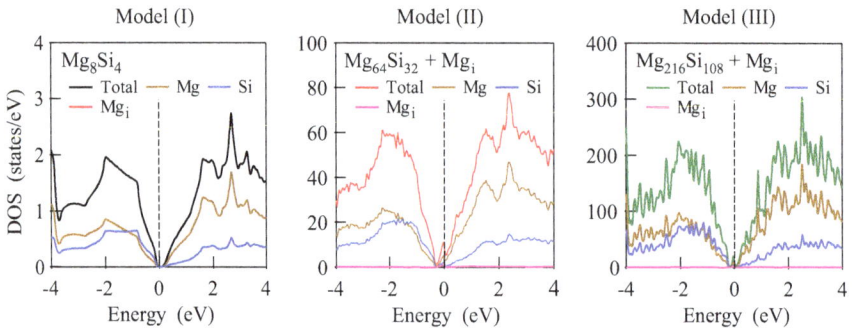

**Figure 5.6:** Electronic density of states (DOS) for three crystal structure models of $Mg_2Si$ with and without $Mg_i$ shown in Figure 5.5. The broken line at 0 eV indicates the Fermi level.

The presence of $Mg_i$ in a $Mg_2Si$ polycrystalline sample has been experimentally revealed using single-crystal X-ray diffraction (XRD) measurements by Kubouchi et al. [66, 67]. Figure 5.8 shows difference Fourier maps at the $z = 0.5$ plane of the $Mg_2Si$ sample with positive (green) and negative (blue) contour levels. Assuming the Mg-deficient $Mg_2Si$ model (Figure 5.8(a)), the residual electron density is obviously high at the $4b$ (1/2 1/2 1/2) site. Figure 5.8(b) shows the calculated difference Fourier map

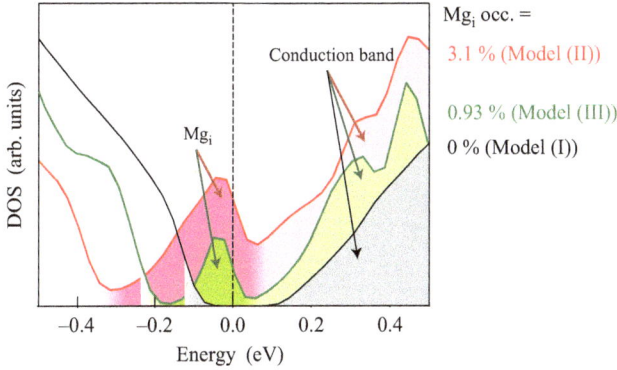

**Figure 5.7:** Enlarged electronic density of states (DOS) around the Fermi level for three crystal structure models Mg$_2$Si with and without Mg$_i$ shown in Figure 5.5.

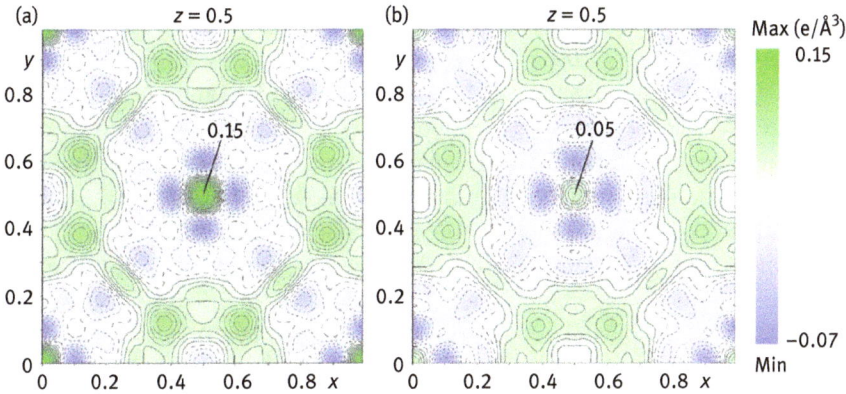

**Figure 5.8:** Difference Fourier maps at $z = 0.5$ plane of a Mg$_2$Si polycrystalline sample assuming (a) Mg-deficient Mg$_2$Si and (b) Mg-deficient Mg$_2$Si with Mg$_i$ models. Green (blue) area corresponds to positive (negative) contour levels of electron density. (Reprinted from Kubouchi et al. [67], Copyright 2015, with permission from The Minerals, Metals & Materials Society.).

assuming the Mg-deficient Mg$_2$Si with Mg$_i$. The residual electron density around the 4$b$ (1/2 1/2 1/2) site is reduced from 0.15 e/Å$^3$ to 0.05 e/Å$^3$ by introducing Mg$_i$. These results verify that not only Mg$_i$ but also V$_{Mg}$ exist in the Mg$_2$Si polycrystalline sample. In other words, there is a possibility that Mg$_i$ and V$_{Mg}$ form a Frenkel defect in Mg$_2$Si. The Mg$_i$ occupancy in polycrystalline samples with the nominal composition of Mg$_{2+x}$Si reported in the previous studies [66, 67] is summarized in Figure 5.9. The Mg$_{2+x}$Si polycrystalline samples are prepared by using Mg$_2$Si (2N), Mg (3N), and Si (4N) powders as starting materials. The Mg$_i$ occupancy increases with increasing $x$ from ~0.6% ($x = 0$) to ~1.2% ($x = 0.3$), and then decreases again above $x = 0.3$. The Mg$_{2+x}$Si polycrystalline samples with $x \leq 0$ also possess Mg$_i$; however, the Mg$_i$ occupancy is nearly constant (~0.5%).

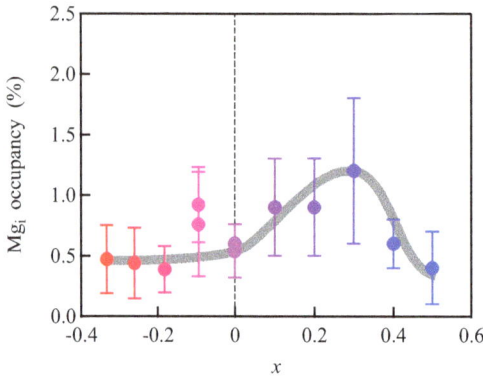

**Figure 5.9:** Occupancy of Mg at the interstitial site ($Mg_i$) of nominal compositions of $Mg_{2+x}Si$ samples. (Data are taken from Kubouchi et al. [66, 67].).

The presence of $Mg_i$ has been also reported for several $Mg_2(Si,Ge,Sn)$ samples. Bi-doped $Mg_2(Si,Ge)$ polycrystalline samples were prepared by Farahi et al. [68]. The starting materials were Mg chips (3N8), Si powder (3N), Ge pieces (6N), and Bi granules (4N). In low-magnification STEM (scanning transmission electron microscopy)-HAADF (high-angle annular dark-field) images of the $Mg_2Si_{0.777}Ge_{0.2}Bi_{0.023}$ sample, there were small-size atoms at $4b$ (1/2 1/2 1/2) sites, that is, $Mg_i$ (highlighted by the red lines in Figure 5.10(a)–(d)). Macario et al. [43] prepared Bi-doped $Mg_2(Si,Sn)$ polycrystalline samples by using Mg chips (3N8), Si powder (3N), Sn granules (3N), and Bi granules (4N) as starting materials. The crystal structure of $Mg_2Si_ySn_{0.97-y}Bi_{0.03}$ samples is investigated by analyzing powder XRD patterns using the Rietveld method. In the $y = 0.30$ sample, not $Mg_i$ but $V_{Mg}$ is present. In contrast, the $y = 0.325$ and $y = 0.35$ samples possess only $Mg_i$. It seems that the $Mg_i$ occupancy increases with increasing $y$ from 0% ($y = 0.30$) to 4% ($y = 0.35$).

The $Mg_i$ and $V_{Mg}$ amounts in the $Mg_2Si$ and $Mg_2(Si,Ge,Sn)$ samples depend on the processing conditions. (i) As mentioned earlier, the formation energy of $Mg_i$ and $V_{Mg}$ is affected by the Mg-rich and Si-rich conditions (Figures 5.3 and 5.4), which, therefore, changes the $Mg_i$ and $V_{Mg}$ amounts. (ii) The higher the processing temperature is, the more $Mg_i$ and $V_{Mg}$ are generated in $Mg_2Si$ [56]. (iii) The purity of the starting materials used for preparing the $Mg_2(Si,Ge,Sn)$ samples is related to the generated $Mg_i$ and $V_{Mg}$ amounts. If the starting materials contain impurities, then so do the $Mg_2(Si,Ge,Sn)$ samples illustrated in previous studies [43, 66–68]. In the case of Si, it was reported that each impurity binds with the lattice defects [69], depending on (1) the charge state and (2) the size of the impurity [70]. (1) An ionic impurity can interact with lattice defects via the Coulomb interaction, and (2) the size difference between an impurity and a Si atom induces local strain around the impurity. These two factors change the formation energy of lattice

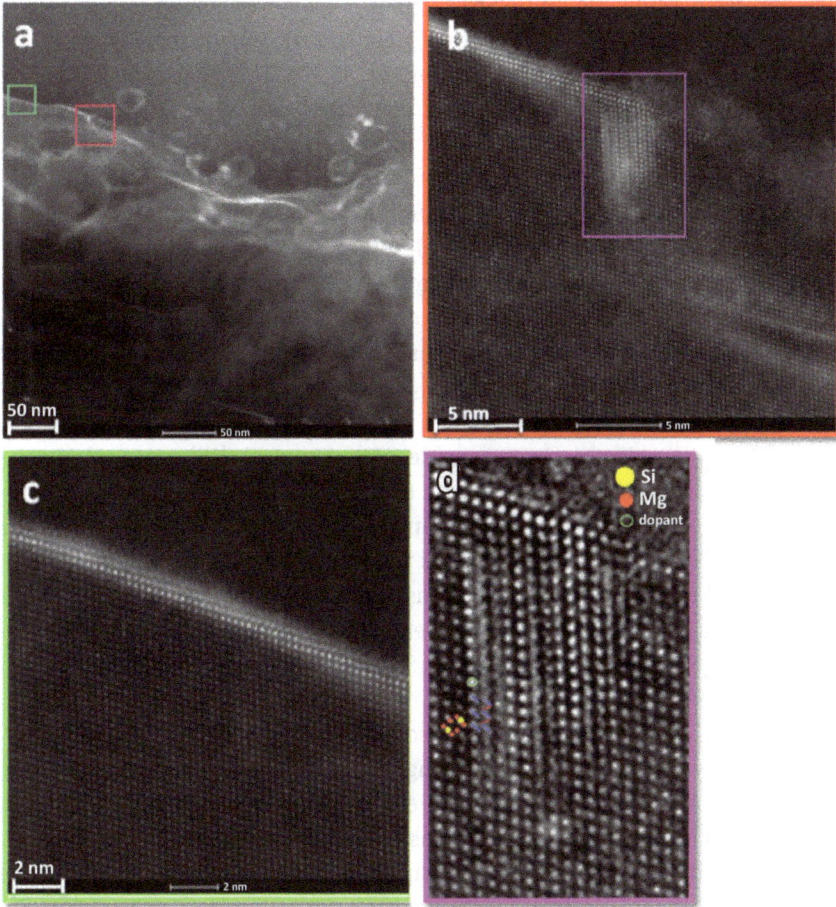

**Figure 5.10:** Low-magnification STEM (scanning transmission electron microscopy)-HAADF (high-angle annular dark-field) images of a Mg$_2$Si$_{0.777}$Ge$_{0.2}$Bi$_{0.023}$ sample. (b) and (c) STEM-HAADF image of the red and green areas in (a). (d) STEM-HAADF image of the purple area in (b). Blue and red lines indicate interstitial Mg and heavy atoms, respectively. (Reprinted from Farahi et al. [68], Copyright 2015, with permission from Elsevier.).

defects in Si, resulting in an increase or decrease of the amount of lattice defects. Detailed discussion about these effects on lattice defects in bismuth-telluride-based materials is surveyed by Zhu et al. [51]. Similarly, the impurities in the Mg$_2$(Si,Ge,Sn) samples are expected to affect the Mg$_i$ and V$_{Mg}$ amounts. Ionized dopant atoms (Sb, Li, Bi, etc.) and neutral substitution atoms (Ge, Sn) in Mg$_2$(Si,Ge, Sn) samples can also be considered impurities; they possess electric charges and are different in size from Mg and Si atoms. Thus, one has to carefully compare the amounts of Mg$_i$ and V$_{Mg}$ in Mg$_2$(Si,Ge,Sn) samples considering the chemical compositions as well as the processing conditions.

Thus far, we describe $Mg_2(Si,Ge,Sn)$ samples actually containing lattice defects such as $Mg_i$ and $V_{Mg}$. It is verified theoretically and experimentally that $Mg_i$ increases in $Mg_2Si$ prepared under the Mg-rich condition [56, 58, 67]. Thus, one expects that the increase in $Mg_i$ can enhance the electrical conductivity of $Mg_2(Si,Ge,Sn)$ samples because $Mg_i$ can act as an electron donor. Other lattice defects, $V_{Mg}$, $Mg_{Si}$, and so on, can also affect electrical conductivity. Liu et al. [14] investigated thermoelectric properties of excess-Mg added $Mg_{2(1+z)}Si_{0.5-y}Sn_{0.5}Sb_y$ polycrystalline samples. The $Mg_{2(1+z)}$ $Si_{0.5-y}Sn_{0.5}Sb_y$ samples in their study were all n-type, and the electrical conductivity of the $y = 0.10$ samples increased with $z$ up to $z = 0.12$ due to an increase in the electron carrier density. The highest $zT$ value of 1.25 at 800 K was obtained for the $y = 0.10$, $z = 0.1$ sample. Therefore, the increase of the electron carriers was attributed to the excess Mg that generated $Mg_i$ and $V_{Si/Sn}$ in the $Mg_{2(1+z)}Si_{0.5-y}Sn_{0.5}Sb_y$ samples. Judging from the actual composition of the $Mg_{2(1+z)}Si_{0.5-y}Sn_{0.5}Sb_y$ samples [14], the presence of $V_{Si/Sn}$ is unlikely. On a separate note, Du et al. [16] prepared $Mg_{2(1+z)}$ $Si_{0.38}Sn_{0.6}Sb_{0.02}$ polycrystalline samples with excess Mg, which showed a lower Si/Sn ratio than those prepared by Liu et al. [14]. In the X-ray photoelectron spectra of the $z = 0.2$ and $z = 0.24$ samples, the Mg $2p$ peak was decomposed into two components, which were attributable to the contributions from $Mg_i$ at the 4b (1/2 1/2 1/2) site and Mg at the 8b (1/4 1/4 1/4) site, respectively. Thus, it is plausible that $Mg_i$ is present in $Mg_{2(1+z)}Si_{0.4}Sn_{0.5}Sb_{0.1}$ and $Mg_{2(1+z)}Si_{0.38}Sn_{0.6}Sb_{0.02}$ polycrystalline samples with excess Mg.

Liu et al. [20] also prepared excess Mg added $Mg_{2(1+z)}(Si_{0.3}Sn_{0.7})_{1-y}Ga_y$ polycrystalline samples, which were of p-type owing to Ga doping. Interestingly, the hole carrier density of the $y = 0.05$ samples increased with increasing $z$ up to $z = 0.12$. This increase could not be explained by assuming the presence of $Mg_i$. Instead, the $Mg_{Si}$ antisite defect, which can act as an acceptor, was presumed. The above hypotheses that different kinds of lattice defects exist in the $Mg_{2(1+z)}Si_{0.5-y}Sn_{0.5}Sb_y$ and $Mg_{2(1+z)}(Si_{0.3}Sn_{0.7})_{1-y}Ga_y$ polycrystalline samples are possible, if Sb and Ga interact with lattice defects differently. To elucidate the effect of excess Mg on the lattice defects further, quantitative analysis of the amounts of lattice defects in these samples is needed.

It should be noted that, in some cases, excess Mg in $Mg_2(Si,Ge,Sn)$ samples results in the precipitation of Mg metals at grain boundaries ($Mg_{GB}$), affecting the electrical conductivity of the samples more than $Mg_i$. Kubouchi et al. [67] and Nieroda et al. [71] reported that excess Mg added $Mg_2Si$ polycrystalline samples with the nominal composition of $Mg_{2+x}Si$ contain $Mg_{GB}$ in the $Mg_2Si$ matrix. In other words, the prepared $Mg_{2+x}Si$ samples are $Mg/Mg_2Si$ composite materials. Furthermore, Kubouchi et al. [67] revealed that the electrical conductivity of $Mg_{2+x}Si$ samples does not correlate with the $Mg_i$ occupancy shown in Figure 5.10, but it tends to increase with increasing area ratio of Mg ($Mg_{GB}$) (Figure 5.11; reprinted from ref. [72].). This is probably because $Mg_{GB}$ generates more electron carriers in the $Mg_{2+x}Si$ samples than $Mg_i$. According to a

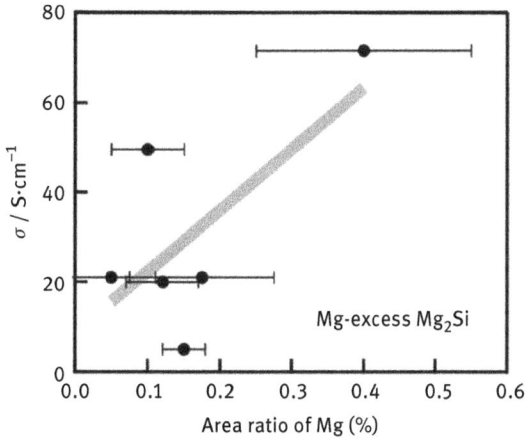

**Figure 5.11:** Electrical conductivity, $\sigma$, of Mg excess Mg$_2$Si polycrystalline samples as a function of the area ratio of Mg in the samples. The measurement temperature is 300 K. Gray line is a guide to the eyes. (Reprinted from Hayashibara et al. [72], Copyright 2018, with permission from The Thermoelectrics Society of Japan.).

simple estimation, the electron concentration generated by Mg$_i$ is ~$8 \times 10^{21}$ cm$^{-3}$, while that of the Mg metal, that is, Mg$_{GB}$, is ~$7 \times 10^{22}$ cm$^{-3}$.

As we have seen in this chapter, lattice defects and secondary phases have the potential to enhance thermoelectric properties, in particular electrical conductivity, of Mg$_2$(Si,Ge,Sn) samples. The ionized lattice defects increase the carrier density. To receive the benefit of this effect, the chemical compositions, processing conditions, and secondary phases of the Mg$_2$(Si,Ge,Sn) samples should be carefully controlled. Adversely, the lattice defects can affect a stable site of a dopant atom. For example, Tobola et al. [73] and Hirayama et al. [74] predicted that a boron (B) atom is likely to substitute for the 4$a$ (0 0 0) Si site in the case of Mg$_2$Si without Mg$_i$. In contrast, Kubouchi et al. [39] reported that the 8$c$ (1/4 1/4 1/4) Mg site or the 4$b$ (1/2 1/2 1/2) interstitial site is preferable for the B substitution site in the case of Mg$_2$Si with Mg$_i$. Regarding the processing conditions, Karkin et al. [75] attempted to form Frenkel defects of Mg$_i$ and V$_{Mg}$ in Al-doped Mg$_2$Si samples by using fast neutron irradiation. Turning to the secondary phases, Hayashibara et al. [72] fabricated Al/Mg$_2$Si composite materials and succeeded in enhancing the electrical conductivity with increasing Al content $x$ (the nominal composition of Mg$_2$Si: Al = 1: $x$; Figure 5.12). The thermoelectric properties of the Mg$_2$(Si,Ge,Sn) samples will be further enhanced by controlling the amounts of lattice defects and secondary phases in combination with the conventional doping and substitution method.

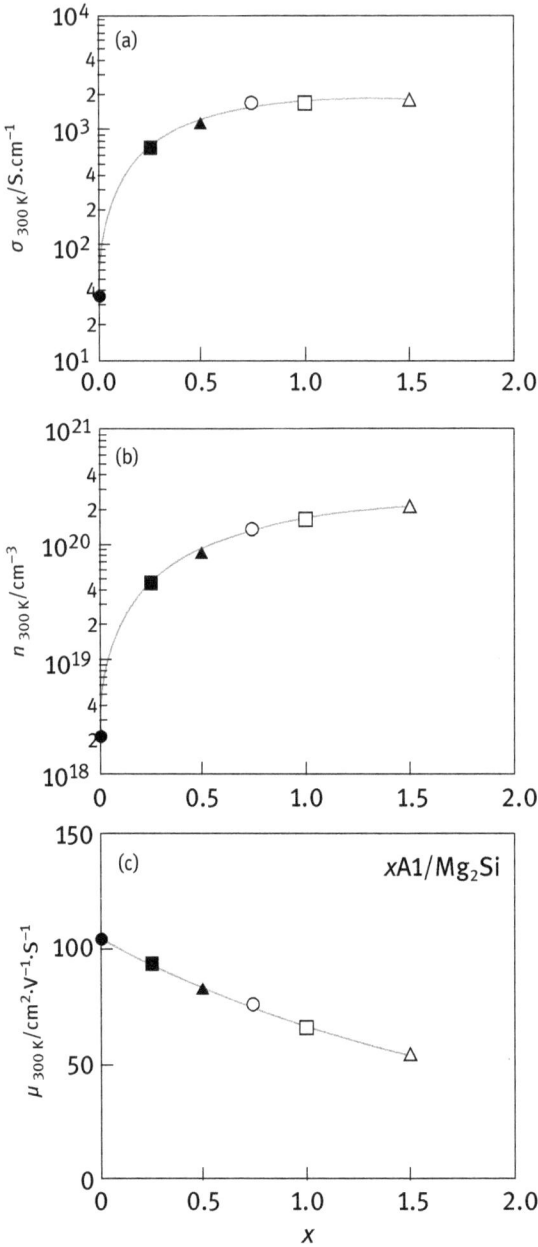

**Figure 5.12:** (a) Electrical conductivity, $\sigma_{300\,K}$, (b) electron carrier density, $n_{300\,K}$, and (c) carrier mobility, $\mu_{300\,K}$, of $Al/Mg_2Si$ composite materials. The nominal ratio of Al: $Mg_2Si$ is equal to $x$: 1. Gray curves are a guide to the eyes. (Reprinted from Hayashibara et al. [72], Copyright 2018, with permission from The Thermoelectrics Society of Japan.).

**Acknowledgments:** K. Hayashi thanks the financial support from the Ministry of Education, Culture, Sports, Science, and Technology (MEXT) of Japan (Grant No. 25289222, 17H03398, 17H05207).

# References

[1] Tani J, Kido, H. Thermoelectric properties of Bi-doped Mg$_2$Si semiconductors, Physica B 2005, 364, 218–224.

[2] Zaitsev V K, Fedorov, M I, Gurieva, E A, Eremin, I S, Konstantinov, P P, Samunin, A Y, Vedernikov, M V. Highly effective Mg$_2$Si$_{1-x}$Sn$_x$ thermoelectrics, Phys. Rev. B 2006, 74, 045207, 1–5.

[3] Tani J, Kido, H. Thermoelectric properties of Sb-doped Mg$_2$Si semiconductors, Intermetallics 2007, 15, 1202–1207.

[4] Zhang Q, Yin, H, Zhao, X B, He, J, Ji, X H, Zhu, T J, Tritt, T M. Thermoelectric properties of n-type Mg$_2$Si$_{0.6-y}$Sb$_y$Sn$_{0.4}$ compounds, Phys. Status Solidi A 2008, 205, 1657–1661.

[5] Mars K, Ihou-Mouko, H, Pont, G, Tobola, J, Scherrer, H. Thermoelectric properties and electronic structure of Bi- and Ag-doped Mg$_2$Si$_{1-x}$Ge$_x$ compounds, J. Electron. Mater. 2009, 38, 1360–1364.

[6] Yang M, Luo, W, Shen, Q, Jiang, H, Zhang, L. Preparation and thermoelectric properties of Bi-doped Mg$_2$Si nanocomposites, Adv. Mater. Res. 2009, 66, 17–20.

[7] Sakamoto T, Iida, T, Kurosaki, S, Yano, K, Taguchi, H, Nishio, K, Kogo, Y, Takanashi, Y. Thermoelectric characteristics of commercialized Mg$_2$Si source doped with Al, Bi, Ag and Cu, J. Electron. Mater. 2010, 39, 1708–1713.

[8] Liu W, Tang, X, Sharp, J. Low-temperature solid state reaction synthesis and thermoelectric properties of high-performance and low-cost Sb-doped Mg$_2$Si$_{0.6}$Sn$_{0.4}$, J. Phys. D: Appl. Phys. 2010, 43, 085406.

[9] Luo W, Yang, M, Chen, F, Shen, Q, Jiang, H, Zhang, L. Preparation and Thermoelectric Properties of Bi-Doped Mg$_2$Si$_{0.8}$Sn$_{0.2}$ Compound, Mater. Trans. 2010, 51, 288–291.

[10] Isoda Y, Tada, S, Nagai, T, Fujiu, H, Shinohara, Y. Thermoelectric properties of p-type Mg$_{2.00}$Si$_{0.25}$Sn$_{0.75}$ with Li and Ag double doping, J. Electron. Mater. 2010, 39, 1531–1535.

[11] You S-W, Park, K-H, Kim, I-H, Choi, S-M, Seo, W-S, Kim, S-U. Solid-state synthesis and thermoelectric properties of Al-doped Mg$_2$Si, J. Electron. Mater. 2011, 41, 1675–1679.

[12] Fedorov M I, Zaitsev, V K, Isachenko, G N. High effective thermoelectrics based on the Mg$_2$Si-Mg$_2$Sn solid solution, Solid State Phenom. 2011, 170, 286–292.

[13] Ihou-Mouko H, Mercier, C, Tobola, J, Pont, G, Scherrer, H. Thermoelectric properties and electronic structure of p-type Mg$_2$Si and Mg$_2$Si$_{0.6}$Ge$_{0.4}$ compounds doped with Ga, J. Alloys Compd. 2011, 509, 6503–6508.

[14] Liu W, Tang, X, Li, H, Sharp, J, Zhou, X, Uher, C. Optimized thermoelectric properties of Sb-doped Mg$_{2(1+z)}$Si$_{0.5-y}$Sn$_{0.5}$Sb$_y$ through adjustment of the Mg content, Chem. Mater. 2011, 23, 5256–5263.

[15] Hayatsu Y, Iida, T, Sakamoto, T, Kurosaki, S, Nishio, K, Kogo, Y, Takanashi, Y. Fabrication of large sintered pellets of Sb-doped n-type Mg$_2$Si using a plasma activated sintering method, J. Solid State Chem. 2012, 193, 161–165.

[16] Du Z, Zhu, T, Chen, Y, He, J, Gao, H, Jiang, G, Tritt, T M, Zhao, X. Roles of interstitial Mg in improving thermoelectric properties of Sb-doped Mg$_2$Si$_{0.4}$Sn$_{0.6}$ solid solutions, J. Mater. Chem. 2012, 22, 6838–6844.

[17]  An T-H, Park, C, Seo, W-S, Choi, S-M, Kim, I-H, Kim, S-U. Enhancement of p-type thermoelectric properties in an $Mg_2Sn$ system, J. Korean Phys. Soc. 2012, 60, 1717–1723.

[18]  Battiston S, Fiameni, S, Saleemi, M, Boldrini, S, Famengo, A, Agresti, F, Stingaciu, M, Toprak, M S, Fabrizio, M, Barison, S. Synthesis and characterization of Al-doped $Mg_2Si$ thermoelectric materials, J. Electron. Mater. 2013, 42, 1956–1959.

[19]  Nemoto T, Iida, T, Sato, J, Sakamoto, T, Hirayama, N, Nakajima, T, Takanashi, Y. Development of an $Mg_2Si$ unileg thermoelectric module using durable Sb-doped $Mg_2Si$ legs, J. Electron. Mater. 2013, 42, 2192–2197.

[20]  Liu W, Yin, K, Su, X, Li, H, Yan, Y, Tang, X, Uher, C. Enhanced hole concentration through Ga doping and excess of Mg and thermoelectric properties of p-type $Mg_{2(1+z)}(Si_{0.3}Sn_{0.7})_{1-y}Ga_y$, Intermetallics 2013, 32, 352–361.

[21]  Liu W, Zhang, Q, Yin, K, Chi, H, Zhou, Y X, Tang, X, Uher, C. High figure of merit and thermoelectric properties of Bi-doped $Mg_2Si_{0.4}Sn_{0.6}$ solid solutions, J. Solid State Chem. 2013, 203, 333–339.

[22]  Khan A, Vlachos, N, Kyratsi, T. High thermoelectric figure of merit of $Mg_2Si_{0.55}Sn_{0.4}Ge_{0.05}$ materials doped with Bi and Sb, Scr. Mater. 2013, 69, 606–609.

[23]  Jiang G, Chen, L, He, J, Gao, H, Du, Z, Zhao, X, Tritt, T M, Zhu, T. Improving p-type thermoelectric performance of $Mg_2(Ge,Sn)$ compounds via solid solution and Ag doping, Intermetallics 2013, 32, 312–317.

[24]  Ioannou M, Polymeris, G, Hatzikraniotis, E, Khan, A U, Paraskevopoulos, K M, Kyratsi, T. Solid-state synthesis and thermoelectric properties of Sb-doped $Mg_2Si$ materials, J. Electron. Mater. 2013, 42, 1827–1834.

[25]  Zhang X, Liu, H, Lu, Q, Zhang, J, Zhang, F. Enhanced thermoelectric performance of $Mg_2Si_{0.4}Sn_{0.6}$ solid solutions by in nanostructures and minute Bi-doping, Appl. Phys. Lett. 2013, 103, 063901, 1–4.

[26]  Tada S, Isoda, Y, Udono, H, Fujiu, H, Kumagai, S, Shinohara, Y. Thermoelectric properties of p-type $Mg_2Si_{0.25}Sn_{0.75}$ doped with sodium acetate and metallic sodium, J. Electron. Mater. 2013, 43, 1580–1584.

[27]  Zhang Q, Cheng, L, Liu, W, Zheng, Y, Su, X, Chi, H, Liu, H, Yan, Y, Tang, X, Uher, C. Low effective mass and carrier concentration optimization for high performance p-type $Mg_{2(1-x)}Li_{2x}Si_{0.3}Sn_{0.7}$ solid solutions, Phys. Chem. Chem. Phys. 2014, 16, 23576–23583.

[28]  Muthiah S, Sivaiah, B, Gahtori, B, Tyagi, K, Srivastava, A, Pathak, B, Dhar, A, Budhani, R. Double-doping approach to enhancing the thermoelectric figure-of-merit of n-type $Mg_2Si$ synthesized by use of spark plasma sintering, J. Electron. Mater. 2014, 43, 2035–2039.

[29]  Farahi N, VanZant, M, Zhao, J, Tse, J, Prabhudev, S, Botton, G, Salvador, J, Borondics, F, Liu, Z, Kleinke, H. Sb- and Bi-doped $Mg_2Si$: location of the dopants, micro- and nanostructures, electronic structures and thermoelectric properties, Dalton Trans. 2014, 43, 14983–14991.

[30]  de Boor J, Compere, C, Dasgupta, T, Stiewe, C, Kolb, H, Schmitz, A, Mueller, E. Fabrication parameters for optimized thermoelectric $Mg_2Si$, J. Mater. Sci. 2014, 49, 3196–3204.

[31]  Gao H L, Zhu, T J, Zhao, X B, Deng, Y. Influence of Sb doping on thermoelectric properties of $Mg_2Ge$ materials, Intermetallics 2015, 56, 33–36.

[32]  Ning H, Mastrorillo, G D, Grasso, S, Du, B, Mori, T, Hu, C, Xu, Y, Simpson, K, Maizza, G, Reece, M J. Enhanced thermoelectric performance of porous magnesium tin silicide prepared using pressure-less spark plasma sintering, J. Mater. Chem. A 2015, 3, 17426–17432.

[33]  Liu W, Kim, H S, Chen, S, Qing Jie, B, Yao, L M, Ren, Z, Opeil, C P, Wilson, S, Chu, C-W, Ren, Z. n-type thermoelectric material $Mg_2Sn_{0.75}Ge_{0.25}$ for high power generation, Proc. Natl. Acad. Sci. USA 2015, 112, 3269–3274.

[34] Nozariasbmarz A, Roy, P, Zamanipour, Z, Dycus, J H, Cabral, M J, LeBeau, J M, Krasinski, J S, Vashaee, D. Comparison of thermoelectric properties of nanostructured Mg$_2$Si, FeSi$_2$, SiGe, and nanocomposites of SiGe–Mg$_2$Si, SiGe–FeSi$_2$, APL Mater. 2016, 4, 104814.

[35] de Boor J, Saparamadu, U, Mao, J, Dahal, K, Müller, E, Ren, Z. Thermoelectric performance of Li doped, p-type Mg$_2$(Ge,Sn) and comparison with Mg$_2$(Si,Sn), Acta Mater. 2016, 120, 273–280.

[36] Yin K, Su, X, Yan, Y, You, Y, Zhang, Q, Uher, C, Kanatzidis, M G, Tang, X. Optimization of the electronic band structure and the lattice thermal conductivity of solid solutions according to simple calculations: a canonical example of the Mg$_2$Si$_{1-x-y}$Ge$_x$Sn$_y$ ternary solid solution, Chem. Mater. 2016, 28, 5538–5548.

[37] Gao P, Davis, J D, Poltavets, V V, Hogan, T P. The p-type Mg$_2$Li$_x$Si$_{0.4}$Sn$_{0.6}$ thermoelectric materials synthesized by a B$_2$O$_3$ encapsulation method using Li$_2$CO$_3$ as the doping agent, J. Mater. Chem. C 2016, 4, 929–934.

[38] Tang X, Wang, G, Zheng, Y, Zhang, Y, Peng, K, Guo, L, Wang, S, Zeng, M, Dai, J, Wang, G, Zhou, X. Ultra rapid fabrication of p-type Li-doped Mg$_2$Si$_{0.4}$Sn$_{0.6}$ synthesized by unique melt spinning method, Scr. Mater. 2016, 115, 52–56.

[39] Kubouchi M, Hayashi, K, Miyazaki, Y. Electronic structure and thermoelectric properties of boron doped Mg$_2$Si, Scr. Mater. 2016, 123, 59–63.

[40] Saparamadu U, de Boor, J, Mao, J, Song, S, Tian, F, Liu, W, Zhang, Q, Ren, Z. Comparative studies on thermoelectric properties of p-type Mg$_2$Sn$_{0.75}$Ge$_{0.25}$ doped with lithium, sodium, and gallium, Acta Mater. 2017, 141, 154–162.

[41] Vlachos N, Polymeris, G S, Manoli, M, Hatzikraniotis, E, Khan, A U, Lioutas, C B, Stefanaki, E C, Pavlidou, E, Paraskevopoulos, K M, Giapintzakis, J, Kyratsi, T. Effect of antimony-doping and germanium on the highly efficient thermoelectric Si-rich-Mg$_2$(Si,Sn,Ge) materials, J. Alloys Compd. 2017, 714, 502–513.

[42] Tang X, Zhang, Y, Zheng, Y, Peng, K, Huang, T, Wang, X L, G, Wang, S, Zhou, X. Improving thermoelectric performance of p-type Ag-doped Mg$_2$Si$_{0.4}$Sn$_{0.6}$ prepared by unique melt spinning method, Appl. Therm. Eng. 2017, 111, 1396–1400.

[43] Macario L R, Cheng, X, Ramirez, D, Mori, T, Kleinke, H. Thermoelectric properties of Bi-doped magnesium silicide stannides, ACS Appl. Mater. Interfaces 2018, 10, 40585–40591.

[44] Wei J, Duan, B, Li, J, Yang, H, Chen, G, Zhai, P. High pressure synthesis of multiple doped Mg$_2$Si-based thermoelectric materials, J. Mater. Sci: Mater. Electron. 2018, 29, 10904–10910.

[45] Li J, Li, X, Cai, B, Chen, C, Zhang, Q, Zhao, Z, Zhang, L, Yu, F, Yu, D, Tian, Y, Xu, B. Enhanced thermoelectric performance of high pressure synthesized Sb-doped Mg$_2$Si, J. Alloys Compd. 2018, 741, 1148–1152.

[46] Tani J, Shinagawa, T, Chigane, M. Thermoelectric properties of impurity-doped Mg$_2$Sn, J. Electron. Mater. 2019. Doi: 10.1007/s11664-19-07093-x

[47] Zaitsev V K, Fedorov, M I, Eremin, I S, Gurieva, E A. Thermoelectrics Handbook: Macro to Nano-Structured Materials, CRC press, New York, 2005, Chap. 29.

[48] Kajikawa T, Shida, K, Sugihara, S, Ohmori, M, Hirai, T, Proceedings of the 16th International Conference on Thermoelectrics (ICT'97) (IEEE, 1997), p. 275.

[49] Tamaki H, Sato, H K, Kanno, T. Isotropic Conduction Network and Defect Chemistry in Mg$_{3+\delta}$Sb$_2$-Based Layered Zintl Compounds with High Thermoelectric Performance, Adv. Mater. 2016, 28, 10182–10187.

[50] Hu L, Zhu, T, Liu, X, Zhao, X. Point defect engineering of high-performance bismuth-telluride-based thermoelectric materials, Adv. Funct. Mater. 2014, 24, 5211–5218.

[51] Zhu T, Hu, L, Zhao, X, He, J. New insights into intrinsic point defects in V2VI3 thermoelectric materials, Adv. Sci. 2016, 3, 1600004.

[52] Guo Q, Luo, S. Improved thermoelectric efficiency in p-type ZnSb through Zn deficiency, Func. Mater. Lett. 2015, 08, 1550028.

[53] Mao J, Wu, Y, Song, S, Zhu, Q, Shuai, J, Liu, Z, Pei, Y, Ren, Z. Defect engineering for realizing high thermoelectric performance in n-type $Mg_3Sb_2$-based materials, ACS Energy Lett. 2017, 2, 2245–2250.

[54] Pei Y, Zheng, L, Li, W, Lin, S, Chen, Z, Wang, Y, Xu, X, Yu, H, Chen, Y, Ge, B. Interstitial point defect scattering contributing to high thermoelectric performance in SnTe, Adv. Electr. Mater. 2016, 2, 1600019.

[55] Wu D, Wu, L, He, D, Zhao, L-D, Li, W, Wu, M, Jin, M, Xu, J, Jiang, J, Huang, L, Zhu, Y, Kanatzidis, M G, He, J. Direct observation of vast off-stoichiometric defects in single crystalline SnSe, Nano Energy 2017, 35, 321–330.

[56] Kato A, Yagi, T, Fukusako, N. First-principles studies of intrinsic point defects in magnesium silicide, J. Phys.: Condens. Matter 2009, 21, 205801, 1–7.

[57] Jund P, Viennois, R, Colinet, C, Hug, G, Fèvre, M, Tédenac, J-C. Lattice stability and formation energies of intrinsic defects in $Mg_2Si$ and $Mg_2Ge$ via first principles simulations, J. Phys.: Condens. Matter 2013, 25, 035403.

[58] Liu X, Xi, L, Qiu, W, Yang, J, Zhu, T, Zhao, X, Zhang, W. Significant roles of intrinsic point defects in $Mg_2X$ ($X$ = Si, Ge, Sn) thermoelectric materials, Adv. Electron. Mater. 2016, 2, 1500284.

[59] Blaha P, Schwarz, K, Madsen, G, Kvasnicka, D, Luitz, J, WIEN2k, An Augmented Plane Wave + Local Orbitals Program for Calculating Crystal Properties (Karlheinz Schwarz, Techn. Universität Wien, Austria, 2001).

[60] Morris R G, Redin, R D, Danielson, G C. Semiconducting Properties of $Mg_2Si$ Single Crystals, Phys. Rev. 1958, 109, 1909–1915.

[61] Hirayama N, Iida, T, Nishino, K, Kogo, Y, Takarabe, K, Hamada, N. Influence of native defects on structural and electronic properties of magnesium silicide, Jpn. J. Appl. Phys. 2017, 56, 05DC05.

[62] Giannozzi P, Baroni, S, Bonini, N, Calandra, M, Car, R, Cavazzoni, C, Ceresoli, D, Chiarotti, G L, Cococcioni, M, Dabo, I, Dal Corso, A, de Gironcoli, S, Fabris, S, Fratesi, G, Gebauer, R, Gerstmann, U, Gougoussis, C, Kokalj, A, Lazzeri, M, Martin-Samos, L, Marzari, N, Mauri, F, Mazzarello, R, Paolini, S, Pasquarello, A, Paulatto, L, Sbraccia, C, Scandolo, S, Sclauzero, G, Seitsonen, A P, Smogunov, A, Umari, P, Wentzcovitch, R M. QUANTUM ESPRESSO: a modular and open-source software project for quantum simulations of materials, J. Phys.: Condens. Matter 2009, 21, 395502.

[63] Giannozzi P, Andreussi, O, Brumme, T, Bunau, O, Buongiorno Nardelli, M, Calandra, M, Car, R, Cavazzoni, C, Ceresoli, D, Cococcioni, M, Colonna, N, Carnimeo, I, Dal Corso, A, de Gironcoli, S, Delugas, P, DiStasio, R A Jr., Ferretti, A, Floris, A, Fratesi, G, Fugallo, G, Gebauer, R, Gerstmann, U, Giustino, F, Gorni, T, Jia, J, Kawamura, M, Ko, H-Y, Kokalj, A, Küçükbenli, E, Lazzeri, M, Marsili, M, Marzari, N, Mauri, F, Nguyen, N L, Nguyen, H-V, Otero-de-la-Roza, A, Paulatto, L, Poncé, S, Rocca, D, Sabatini, R, Santra, B, Schlipf, M, Seitsonen, A P, Smogunov, A, Timrov, I, Thonhauser, T, Umari, P, Vast, N, Wu, X, Baroni, S. Advanced capabilities for materials modelling with Quantum ESPRESSO, J. Phys.: Condens. Matter 2017, 29, 465901.

[64] Imai M, Mori, Y, Nakamura, S, Takarabe, K. Consideration about the synthesis pressure effect on lattice defects of $Mg_2Si$ using 1st principle calculations, J. Alloys Compd. 2016, 644, 369–377.

[65] Payne M C, Teter, M P, Allan, D C, Joannopoulos, J D. Iterative minimization techniques for ab initio total-energy calculations: molecular dynamics and conjugate gradients, Rev. Mod. Phys. 1992, 64, 1045–1097.

[66] Kubouchi M, Hayashi, K, Miyazaki, Y. Quantitative analysis of interstitial Mg in $Mg_2Si$ studied by single crystal X-ray diffraction, J. Alloys Compd. 2014, 617, 389–392.

[67] Kubouchi M, Ogawa, Y, Hayashi, K, Takamatsu, T, Miyazaki, Y. Effect of interstitial Mg in $Mg_{2+x}Si$ on electrical conductivity and Seebeck coefficient, J. Electron. Mater. 2016, 45, 1589–1593.

[68] Farahi N, Prabhudev, S, Botton, G A, Zhao, J, Tse, J S, Liu, Z, Salvador, J R, Kleinke, H. Local structure and thermoelectric properties of $Mg_2Si_{0.977-x}Ge_xBi_{0.023}$ $(0.1 \leq x \leq 0.4)$, J. Alloys Compd. 2015, 644, 249–255.

[69] Watkins G D. Intrinsic defects in silicon, Mater. Sci. Semicond. Proc. 2000, 3, 227–235.

[70] Sueoka K, Kamiyama, E, Vanhellemont, J. Density functional theory study on the impact of heavy doping on Si intrinsic point defect properties and implications for single crystal growth from a melt, J. Appl. Phys. 2013, 114, 153510.

[71] Nieroda P, Kolezynski, A, Leszczynski, J, Nieroda, J, Pasierb, P. The structural, microstructural and thermoelectric properties of $Mg_2Si$ synthesized by SPS method under excess Mg content conditions, J. Alloys Compd. 2019, 775, 138–149.

[72] Hayashibara Y, Hayashi, K, Ando, I, Kubouchi, M, Ogawa, Y, Saito, W, Miyazaki, Y. Fabrication and thermoelectric properties of $Al/Mg_2Si$ composite materials, Mater. Trans. 2018, 59, 1041–1045.

[73] Tobola J, Zwolenski, P, Kaprzyk, S. Theoretical search for p-type dopants in $Mg_2X$ $(X = Si, Ge)$ semiconductors for thermoelectricity, Solid State Phenom. 2013, 194, 266–271.

[74] Hirayama N, Iida, T, Funashima, H, Morioka, S, Sakamoto, M, Nishio, K, Kogo, Y, Takanashi, Y, Hamada, N. Theoretical analysis of structure and formation energy of impurity-doped $Mg_2Si$: Comparison of first-principles codes for material properties, Jpn, J. Appl. Phys. 2015, 54, 07JC05.

[75] Karkin A E, Voronin, V I, Morozova, N V, Ovsyannikov, S V, Takarabe, K, Mori, Y, Nakamura, S, Shchennikov, V V. Unconventional electronic properties of $Mg_2Si$ thermoelectrics revealed by fast-neutron irradiation doping, J. Phys. Chem. C 2016, 120, 96929701.

Supree Pinitsoontorn

# 6 Thermoelectric oxides

**Abstract:** This chapter presents a comprehensive review of the high-performance p-type thermoelectric (TE) oxide materials. The benefits of using oxides for TE applications include high stability and durability at elevated temperature, abundancy, low-cost, and non-toxicity. The research on TE oxides has increased substantially in the last two decades. The most popular candidates for p-type TE oxides include $Na_xCoO_2$, $Ca_3Co_4O_9$, delafossite oxides, and BiCuSeO. They are the main subject of this chapter. There are reports on many aspects of these oxides, for example, crystal structure, defect chemistry, microstructure, electronic structure, fabrication process, or other properties. Hence, in this chapter, we only focus on the TE properties at elevated temperature of high-performance $p$-type oxides, with large dimensionless figure-of-merit ($ZT$). The details of the results are summarized for the two main objectives: why and how such oxides exhibit high $ZT$. Polycrystalline $Na_xCoO_2$ shows the relatively large $ZT$ values of 0.4–0.8, which can be enhanced to >0.9 with doping. The $ZT$ of the pristine $Ca_3Co_4O_9$ is relatively lower ($ZT < 0.4$), but with proper doping, the recorded $ZT$ value of 0.74 at 800 K for this oxide is achieved. Delafossite oxides are another class of materials that have been extensively studied. However, the $ZT$ of the delafossite oxides is still much lower than other $p$-type oxides. The thermoelectric properties of BiCuSeO were firstly reported in 2010, and since then, the compound has been intensively studied. Recently, the $ZT$ value of BiCuSeO has climbed up exponentially to 1.5, which is comparable to or even better than the current state-of-the-art thermoelectric materials, This makes BiCuSeO a very promising $p$-type thermoelectric oxide for high-temperature applications.

**Keywords:** Thermoelectric oxides, $Na_xCoO_2$, $Ca_3Co_4O_9$, Delafossite, BiCuSeO

## 6.1 Introduction

In the past, oxides were not considered to be a good candidate for high-performance thermoelectric (TE) materials because their characteristics are out of the criteria for TE materials selection. These criteria are based on classical semiconductor theories, which state that the candidate for good TE materials should be found in degenerate semiconductors or semimetals with carrier concentration around $10^{18}$–$10^{20}$ cm$^{-3}$ with

**Supree Pinitsoontorn,** Department of Physics, Faculty of Science, Khon Kaen University, Khon Kaen, Thailand; Institute of Nanomaterials Research and Innovation for Energy (IN-RIE), Research Network of NANOTEC-KKU (RNN), Khon Kaen University, Khon Kaen, Thailand

https://doi.org/10.1515/9783110596526-006

a band gap of ~$10k_BT$, where $k_B$ is the Boltzmann constant and $T$ is the absolute temperature. Moreover, materials with large effective mass ($m^*$) and high carrier mobility ($\mu$) are required, which are normally found in materials with high-symmetry crystal structure and similar electronegativity between different constituents. Furthermore, heavy elements or complex unit cells are favored for suppressing lattice thermal conductivity ($\kappa_l$). These guidelines show that most of the current state-of-the-art TE materials fall in such categories, except oxides.

Typically, oxides are inorganic compounds consisting of metal cations and oxygen anions with large difference in their electronegativity, leading to ionic bonding. As a result, conducting charge carriers tend to be localized by the polarized cations or anions, making the carrier mobility much lower than the other semiconducting TE materials. In addition, the small atomic mass of oxygen in oxides and the large bonding energy result in a large sound velocity which yields a large $\kappa_l$. Although oxides were believed to make poor TE materials at first, the discovery of $Na_xCoO_2$ in 1997 [1] with the power factor ($S^2\sigma$) comparable to $Bi_2Te_3$ has stimulated a number of subsequent studies on TE oxides ($S$ = Seebeck coefficient or thermopower, and $\sigma$ is electrical conductivity). A single crystal of $Na_xCoO_2$ was later shown to possess the dimensionless figure of merit ($ZT$) exceeding 1, which suggested that oxide materials could be very promising for use in high-temperature thermoelectric power generation [2].

The benefits of using oxide materials for TE applications over heavy element semiconductor alloys include the stability and durability at high temperature, high abundancy, and the constituents consisting of relatively low-cost elements and nontoxic. The research on TE oxides has gained popularity from many research groups around the world, considering the number of publications in the subject was less than 50 papers in 1999 but became more than 420 publications in 2018 (Figure 6.1). A recent report by

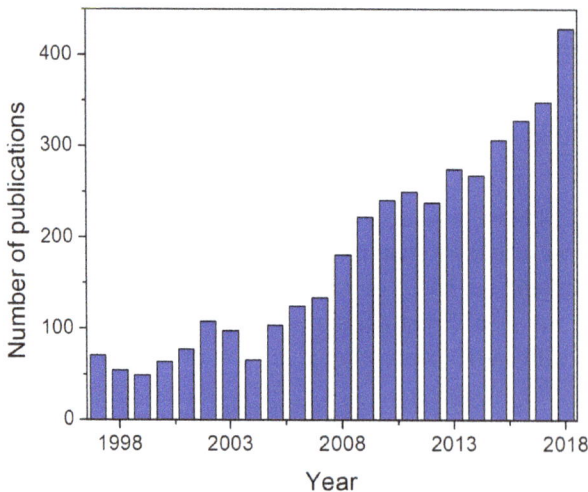

**Figure 6.1:** Number of publications for thermoelectric oxides during 1997–2018.

Gaultois et al. collected data from published literatures for many classes of TE materials, and not only considered TE properties but also production, reserve and scarcity of materials, using Herfindahl–Hirschman index (HHI) [3]. Their analysis showed that oxides are one of the promising TE material families for large-scale applications [3]. So far, there are a number of reports on numerous types of TE oxides in many aspects, for example, crystal structure, defect chemistry, microstructure, electronic structure, fabrication process, or other properties. However, the high-performance oxides for TE conversion (large *ZT*) are limited to not so many species, namely $Na_xCoO_2$, $Ca_3Co_4O_9$, $CuAlO_2$ (delafossite), BiCuSeO for *p*-type materials, and $SrTiO_3$, $CaMnO_3$, ZnO, $In_2O_3$ for *n*-type materials. However, the main subject of this chapter focuses on the TE properties of *p*-type oxide materials of which the crystal structures are shown in Figure 6.2. For *n*-type oxides as well as more details on TE oxide materials, interested readers can find several excellent comprehensive reviews [4–12].

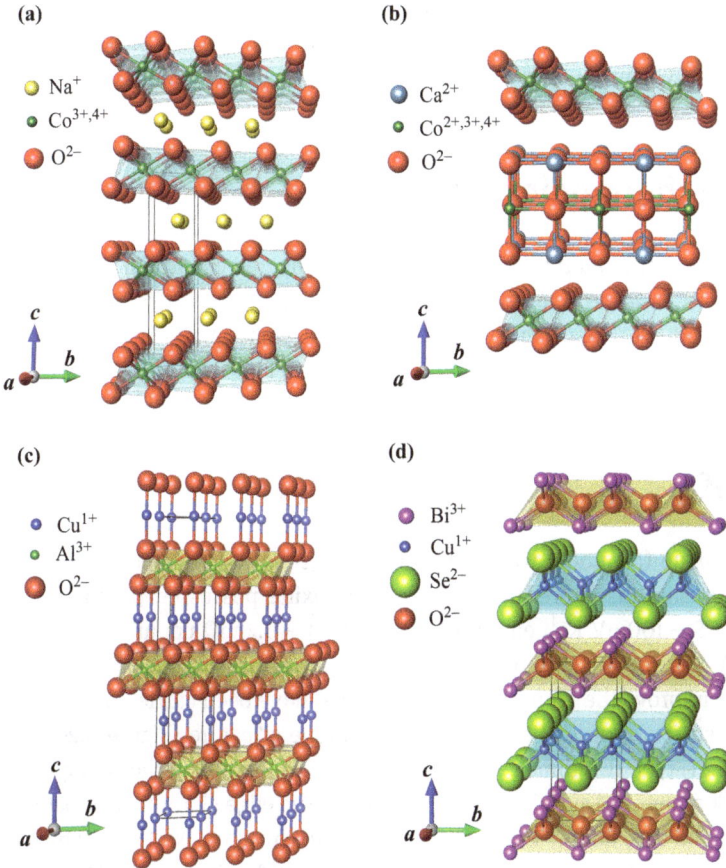

**Figure 6.2:** The crystal structure of *p*-type thermoelectric oxides: (a) $Na_xCoO_2$, (b) $Ca_3Co_4O_9$, (c) $CuAlO_2$ (delafossite), and (d) BiCuSeO.

Although single crystals or superlattice thin films have been reported for a very large $ZT$, the fabrication cost and mass production are the main obstacles for utilizing such materials in TE devices. Therefore, though with lower TE conversion efficiency, polycrystalline oxides in bulk form are the focus of this chapter. Also, in many publications, not all TE parameters, i.e., Seebeck coefficient ($S$), electrical conductivity ($\sigma$), thermal conductivity ($\kappa$) are reported, and thus the dimensionless $ZT$ cannot be calculated. Those publications are not generally considered here. Many researches can show an improvement in power factor without reporting $\kappa$ but normally $\kappa$ tends to increase with $\sigma$, and in fact no improvement in $ZT$ is achieved. In addition, many unsuccessful attempts in enhancing $ZT$ by various means, for example, doping, composite, pressing, or sintering, can be found in literature. The $ZT$ in those works are not improved or even worsened and thus are not considered here. Furthermore, TE oxides are useful for TE applications with temperature ranging from middle- to high-temperature regime (500–900 °C); hence, the low-temperature studies (<300 K) are thus not included. Hence, in this chapter, we mainly discuss the TE properties of high-performance $p$-type oxides as shown in Figure 6.2. The details of the results will be summarized for the two main objectives: why and how such oxides exhibit high $ZT$.

## 6.2 $p$-Type thermoelectric oxides

The most extensively studied $p$-type TE oxides are the layered cobalt oxides such as $Na_xCoO_2$ and $Ca_3Co_4O_9$. The crystal structures of these materials are shown in Figure 6.2(a) and (b). The other related oxide in the same family is the BiSrCoO but the TE performance is much lower than the other two types.

## 6.3 $Na_xCoO_2$

Sodium cobalt oxide ($Na_xCoO_2$) was originally researched for its superconducting behavior but Terasaki discovered in 1997 that this oxide possessed promising TE properties (with $S \sim 100$ μV K$^{-1}$, electrical resistivity $\rho \sim 0.2$ mΩ cm and the power factor of 50 μW K$^{-2}$ cm$^{-1}$) [1], comparable to $Bi_2Te_3$. Since then, a lot of efforts to understand and improve the TE properties of $Na_xCoO_2$ have been carried out.

The crystal structure of $Na_xCoO_2$ is the layered-type structure (Figure 6.2(a)). It consists of the CdI$_2$-type two-dimensional hexagonal CoO$_2$ layer and the Na layer alternatingly stack along its $c$-axis. The CoO$_2$ layer forms the edge-sharing CoO$_6$ octahedra, which acts like the conduction path of electron, whereas the Na layer, which is normally nonstoichiometric, works as a low thermal conductivity path. Therefore, $Na_xCoO_2$ can be considered as a subnanostructuring approach, where the electrical conducting layer of CoO$_2$ is combined with the thermal barrier layer of Na and behave in a class of TE materials, the so-called phonon glass/electron crystal. Moreover, the $CoO_2$ layer was found to be a strongly correlated electron system [13]. The Co ions are the mixed valence states between $Co^{3+}$ and $Co^{4+}$ with high spin degeneracy [14, 15]. These factors are believed to be the contribution of the large thermopower, and the Heikes formula, $S = -\frac{k_B}{e} \ln\left(\frac{g_3}{g_4}\frac{x}{1-x}\right)$, is generally used for determination of $S$, where $g_3$ and $g_4$ are the degeneracies of $Co^{3+}$ and $Co^{4+}$, and $x$ is the concentration of $Co^{4+}$ [15]. The Na content in $Na_xCoO_2$ can be varied from 0.3 to 1.0 but the most common for TE studies are $x = 0.5$–$0.7$.

The TE dimensionless $ZT$ values for the high-performance polycrystalline $Na_xCoO_2$ bulk compounds from room to high temperature are illustrated in Figure 6.3. Figure 6.3(a) presents the research for the pristine $Na_xCoO_2$, whereas Figure 6.3(b) shows the results from doping. For an ease of comparison, both figures use the same scales in $x$- and $y$-axes. The maximum $ZT$ ranges from 0.28 to 0.80. Most of the researches are from the early studies in the late 1990s and early 2000s. Yakabe et al. reported the TE properties of polycrystalline $Na_{0.5}CoO_2$ in 1997 [16]. The oxide powders were synthesized by a standard solid state reaction (SSR) method, and the pellet was formed by cold pressing and sintering, or hot pressing. By cold pressing, the values of $\rho$ and $S$ were 2 m$\Omega$ cm and 100 µV K$^{-1}$, respectively. Hot pressing improved the crystallinity, reduced $\rho$ of about 40%, and boosted $S$ of up to 1.4 times with only a slight increase in $\kappa$. The maximum $ZT$ in this work was found to be 0.36 at 673 K. Ohtaki studied the variation of $x$ in $Na_xCoO_2$ ($x = 0.29$–$1.0$) on the TE properties [17]. The decrease of $x$ tended to increase $\sigma$ due to the formation of positive holes from the Na deficiency, but $S$ was insensitive to $x$. On the other hand, $\kappa$ was lowest at $x \approx 0.5$ due to the most randomness of the Na vacancies. As a result, the highest $ZT$ was observed at $x = 0.46$, $Na_{0.46}CoO_2$, with the maximum value of ~0.5 at 973 K. In another work, Matsubara et al. investigated the effect of Na content in the spark plasma sintered $Na_xCoO_2$ [18]. The best composition was found in $Na_{0.5}CoO_2$. Spark plasma sintering (SPS) technique helped consolidating the oxide powders, leading to very dense samples (~98% of theoretical density) with high texture. Consequently, $\sigma$ was improved but with the cost of higher $\kappa$, while $S$ was not so sensitive to density or grain alignment. The maximum $ZT$ in this work was found to be 0.3 at 973 K.

The recorded high value of $ZT$ in the polycrystalline $Na_xCoO_4$ was published by Ito et al. in 2003 [19]. The $Na_xCoO_2$ flakes were synthesized by the polymerized complex (PC) method. After normal compacting and pressing, very dense samples with a high texture along the pressing direction were obtained. Compared to the samples prepared by SSR, the grain size from the PC method was significantly smaller (leading to lower $\kappa$), and the composition homogeneity was better for the overall area of the samples (resulting in larger $S$). With these reasons, the $ZT$ of the $Na_{0.55}CoO_2$ prepared by PC method showed a substantial improvement with the maximum $ZT$ of 0.8 at 955 K. A later work by Wu et al. studied the TE properties of $Na_{0.5}CoO_2$ synthesized by different techniques, i.e., SSR, sol–gel, and molten salt (MSS) [20]. The Seebeck coefficients are

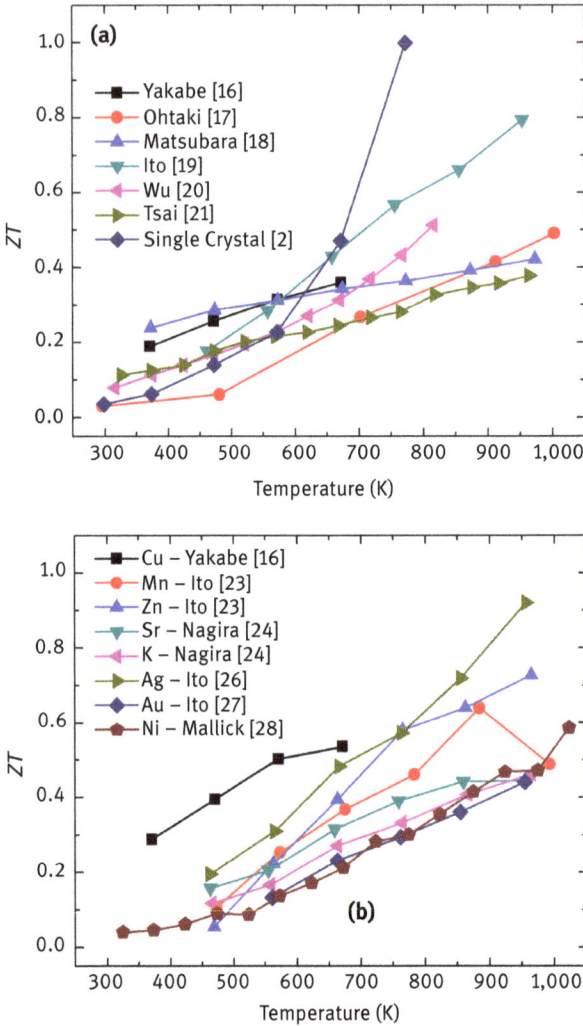

**Figure 6.3:** The *ZT* plots of (a) pristine Na$_x$CoO$_2$ and (b) doped Na$_x$CoO$_2$.

not much different but $\sigma$ are highest for MSS and SSR techniques. The SSR sample also showed the lowest $\kappa$ due to its finest microstructure, leading to the maximum *ZT* value of 0.51 at 814 K. In addition, in a strongly correlated system like Na$_x$CoO$_2$, the thermopower depends on the ratio of Co$^{3+}$ and Co$^{4+}$, which can be changed by varying Na concentration $x$, or adjusting oxygen vacancy. Thus, Tsai et al. did manipulate the oxygen vacancy in Na$_{0.73}$CoO$_{2-\delta}$ ($\delta$ = 0.09, 0.12, 0.14, 0.16) by adjusting oxygen partial pressure during heat treatment [21]. Their results showed that by increasing concentration of oxygen vacancy ($\delta$), $\rho$ and $S$ simultaneously increased due to the recombination of electrons and holes and the change in spin states of Co ions, respectively. $\kappa$ was reduced with decreasing $\delta$. The maximum $ZT$ of 0.38 was found at 963 K for $Na_{0.73}CoO_{1.86}$.

In Figure 6.3(a), the $ZT$ of the single crystal $Na_xCoO_2$ is also plotted. It can be seen that although many efforts have been done to improve $ZT$ of the polycrystalline $Na_xCoO_2$, none of the results can compete with the single crystal $Na_xCoO_2$ ($ZT \sim 1.0$) [2]. This is mainly due to the fact that $\sigma$ of polycrystals is about 1 order of magnitude lower than that of single crystals but their $\kappa$ values at high temperature is not so much reduced since the Umklapp phonon–phonon scattering dominates in this regime. The development of nanograins or nanocomposites could further suppress thermal conductivity without disturbing much of $\sigma$. However, there was a recent study of the $Na_xCoO_2$ ceramic, prepared by an SSR followed by a conventional pressing and sintering, which reported the $ZT$ value of 1.57 at 1,100 K [22]. This value is unbelievably high. In fact, it is too high to be true and need verification by other research groups.

Figure 6.3(b) shows the $ZT$ plots for the doped $Na_xCoO_2$. A number of elements have been substituted either at Na or Co sites in an attempt to improve TE performance. One of the early works was carried out by Yakabe in 1997, where Ba or Bi was partially substituted for Na, and Cu or Bi was substituted for Co [16]. However, the only improvement was observed in the Cu-doped system $Na_{0.5}(Co_{0.95}Cu_{0.05})O_2$ with the maximum $ZT$ of 0.54 at 673 K. The main reasons for the improvement were from the suppression of $\kappa$ simultaneously with an increase of $S$. Ito et al. studied a partial substitution of 3d transition metals (Cr, Mn, Fe, Ni, and Zn) for Co in $Na_xCo_{1-y}M_yO_2$ [23]. For Cr, Fe, and Ni substitution, a relatively large portion of second phases were found and led to the deterioration of TE properties. The Mn-doped sample possessed a single phase with fine microstructure and many pores, which were effective for reducing $\kappa$, yielding a peak $ZT$ of 0.64 at 873 K [23]. The maximum $ZT$ in the series (0.72 at 965 K) was observed in the Zn-doped sample [23]. In this particular sample, the small amount of the second phase (NaO) was observed but actually advantageous since it assisted in decreasing $\kappa$, whereas $S$ and $\sigma$ were affected only slightly. In 2003, Nagira et al. published an article on the partial substitution of Na with several elements in $(Na_{1-y}M_y)_xCoO_2$ (M = K, Sr, Y, Nd, Sm, and Yb) [24]. The doping caused the changes in TE properties ($S$, $\sigma$, $\kappa$). However, the significant improvement of TE $ZT$ was not observed. The Sr-doped sample showed a slight $ZT$ enhancement at temperature <800 K. The K- and Nd-doped samples showed similar $ZT$ values with the undoped samples, whereas the other samples exhibited poorer TE performance. The reasons for the non-enhancement of $ZT$ could be due to the existence of the second phases (other oxide compounds) in all doped samples. If pure phase could be synthesized, the TE properties could be enhanced.

Silver (Ag) is another choice of materials that have been added to the $Na_xCoO_2$ system by Ito et al. [25–27]. The presence of Ag in the $Na_xCoO_2$/Ag composite phases prepared by SSR changed the carrier concentration and led to the significant increase in power factor. Nonetheless, the high $\kappa$ of the Ag phase increases the total $\kappa$ of the composite. As a consequence, the $ZT$ value was not different from the sample without Ag [25, 27]. However, if the $Na_xCoO_4$/Ag composite was fabricated by the PC route, a slight increase in $ZT$ was observed [26]. This was due to the fine Ag particles and their better homogeneous distribution by the PC method. The maximum $ZT$ of 0.92 at 956 K was recorded, which was the highest value for the polycrystalline $Na_xCoO_2$ [26]. Another noble metal (Au) was introduced to form the $Na_xCoO_2$/Au composite, but $ZT$ was relatively the same as the pure $Na_xCoO_2$ [27]. A recent research by Mallick and Vitta partially substituted Co with Ni in $Na_{0.7}Co_{1-x}Ni_xO_2$ compound [28]. Ni substitution changed the carrier concentration and the ratio of $Co^{3+}/Co^{4+}$, and thus enhance the power factor for the compound $Na_{0.7}Co_{0.9}Ni_{0.1}O_2$. Although small amount of NiO was presented, it helped in suppressing $\kappa$, and led to the 57% improvement in $ZT$ (0.58 at 1,023 K). In summary, for the doped $Na_xCoO_2$, not much enhancement of $ZT$ could be found in literature, except the addition of Ag where the recorded value of 0.92 was obtained. For other elemental doping, the TE properties are very similar, if not worse, compared to the pristine $Na_xCoO_2$.

A big disadvantage of $Na_xCoO_2$ is the stability of the compound. When $Na_xCoO_2$ is exposed to ambient atmosphere for a long time, Na reacts with moisture in air and cause hydration. After a while, white scale is normally observed on the surface of the samples. The loss of Na lead to degradation of the compound and its TE performance. With this reason, though with attractive TE properties, many research groups chose not to focus on $Na_xCoO_2$ but concentrate on the related oxide, namely $Ca_3Co_4O_9$.

## 6.4 $Ca_3Co_4O_9$

The structure of $Ca_3Co_4O_9$ (Figure 6.2(b)) is closely related to $Na_xCoO_2$ such that both oxides consist of $CoO_2$ layers, but the Na layer is replaced with the distorted CaO–CoO–CaO ($Ca_2CoO_3$) rock-salt-type layer. Moreover, the alternating $CoO_2$ layer and $Ca_2CoO_3$ layer share the lattice parameters $a$, $c$, and $\beta$ in the monoclinic unit cell but the lattice parameters $b$ are different. The ratio of $b_1$ ($Ca_2CoO_3$) and $b_2$ ($CoO_2$) is the golden ratio ($\approx 1.62$). Thus, $Ca_3Co_4O_9$ is considered as a misfit layered structure, and the actual formula should be written as $(Ca_2CoO_3)(CoO_2)_{1.62}$. The detailed studies for the crystal structure and electronic structure of $Ca_3Co_4O_9$ using X-ray diffraction and photoemission spectroscopy can be found elsewhere [29–31].

Similar to $Na_xCoO_2$, $Ca_3Co_4O_9$ is a strongly electron correlated system [32], and the large thermopower can also be explained from the Heikes formula [15]. For a single crystal $Ca_3Co_4O_9$, $\rho$, $S$, and $\kappa$ are in the range of $2$–$3 \times 10^{-3}$ $\Omega$ cm, $100$–$250$ $\mu V$ $K^{-1}$, and $3$–$4$ W $m^{-1}$ $K^{-1}$, respectively, from room temperature to 973 K [33]. As a result, $ZT$ of the single crystal $Ca_3Co_4O_9$ increases with temperature and shows the maximum of 0.87 at 973 K, as shown in Figure 6.4(a). This value is only slightly below the $ZT$ of $Na_xCoO_2$. However, TE properties of polycrystalline $Ca_3Co_4O_9$ are much inferior to its

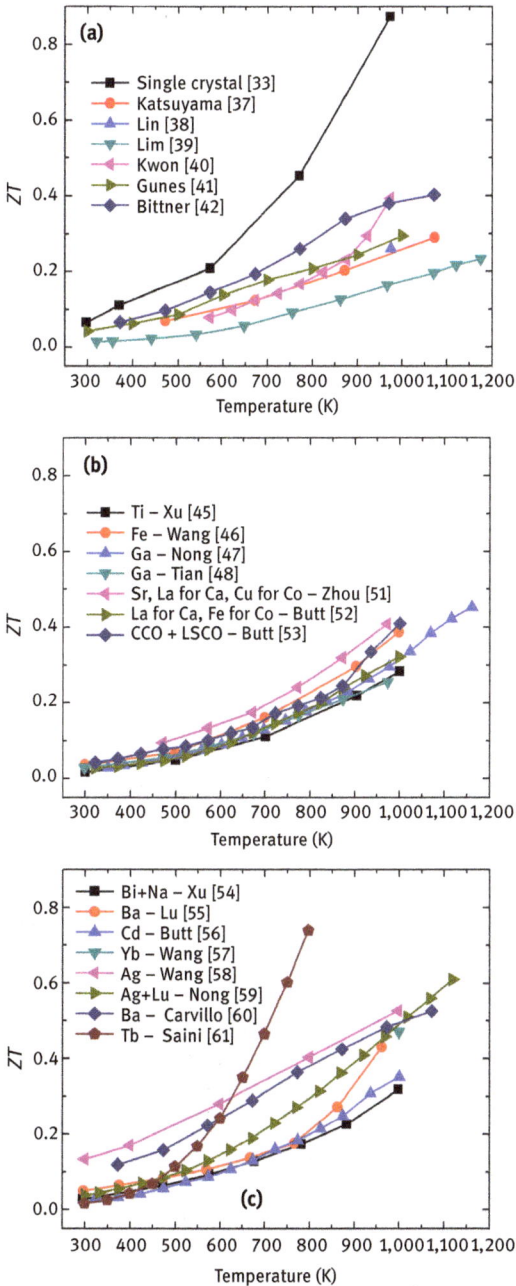

**Figure 6.4:** The *ZT* plots of (a) pristine $Ca_3Co_4O_9$, (b) $Ca_3Co_4O_9$ doped at Co site or both sites, and (c) $Ca_3Co_4O_9$ doped at Ca site.

single crystal, mainly due to a substantial decrease of $\sigma$, nearly an order of magnitude lower. The $ZT$ of polycrystalline $Ca_3Co_4O_9$ was generally found to be much lower ($ZT < 0.2$) [34–36]. The TE properties of $Ca_3Co_4O_9$ polycrystals can be improved by adjusting processing techniques or by doping. Figure 6.4(a) shows the high TE performance of $Ca_3Co_4O_9$ polycrystals with relatively large $ZT$ around 0.2–0.4. Katsuyama et al. synthesized $Ca_3Co_4O_9$ powders by the PC method and consolidated the powders to form a bulk pellet by hydrothermal hot-pressing (HHP) technique [37]. The PC method could assist in refining the microstructure and the HHP technique with high pressure and operating temperature increased the sample density, which significantly improved electrical conductivity with small increase in $\kappa$. The $ZT$ of 0.29 at 1,073 K was achieved.

Increasing the anisotropic behavior of $Ca_3Co_4O_9$ is another approach for the enhancement of TE properties. Lin et al. employed SPS combined with a dynamic forging process in fabrication of $Ca_3Co_4O_9$ ceramics [38]. Highly (00$l$) oriented grains were obtained, lowering $\rho$ and enhancing power factor with a little effect on $S$. The maximum $ZT$ of 0.23 at 975 was observed. Similarly, Lim et al. prepared the $Ca_3Co_4O_9$ ceramics using an SPS process and obtained a highly textured ceramic [39]. The TE properties were improved along the direction perpendicular to the SPS press axis, with the $ZT$ of 0.22 at 1,173 K. Another research by Kwon et al. fabricated a highly textured $Ca_3Co_4O_9$ ceramic by a multisheet cofiring technique [40]. First, 10-mm disk-shaped specimens were cut out of the $Ca_3Co_4O_9$ sheet prepared by uniaxial pressing. Those disks were stacked together and sintered under uniaxial pressure. This cost-effective method was proved to reduce $\rho$ but nearly unchanged $\kappa$, and hence a large $ZT$ of 0.4 at 973 K was observed [40]. On the other hand, Gunes and Ozenbas studied the effect of grain size and porosity on phonon scattering enhancement of $Ca_3Co_4O_9$ synthesized using the citrate sol–gel method [41]. The average grain sizes and densities were varied from 18 nm to 2 µm and 66.0% to 79.5%, respectively. For the smaller grain size and lower density, lattice thermal conductivity was substantially reduced due to grain boundary scattering and nanopore scattering. Despite the smaller power factor, the specimen with the smallest grain and density exhibited the largest $ZT$ of 0.29 at 1,000 K [41]. This work showed that the nanostructuring approach combined with high porosity is fairly effective for improving $ZT$ of $Ca_3Co_4O_9$. Later, Bittner et al. applied the principle of porous $Ca_3Co_4O_9$ in their research and could produce extremely low $\kappa$ of 0.63 W m$^{-1}$ K$^{-1}$ for the porous ceramic with relatively low density (67.7%) [42]. The maximum $ZT$ in their work reached 0.4 at 1,073 K, which is the highest reported value for the undoped polycrystalline $Ca_3Co_4O_9$.

Another approach to improve the TE properties of $Ca_3Co_4O_9$ is via doping. The dopants can be substituted at either Ca site or Co site. It has been argued that the elemental doping at Co site should be more effective since the density functional theory calculation showed that the electron contribution near the Fermi level was mostly limited to the $CoO_2$ layer [43, 44]. Therefore, a few researches have attempted by partially replacing Co with some other constituents. For example, Xu et al. studied the $Ca_3Co_{4-x}Ti_xO_9$ system and found that Ti doping decreased the carrier concentration [45]. Since the valence of Ti ion is higher than the valence of Co in $Ca_3Co_4O_9$, the substitution of Ti for Co is considered as electron doping. Consequently, $\rho$ and $S$ increased simultaneously with higher Ti concentration. Moreover, $\kappa$ was also reduced with Ti content from the disorder and structural distortion; therefore, the $ZT$ of 0.29 was achieved at 1,000 K [45]. The higher $ZT$ of 0.4 at 1,000 K was observed in the Fe doping system ($Ca_3Co_{4-x}Fe_xO_9$) in the research by Wang et al. who fabricated $Ca_3Co_{4-x}Fe_xO_9$ (M = Fe, Mn, and Cu) ceramics by cold high-pressure compacting technique [46]. By using this technique, high pressure of 3 GPa was uniaxially applied using a special steel die leading to high compaction with a large texture. As a consequence, $\rho$ was significantly reduced, compared to a conventional pressing, but $S$ was not changed. The $ZT$ was improved for all dopants but the maximum value was found for Fe doping. Ga is another element, which was substituted at Co-site for TE enhancement [47, 48]. Ga ion charge state is 3+, which is lower than the average valence of Co ion. Therefore, the substitution of Ga for Co introduced more hole carriers, which decreased $\rho$ but showed a small effect on $S$. $\kappa$ was also slightly reduced due to increased phonon scattering by point defects. The maximum $ZT$ for $Ca_3Co_{3.95}Ga_{0.05}O_9$ was found to be 0.45 at 1,200 K for the hot-pressed sample [47] and 0.26 at 973 K for the SPS sample [48].

The $ZT$ plots of $Ca_3Co_4O_9$ doped at Co sites are summarized in Figure 6.4(b). It can be seen that only slight TE enhancement can be observed compared to pristine $Ca_3Co_4O_9$ (Figure 6.4(a). The reason for the lower TE enhancement than expected is because the dopants are not substituted for Co in the $CoO_2$ layer but are preferentially replaced Co in the rock-salt $Ca_2CoO_3$ layer. X-ray absorption spectroscopy (XAS) provided direct evidence that Fe and Ga atoms are more likely to be located in the RS layer rather than in the $CoO_2$ layer as presumed previously [48–50]. With this reason, the electronic contribution near Fermi level from the $CoO_2$ layer is thus not strongly affected. Alternatively, some research have partially substituted both Ca and Co ions at the same time. Zhou et al. fabricated the $(Ca_{2.7}Sr_{0.2}La_{0.1})(Co_{3.9}Cu_{0.1})O_9$ ceramics by the magnetic alignment technique and SPS [51]. Dual sites substitution of Sr, La, and Cu did not appreciably change $\sigma$ and $S$ but significantly reduced $\kappa$, which mainly due to the heavier ions of $Sr^{2+}$ and $La^{3+}$, compared to $Ca^{2+}$. The remarkable $ZT$ of 0.41 was achieved at 973 K [51]. A recent article by Butt et al. dually doped La and Fe in $Ca_{3-x}La_xCo_{4-y}Fe_yO_9$ ceramics prepared by the sol–gel method followed by SPS [52]. The power factor was enhanced by the effect of dual-site doping. More importantly, the doping affected lattice distortion due to the difference in ionic radii, and thus increased phonon scattering. The $ZT$ of 0.32 at 1,000 K was observed for a small amount of La and Fe doping ($Ca_{2.8}La_{0.1}Co_{3.9}Fe_{0.2}O_9$) [52]. A year later, Butt et al. published an article on TE properties of the oxide composite between $Ca_3Co_4O_9$ and $La_{0.8}Sr_{0.2}CoO_3$ (LSCO) [53]. The mixed oxide with 25 wt.% LSCO showed the maximized $ZT$ of 0.41 at 1,000 K, owing to the hierarchical mesoscopic structure with nanoscale LSCO particles and submicrometer-scale grain boundaries. The $ZT$ plots of dual-site doping and the oxide composite are also included in Figure 6.4(b).

A larger number of researches have been focusing on partial substitution at Ca site of $Ca_3Co_4O_9$. Figure 6.4(c) shows the $ZT$ plots of those studies. It should be noted that the scale in $x$- and $y$-axes in Figure 6.3(a)–(c) is equal for the ease of comparison. One of the early works by Xu et al. explored $Bi^{3+}$ and $Na^+$ double substitution for $Ca^{2+}$ in $Ca_{2.4}Bi_{0.3}Na_{0.3}Co_4O_9$ [54]. The double substitution caused the increase in both $\sigma$ and $S$ simultaneously due to the grain size and orientation, the increase of carrier concentration, and the change of energy dependence mobility. Moreover, the double substitution of $Bi^{3+}$ and $Na^+$ greatly suppressed $\kappa_{l,}$ which led to the enhanced $ZT$ of 0.32 at 1,000 K [54]. Lu et al. synthesized $Ca_{3-x}Ba_xCo_4O_9$ by sol–gel and SPS method [55]. They showed that with a certain amount of Ba doping ($x$ = 0.2), $\rho$ and $\kappa$ were reduced with a slight change in $S$. Also, with a prolonged sintering time of 2 h, the density, grain size, and grain orientation increased, which resulted in the optimized $ZT$ of 0.43 at 973 K [55]. Butt et al. partially replaced Co with Cd but Cd in fact occupied the Ca site, confirmed by XAS analysis [56]. They showed the increase of charge carrier concentration and distinctive charge redistribution, which induced a spin-entropy enhancement, leading to the enhancement of $\sigma$ and $S$. The excess Cd formed CdO nanoinclusions serving as phonon scattering centers, which reduced $\kappa$ and resulted in $ZT$ of 0.35 at 1,000 K [56].

Remarkable TE enhancement of $Ca_3Co_4O_9$ by Ca site substitution can be found in the following research [57–61]. Wang et al. studied the TE properties of $Ca_{3-x}Ln_xCo_4O_9$, where Ln is the rare-earth element from La to Yb [57]. The doped Ln ions have noticeable influences on the TE characteristics; the substitutions of $Ln^{3+}$ for $Ca^{2+}$ change hole concentration and also introduce strong point-defect phonon scattering. Among different Ln ions, the $Ca_3Co_4O_9$ doped with Yb exhibited the strongest structural distortion, electron correlation, and ionic mass fluctuation, and thus the most enhanced TE properties. Furthermore, highly textured nanosize grain ceramics, compacted under a high pressure of 3.5 GPa using a special die, double the $ZT$, compared with the specimens prepared by a conventional pressing and sintering. The impressive $ZT$ of 0.47 at 1,000 K was observed [57]. The same research group also studied the effect of Ag doping and Ag addition in $Ca_3Co_4O_9$ on the TE properties [58]. Electrical conductivity of the Ag-doped samples increased attributed to the substitution of univalent $Ag^+$ for divalent $Ca^{2+}$, creating more hole carriers, whereas in the Ag-added samples, Ag metals form a well connection between $Ca_3Co_4O_9$ grains, reducing carrier scattering at grain boundaries, and thus improve $\sigma$. The $S$ of the Ag-doped or Ag-added samples only changed slightly. Although Ag doping and Ag addition led to an increase in $\kappa$, it was well compensated by significant improvement of $\sigma$; hence, the remarkable $ZT$ of 0.52 at 1,000 K was achieved [58]. By pursuing the same principle, Nong et al. dually doped $Ag^+$ and $Lu^{3+}$ at Ca site of $Ca_3Co_4O_9$ [59]. Heavy $Lu^{3+}$ ion changed the carrier concentration as well as behaved as point defects for phonon scattering, while Ag formed metallic nanosized inclusions within the grain interiors, leading to an effective way of improving $S$

and $\sigma$ and yet suppressing $\kappa$. The very high $ZT$ of 0.61 at 1,118 K was obtained, and was further improved after subsequent heat treatment, and maintained even after further heating [59]. In addition, Carvillo reported the enhanced $ZT$ of $Ca_3Co_4O_9$ through Ba grain boundary segregation [60]. $Ca_3Ba_xCo_4O_9$ ceramics were prepared by sol–gel and then cold pressing and sintering. The doped Ba was found to segregate to grain boundaries, which helped in enhancing $S$ and $\sigma$. The Ba-rich grain boundaries acted as an energy barrier filters low-energy carriers out, leading to increased $S$ from energy filtering effect. Simultaneously, Ba grain boundary segregation lowers grain boundary energy and improves crystallographic texture, which promotes carrier mobility. Also, the presence of heavy element Ba at grain boundaries enhanced phonon scattering effect, and the $ZT$ of 0.52 at 1,073 K was observed for $Ca_3Ba_{0.05}Co_4O_9$ [60]. Last but not least, Saini et al. recently reported the recorded value of $ZT$ (0.74 at 800 K) for polycrystalline bulk $Ca_3Co_4O_9$ [61]. The heavy element terbium (Tb) was partially substituted for Ca in the $Ca_{3-x}Tb_xCo_4O_9$ system. $S$ and $\rho$ increased with Tb content up to $x = 0.5$ due to the charge difference between $Tb^{3+}/Tb^{4+}$ and $Ca^{2+}$, which suggested a decrease in hole carrier concentrations. Above $x = 0.5$, the second phase was found, which degraded TE properties. The use of heavy Tb ions suppressed $\kappa$ significantly, and the maximum $ZT$ of 0.74 at 800 K was found for $Ca_{2.5}Tb_{0.5}Co_4O_9$ [61].

## 6.5 Delafossite oxides

Delafossite is a common name of the crystal structure, where the general chemical formula is written as $ABO_2$, where A = Cu, Ag, Pd, or Pt and B = Cr, Fe, Co, Al. As illustrated in Figure 6.2(c), the delafossite structure is another layered-type structure formed by alternatively stacking the O–A–O dumbbell layers with the $BO_2$ edge-shared octahedral layers along the $c$-axis. The most commonly studied compounds are, for instance, $CuAlO_2$, $CuFeO_2$, and $CuCrO_2$. Originally, the delafossite oxides were interested for their electrical and optical properties and were mainly focused for the use as $p$-type transparent conducting thin film. $CuAlO_2$ thin film was the key compound to achieve high optical transparency in the visible region [62] and it remains to be an active material for research in optoelectronic transparent conductive oxide films. Subsequently, Banerjee et al. investigated the TE properties of $CuAlO_2$ thin film in 2005 and found relatively large $\sigma$ and $S$ [63]. Since then, the delafossite oxides were considered as promising TE oxides and a number of related compounds were studied for their TE properties.

Despite the fact that a number of compounds in the delafossite families have been investigated, including $CuAlO_2$, $CuFeO_2$, $CuMgO_2$, $CuCrO_2$, $CuCoO_2$, $CuBO_2$, in both film and bulk forms, the $ZT$ of these oxides are still low. For example, a recent work showed the $ZT$ of $CuAlO_2$ in the order of $10^{-2}$ [64] and in fact the $ZT$ values for many delafossite oxides found in literatures are around this order of magnitude or even lower. Doping in delafossite has shown a significant effect in tuning TE properties since it can change $\sigma$ by several orders of magnitude, turning insulators to semiconductors. However, the $ZT$ greater than 0.2 of delafossite oxides have never been reported. Here, we collected the studies of TE delafossite oxides with a relatively high $ZT$ (>0.04) among other compounds in the same family.

The $ZT$ plots for delafossite oxides are shown in Figure 6.5. One of the early works was published by Kuriyama et al. in 2006, where they reported the high-temperature TE properties of delafossite oxide $CuRh_{1-x}Mg_xO_2$ [65]. The parent compound $CuRhO_2$ is an insulator but substitution of $Mg^{2+}$ for $Rh^{3+}$ increases hole concentration, leading to the metallic conductivity of $CuRh_{0.9}Mg_{0.1}O_2$ with a large thermopower of +270 μV K$^{-1}$. As a result, the $ZT$ of 0.15 was realized above 1,000 K, which was the highest record for the delafossite family [65]. Subsequently, Hayashi et al. studied a delafossite compound $CuCrO_2$ and the effect of doping on TE properties of the oxide [66]. Single and dual doping using several dopants (Mg, Zn, Ca, Ni, Co) have been substituted at Cr site. For the single doping, the maximum power factor was obtained for the Mg-doped sample, whereas for the dual doping, the combined Mg and Ni substitution for Cr showed the most improvement in TE properties due to the overlapping integral between Cu 3d and O 2p orbitals and the increase in the hole concentration. The resultant $ZT$ was found to be 0.10 at 1,100 K for the $CuCr_{0.93}Mg_{0.03}Ni_{0.04}O_2$ compound [66]. Furthermore, the effect of Mg substitution at Cr site in $CuCr_{1-x}Mg_xO_2$ on structure and TE properties was investigated by Ono et al [67]. The values of $\rho$ decreased with the amount of Mg content from the hole doping effect by substituting $Cr^{3+}$ with $Mg^{2+}$. The change of $S$ with $Mg^{2+}$ can be explained by Heikes formula considering spin entropy. The highest $ZT$ value of 0.04 at 950 K was observed [67].

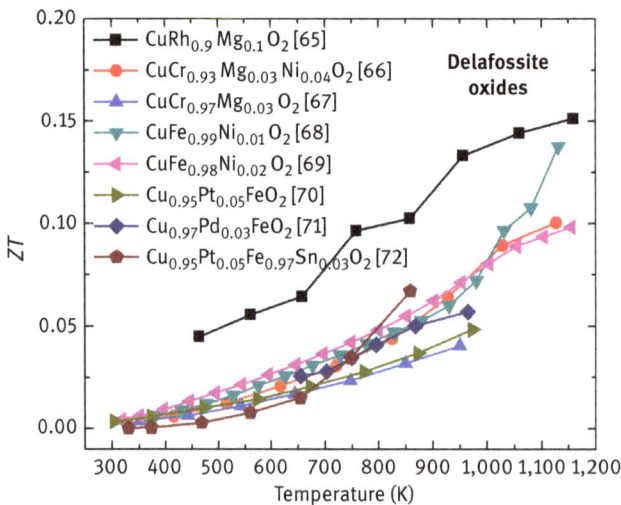

Figure 6.5: The $ZT$ plots of delafossite oxides.

TE properties of the delafossite oxide $CuFeO_2$ were studied by several research groups. Hayashi et al. studied the high-temperature TE properties of $CuFe_{1-x}Ni_xO_2$ and found that Ni doping increased hole concentration and thus enhanced $\sigma$ [68]. Although $S$ decreased by Ni doping, the high value of $S$ of >250 $\mu V\ K^{-1}$, and relatively low $\kappa$ were obtained, resulting in the maximum $ZT$ of 0.14 at 1,100 K. Later, Nozaki et al. investigated the chemical composition and electronic structure of the same system using X-ray photoelectron spectroscopy (XPS) [69]. The hole carrier density in the Cu and $FeO_2$ layers of the Ni-doped sample was found to be different from that of $CuFeO_2$, leading to the enhancement of electrical conductivity by Ni substitution. The maximum $ZT$ in this work was found to be 0.10 at 1,150 K [69]. Ruttanapun et al. published a few works related to the TE properties of $CuFeO_2$. Unlike most research reported previously, the substitution was taken at A-site, that is, partial replacement of Cu ions. In the $Cu_{1-x}Pt_xFeO_2$ delafossite system [70], increasing Pt concentration resulted in the enhancement of $\sigma$ but no obvious effect on $S$. Although $\kappa$ also increased with Pt, the $ZT$ was still improved with Pt substitution with the highest $ZT$ of 0.05 at 960 K [70]. In another system, $Cu_{1-x}Pd_xFeO_2$ compound, Pd partial substitution for Cu led to an enhanced power factor compared to the undoped sample [71]. Also, the large atomic mass of Pd helped suppressing $\kappa_l$, resulting in the maximum $ZT$ of 0.055 at 950 K. Finally, Ruttanapun et al. simultaneously substitute at Cu site and Fe site in the delafossite oxide $Cu_{0.95}Pt_{0.05}Fe_{0.97}Sn_{0.03}O_2$ [72]. By combining the effect of $\sigma$ enhancement by Pt doping, $S$ enhancement by Sn doping, and $\kappa_l$ reduction by employing complex structure, the studied compound showed an improved TE properties with maximum $ZT$ of 0.07 at 860 K [72].

In summary, even though a number of effects have been carried out to improve the TE properties of delafossite oxides, the $ZT$ values are still very low, much lower than $Na_xCoO_2$, $Ca_3Co_4O_9$, or BiCuSeO. It cannot be seen in the near future that the delafossite compounds will be widely utilized for TE applications.

## 6.6 BiCuSeO

BiCuSeO is a relatively new class of TE oxides. The TE properties of BiCuSeO were first reported by Zhao et al. in 2010, with the $ZT$ of ~0.4 at 873 K for the pristine compound [73]. Since then, it has become one of the attractive materials for TE studies. With proper doping, the carrier concentration can be optimized but the power factor of BiCuSeO is usually less 0.6 mW $m^{-1}\ K^{-1}$ at 923 K [74]. This value is much lower than those observed in the state-of-the-art TE materials or other layered oxides. However, it is more than compensated by an extremely low $\kappa$ of 0.40 at 923 K [75]. The intrinsically low $\kappa$ of BiCuSeO, much lower than the values observed for most TE materials, is thought to originate from its strong crystal anharmonicity, weak chemical bond, and low Debye temperature [76]. The $ZT$ of BiCuSeO has been continuously increased and reached the recorded value of 1.5 at 873 K in recent years [77, 78].

BiCuSeO is in an oxychalcogenide group, with a general formula RMChO, where R = Bi, Ce to Dy, M = Cu, Ag, and Ch = S, Se, Te, commonly called 1111-type compounds. The crystal structure of BiCuSeO is the ZrCuSiAs-type tetragonal structure with the space group P4/nmm [75]. Interestingly, the structure of BiCuSeO is also an alternating layered structure, consisting of $Bi_2O_2$ layers and $Cu_2Se_2$ layers stacking along the $c$-axis, forming a natural superlattice (Figure 6.2(d)). The electrical transport pathway is mainly limited to the $Cu_2Se_2$ layers and $Bi_2O_2$ layers, whereas the $Bi_2O_2$ layers behave as a charge reservoir [79]. Similar to $NaCoO_2$, $Ca_3Co_4O_9$, and delafossite oxides, BiCuSeO exhibits anisotropic transport behaviors due to its layered structure. A simple ionic model is commonly applied to describe the structure of BiCuSeO. With this model, to keep the charge balance, the ionic states of each element are assigned: $Bi^{3+}$, $Cu^{1+}$, $Se^{2-}$, $O^{2-}$. Hence, each layer is usually written as $(Bi_2O_2)^{2+}$ and $(Cu_2Se_2)^{2-}$. This simplified model is useful to understand the doping behavior in this compound [75].

In its pristine form, BiCuSeO is a semiconductor with a moderately large band gap ($E_g \sim 0.8$ eV). However, the TE properties of BiCuSeO are strongly dependent on the synthesis conditions since the presence of Cu vacancies is often found, which leads to a significantly increased hole concentration. Room temperature $\sigma$ can vary up to an order of magnitude depending on the synthesis conditions [74]. Figure 6.6 (a) summarizes the $ZT$ plots of the pristine BiCuSeO. Since the discovery of BiCuSeO by Zhao et al. in 2010 ($ZT \sim 0.4$ at 873 K) [73], there have been several efforts to boost up the $ZT$ of pristine BiCuSeO. The early effort was carried out by Liu et al. in 2011 [80]. In their work, the TE performance of BiCuSeO was remarkably enhanced by intentional inducing Cu deficiencies. The electrical conduction path in BiCuSeO is mainly confined in the $(Cu_2Se_2)^{2-}$ layer. Thus, an introduction of Cu vacancies in the $(Cu_2Se_2)^{2-}$ layers increased hole concentration and improved $\sigma$ for more than an order of magnitude. This method still maintained a relatively high $S$ and low $\kappa$. As a result, the high $ZT$ of 0.81 at 923 K was achieved, nearly two times larger than that of the pristine BiCuSeO [80]. In the subsequent research, instead of monovacancy, 2.5% Bi and Cu dual vacancies were introduced in the insulating $(Bi_2O_2)^{2+}$ layers and the conductive $(Cu_2Se_2)^{2-}$ layers, respectively [81]. The dual vacancies were proved to effectively decrease $\kappa_l$ in the $Bi_{0.975}Cu_{0.975}SeO$ compound. More importantly, there existed the interlayer charge transfer between those Bi/Cu vacancies, which led to a significant increase in $\sigma$ with relatively high $S$. The maximum $ZT$ in this work was found to be 0.84 at 750 K. The findings also provide a new strategy for rational design of high-performance TE materials [81].

As mentioned earlier, the TE properties of BiCuSeO strongly depend on synthesis conditions; thus, Li et al. explored the fabrication of polycrystalline BiCuSeO by a high-energy ball milling process before SPS [82]. The ball milling promoted composition homogeneity and grain size refining, which reduced the charge carrier mobility due to

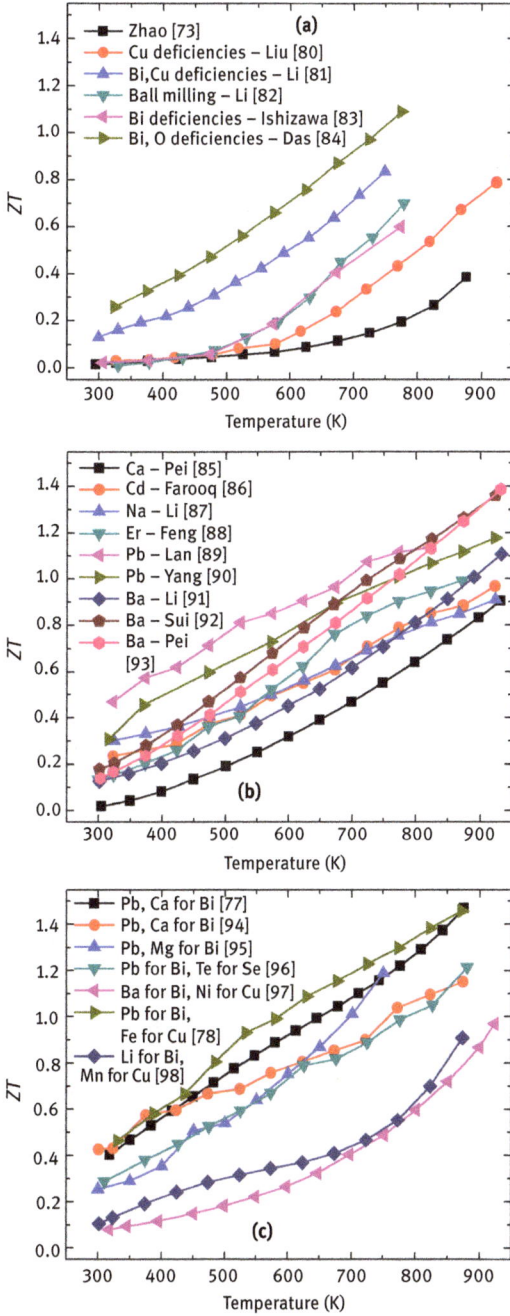

**Figure 6.6:** The *ZT* plots of (a) pristine BiCuSeO, (b) single-doped, and (c) dual-doped BiCuSeO.

the increased grain boundary scattering. However, carrier concentration increased by ball milling, which was believed to be a result of Cu vacancy formed during the synthesis process. Consequently, $\sigma$ of the ball-milled sample increased several folds compared to a conventional synthesis. The $ZT$ was improved to 0.70 at 773 K, 3.5 times higher than that of the pristine sample [82]. More recent works for the TE properties of pristine BiCuSeO were reported by Ishizawa et al. [83] and Das et al. [84]. Ishizawa et al. fabricated BiCuSeO with Bi or Se deficiencies in the nominal composition. However, Bi or Se deficiencies in the starting composition caused the decomposition of the compounds as the pure BiCuSeO matrix phase plus impurity phases, resulting in the maximum $ZT$ of 0.60 at 773 K for the $Bi_{0.95}CuSeO$ compound. In addition, this work employed the XPS analysis, which confirmed the valence states of each ion to be close to $Bi^{3+}$, $Cu^{1+}$, $Se^{2-}$, and $O^{2-}$ [83]. The maximum $ZT$ of 1.09 for the pristine BiCuSeO was reported by Das et al. [84]. In their study, the $CuSe_2$ secondary phase was found in the Bi- and O-deficient BiCuSeO matrix. These defects led to the enhanced TE properties by several means. First, Bi and O deficiency increased the carrier concentration, and thus increased $\sigma$. Second, Bi and O vacancies as well as the presence of $CuSe_2$ impurity suppressed $\kappa_l$ due to the all-length-scale phonon scattering. These combined properties resulted in the high TE performance.

Since $\kappa$ of pristine BiCuSeO is intrinsically low, the main focus of many researches is to enhance the power factors, especially to increase $\sigma$ by optimizing carrier concentration (Figure 6.6(b)). The simplest way to do this is to partially dope in the $(Bi_2O_2)^{2+}$ layers, particularly doping 2+ or 1+ ions for $Bi^{3+}$ (hole doping). Doping in the $(Bi_2O_2)^{2+}$ layers can increase carrier concentration but still can maintain large mobility since the main conduction path is confined in the $(Cu_2Se_2)^{2-}$ layers. For example, Pei et al. partially substituted $Ca^{2+}$ for $Bi^{3+}$ to increase carrier concentration, and significantly improve $\sigma$ [85]. The $S$ was still large (~240 µV K$^{-1}$ at 923 K) and $\kappa_l$ was still low (~0.45 W m$^{-1}$ K$^{-1}$ at 923 K). Therefore, the high $ZT$ of ~0.9 at 923 K was achieved for $Bi_{0.925}Ca_{0.075}CuSeO$. Similarly, $Cd^{2+}$ was doped at Bi site in BiCuSeO to improve carrier concentration and hence $\sigma$ [86]. BiCuSeO with 5% Cd doping was proved to possess large $S$ and high $\sigma$ up to 923 K, resulting in the largest $ZT$ (0.98 at 923 K). In another work, instead of using 2+ ions, Li et al. partially substituted $Na^+$ for $Bi^{3+}$ [87]. The $Na^+$ ion has a similar ionic size to $Bi^{3+}$ ion, and two extra holes can be generated by one $Na^+$ substitution. Thus, optimization of carrier concentration is achieved at lower doping levels. As a result, the high $ZT$ of 0.91 at 923 K was found for a low concentration of $Na^+$ doping ($Bi_{0.985}Na_{0.015}CuSeO$). Recently, Feng et al. reported the enhanced TE performance in BiCuSeO ($ZT$ of 0.99 at 873 K) via Er and 3D modulation doping [88]. The Er-doped samples showed lower carrier concentration, but higher carrier mobility. Therefore, in modulation doping, the heavily Er-doped BiCuSeO powders were ball milled with the undoped BiCuSeO before sintering. The undoped grains created the higher carrier concentration region for which the charge carriers were transferred to the doped grains with lower carrier concentration but higher carrier mobility. These effects provided fast channels for carrier transporting and nearly doubled $\sigma$ as compared to the uniformly doped samples.

The most effective dopants for TE enhancement of BiCuSeO belong to Pb and Ba ions. Lan et al. fabricated Pb-doped BiCuSeO ceramics with fine grain structure containing nanodots that effectively suppress $\kappa_l$ [89]. Moreover, Pb doping was found to be more effective than other dopants for enhancing weighted mobility, $\mu(m^*)^{3/2}$, where $m^*$ is the density of states (DOS) effective mass and $\mu$ is the charge carrier mobility. The weighted mobility is an important parameter for a simultaneous increase of $\sigma$ and $S$. The enhanced weighted mobility is due to the fact that the doped Pb created the partial DOS from 6s components close to the valence band maximum. Such delocalized 6s orbitals from Pb substitution are believed to enhance the hole mobility. Thus, Pb doping in BiCuSeO increased $\sigma$ more effectively than other dopants and raised the $ZT$ to 1.14 at 823 K [89]. Yang et al. also doped Pb in the BiCuSeO system, which was synthesized by self-propagating high-temperature synthesis (SHS) [90]. This rapid process was proved to synthesize the single phase of BiCuSeO in a few seconds. By using the combined X-ray diffraction, differential scanning calorimetry, and quenching experiments, the reactions and the products from each step of the SHS process were elucidated. The $Bi_{0.94}Pb_{0.06}CuSeO$ compound synthesized by the SHS process showed the high $ZT$ value of 1.2 at 923 K [90]. Ba is another dopant which can effectively improve the TE properties of BiCuSeO. For instance, Li et al. fabricated the Ba heavily doped BiCuSeO by a two-step SSR route before SPS [91]. Doping $Ba^{2+}$ for $Bi^{3+}$ improved $\sigma$ due to the increased carrier concentration. Moreover, $\kappa_l$ was significantly reduced by point defect scattering through Ba doping, which originated from both the mass and size differences between $Ba^{2+}$ and $Bi^{3+}$. The maximum $ZT$ value of 1.1 at 923 K was obtained for the compound $Bi_{0.875}Ba_{0.125}CuSeO$ [91]. The TE properties of Ba-doped BiCuSeO was further boosted via texturation [92] and 3D modulation doping [93]. Sui et al. utilized a hot-forging process in fabricating $Bi_{0.875}Ba_{0.125}CuSeO$ ceramics [92]. The carrier mobility along the direction perpendicular to the pressing direction was significantly increased, resulting in an increase in $\sigma$, power factor, and $ZT$ (1.4 at 923 K). On the other hand, Pei et al. improved the carrier mobility of Ba-doped BiCuSeO by modulation doping approach [93]. In doing so, the pristine BiCuSeO powders (lower carrier concentration but higher carrier mobility) and $Bi_{0.75}Ba_{0.25}CuSeO$ powders (higher carrier concentration but lower carrier mobility) were mixed via ball milling, followed by SPS. Since the Fermi levels of the two phases are different, charge carriers diffuse from the doped phase to the undoped phase, resulting in overall enhanced carrier mobility in comparison to uniform doping. As a result, the modulation-doped sample possessed similar $S$ but higher $\sigma$, compared to the uniformly doped sample with the same carrier concentration. The $ZT$ was enhanced from 1.1 (uniformly doping) to 1.4 (modulation doping) at 923 K [93]. The $ZT$ values of the enhanced TE performance BiCuSeO by doping are summarized in Figure 6.6(b).

The effect of dual doping on the TE properties enhancement of BiCuSeO was investigated by several research groups, and the $ZT$ of those studies are plotted in

Figure 6.6(c). Since Pb and Ba are the most effective dopants for enhancing the TE properties of BiCuSeO, they are chosen as the main dopants, and then the other elements were added for further TE improvement. The remarkable work by Liu et al. used this strategy for synergistically optimizing electrical and thermal transport properties of BiCuSeO [77]. The closely packed micrograins of the $Bi_{1-2x}Ca_xPb_xCuSeO$ compounds were fabricated by dual-doping $Pb^{2+}$ and $Ca^{2+}$ for $Bi^{3+}$. Pb was evidently identified to replace Bi and remain in solid solution, whereas Ca did not dissolve in the matrix but formed $CaO_2$ nanoclusters. Moreover, the microstructural investigation revealed the evidence of Bi-rich nanoinclusions. This combined effect led to the enhanced $\sigma$ due to increased carrier concentration from Pb doping as well as the suppressed $\kappa_l$ from the presence of nanostructure phases. The recorded value of $ZT$ (1.5 at 873 K) was achieved for the $Bi_{0.88}Ca_{0.06}Pb_{0.06}SeO$ composition [77]. Subsequently, similar work was carried out for the $Bi_{0.88}Ca_{0.08}Pb_{0.08}SeO$ compound [94]. By optimizing processing conditions (ball milling time and sintering process), the maximum $ZT$ value of 1.15 at 873 K was obtained. Sun et al. modified the compound slightly by replacing $Ca^{2+}$ with $Mg^{2+}$ [95]. It is argued that by changing $Ca^{2+}$ to $Mg^{2+}$, the larger mass and size difference could be produced and might enhance the TE properties of BiCuSeO even further. The maximum $ZT$ of 1.19 at 750 K was observed, which was a very high value at such temperature [95].

Another strategy for dual doping is to dope not only at the Bi site but also at Cu or Se sites. For example, Ren et al. prepared the Pb- and Te-dually doped BiCuSeO by SHS process and found significantly enhanced TE properties [96]. Pb doping at Bi site, as discussed earlier, had a role in optimizing carrier concentration, thus increased $\sigma$, and in suppressing $\kappa_l$ due to point defect scattering or nanoinclusion scattering. On the other hand, Te was substituted for Se in the conductive $(Cu_2Se_2)^{2-}$ layers; thus, it had a direct impact on the electrical transport. Te is less electronegative than Se ($\chi_{Se}$ ~ 2.55, $\chi_{Te}$ ~ 2.1), and its electronegativity is closer to Cu ($\chi_{Cu}$ ~ 1.90). Therefore, doping Te for Se improved chemical bond covalency, resulting in improved carrier mobility through the weakening of carrier–phonon coupling. The Te-doped samples thus retained large $\sigma$ and large power factor values even with decreased carrier concentration. The maximum $ZT$ value of 1.2 at 873 was achieved in $Bi_{0.96}Pb_{0.04}CuSe_{0.95}Te_{0.05}O$, 2.4 times higher than that of the pristine sample [96]. Wen et al. also utilized the dual-doping strategy in enhancing the TE performance of BiCuSeO [97]. They partially substituted Ba for Bi and Ni for Cu and found the enhanced $\sigma$. Meanwhile, the introduction of spin entropy by magnetic Ni ions resulted in increased $S$. The enhanced power factor and low $\kappa$ in this work led to the high $ZT$ of 0.97 at 923 K for $Bi_{0.875}Ba_{0.125}Cu_{0.85}Ni_{0.15}SeO$ [97]. In another research, the very high $ZT$ of 1.5 at 873 K was recorded for the dually doped $Bi_{0.94}Pb_{0.06}Cu_{0.99}Fe_{0.01}SeO$ compound [78]. The partial substitution of Fe for Cu in $Bi_{0.94}Pb_{0.06}CuSeO$ enhanced both $S$ and $\sigma$ simultaneously. The presence of Fe reduced carrier concentration resulting in increased $S$. On the other hand, the carrier mobility increased for low Fe concentration doping, leading to increased $\sigma$. However, the reason for the increased mobility was not totally clear but was thought to originate from the reduction of Cu vacancies due to Fe doping, allowing fast transport of holes in the $Cu_2Se_2$ layers [78]. Lastly, Tang et al. demonstrated that light element doping can be effective for TE enhancement of BiCuSeO [98]. In their work, Li was doped at Bi site, which boosted carrier mobility for about twofold due to largely reduced carrier scattering. Moreover, the $S$ was remarkably increased by Mn doping at the Cu site in $Bi_{1-x}Li_xCuSeO$, contributing to induced spin entropy. Additionally, nanoprecipitates and dual-atom point defect led to a significant reduction in $\kappa_l$. The high $ZT$ value of 0.9 at 873 K was achieved in this work for $Bi_{0.9}Li_{0.1}Cu_{0.9}Mn_{0.1}SeO$.

## 6.7 Conclusion

Oxides were not originally considered as candidates for good thermoelectric materials but the discovery of $Na_xCoO_2$ with a large power factor in 1997 has sparked a number of research groups to reconsider them. This chapter has presented an extensive review on the high-performance $p$-type TE oxides, which have been continuously researched in literature, namely, $Na_xCoO_2$, $Ca_3Co_4O_9$, delafossite oxides, and BiCuSeO. All of these oxides possess the layered-type structure, indicating that such structure is important for high thermoelectric performance. The layered oxides consist of the electrical conducting layer and the thermal insulating layer, thus behaving as "phonon glass/electron crystal." The research on polycrystalline $Na_xCoO_2$ showed the relatively large $ZT$ values of ~0.4–0.8. By doping, the $ZT$ was enhanced to >0.9 at 956 K. However, the $Na_xCoO_2$ compound is sensitive to moisture, which causes hydration and decomposition, making the stability of $Na_xCoO_2$ a problematic issue. The related compound $Ca_3Co_4O_9$ is more stable in an ambient environment but the $ZT$ of the pristine $Ca_3Co_4O_9$ is relatively lower ($ZT < 0.4$). Doping at the Co site of $Ca_3Co_4O_9$ did not show much TE improvement compared to the pristine form, but doping at Ca site is more promising. The recorded $ZT$ value of 0.74 at 800 K was found in the Tb-doped $Ca_3Co_4O_9$. Delafossite oxides are another class of materials that have been extensively studied for their thermoelectric properties. Despite that fact, the thermoelectric performance of the delafossite oxides ($ZT < 0.2$) is still much lower than other oxides. The most promising oxide for thermoelectric applications belongs to BiCuSeO. Though only discovered in 2010, intensive studies and numerous publications have been found in the recent years. The $ZT$ value of BiCuSeO has exponentially climbed up from 0.4 to 1.5 in only 7 years. Several strategies such as carrier concentration optimization, carrier mobility enhancement, and all-length-scale phonon scattering have been adopted for such development. The thermoelectric performance of this material is comparable to or better than the current state-of-the-art thermoelectric materials, causing BiCuSeO as a very promising $p$-type thermoelectric oxide for high-temperature applications.

**Acknowledgments:** This work was funded by the following: (1) Thailand Research Fund (TRF) in cooperation with Synchrotron Light Research Institute (public organization) and Khon Kaen University (RSA6280020). (2) Research Network NANOTEC (RNN) program of the National Nanotechnology Center (NANOTEC), NSTDA, Ministry of Higher Education, Science, Research and Innovation, and Khon Kaen University, Thailand.

# References

[1]   Terasaki I, Sasago, Y, Uchinokura, K. Large thermopower in a layered oxide $NaCo_2O_4$, Phys. Rev. B. 1997, 56(20), R12685–R7.

[2]   Fujita K, Mochida, T, Nakamura, K. High-temperature thermoelectric properties of $Na_xCoO_{2-\delta}$ single crystals, Jpn. J. Appl. Phys. 1 2001, 40(7), 4644–4647.

[3]   Gaultois M W, Sparks, T D, Borg, C K H, Seshadri, R, Bonificio, W D, Clarke, D R. Data-Driven Review of Thermoelectric Materials: Performance and Resource Considerations, Chem. Mater. 2013, 25(15), 2911–2920.

[4]   Koumoto K, Terasaki, I, Funahashi, R. Complex oxide materials for potential thermoelectric applications, MRS Bulletin 2006, 31(3), 206–210.

[5]   Ohta H, Sugiura, K, Koumoto, K. Recent progress in oxide thermoelectric materials: p-type $Ca_3Co_4O_9$ and n-type $SrTiO_3$-, Inorg. Chem. 2008, 47(19), 8429–8436.

[6]   Ohtaki M. Recent aspects of oxide thermoelectric materials for power generation from mid-to-high temperature heat source, J. Ceram. Soc. Jpn. 2011, 119(1395), 770–775.

[7]   He J, Liu, Y F, Funahashi, R. Oxide thermoelectrics: The challenges, progress, and outlook, J. Mater. Res. 2011, 26(15), 1762–1772.

[8]   Fergus J W. Oxide materials for high temperature thermoelectric energy conversion, J. Eur. Ceram. Soc. 2012, 32, 525–540.

[9]   Walia S, Balendhran, S, Nili, H, Zhuiykov, S, Rosengarten, G, Wang, Q H. et al. Transition metal oxides – Thermoelectric properties, Prog. Mater. Sci. 2013, 58(8), 1443–1489.

[10]  Ren G K, Lan, J L, Zeng, C C, Liu, Y C, Zhan, B, Butt, S. et al. High Performance Oxides-Based Thermoelectric Materials, JOM-US 2015, 67(1), 211–221.

[11]  Hebert S, Berthebaud, D, Daou, R, Breard, Y, Pelloquin, D, Guilmeau, E. et al. Searching for new thermoelectric materials: some examples among oxides, sulfides and selenides, J. Phys.: Condens. Matter 2016, 28(1), 013001.

[12]  Korotcenkov G, Brinzari, V, Ham, M H. $In_2O_3$-Based Thermoelectric Materials: The State of the Art and the Role of Surface State in the Improvement of the Efficiency of Thermoelectric Conversion, Crystals 2018, 8(1), 14.

[13]  Ando Y, Miyamoto, N, Segawa, K, Kawata, T, Terasaki, I. Specific-heat evidence for strong electron correlation in the thermoelectric materials $(Na,Ca)Co_2O_4$, Phys. Rev. B. 1999, 60, 10580.

[14]  Wang Y Y, Rogado, N S, Cava, R J, Ong, N P. Spin entropy as the likely source of enhanced thermopower in NaxCo2O4, Nature 2003, 423(6938), 425–428.

[15]  Koshibae W, Tsutsui, K, Maekawa, S. Thermopower in cobalt oxides, Phys. Rev. B. 2000, 62 (11), 6869–6872.

[16]  Yakabe H, Kikuchi, K, Terasaki, I, Sasago, Y, Uchinokura, K Thermoelectric Properties of Transition-Metal Oxide $NaCo_2O_4$ System. 16th International Conference on Thermoelectrics. Dresden, Germany 1997.

[17]  Ohtaki M Thermoelectric Properties and Crystal Chemistry of Promising Oxide Candidate $Na_xCoO_2$. 18th International Conference on Thermoelectrics. Baltimore, Maryland, USA: IEEE; 1999.

[18]  Matsubara I, Zhou, Y Q, Takeuchi, T, Funahashi, R, Shikano, M, Murayama, N. et al. Thermoelectric properties of spark-plasma-sintered Na1+xCo2O4 ceramics, J. Ceram. Soc. Jpn. 2003, 111(4), 238–241.

[19]  Ito M, Nagira, T, Furumoto, D, Katsuyama, S, Nagai, H. Synthesis of NaxCo2O4 thermoelectric oxides by the polymerized complex method, Scripta. Mater. 2003, 48(4), 403–408.

[20]  Wu Y, Wang, J, Yaer, X B, Miao, L, Zhang, B Y, Guo, F. et al. Effects of preferred orientation and crystal size on thermoelectric properties of sodium cobalt oxide, Funct. Mater. Lett. 2016, 9(1), 1650010.

[21]  Tsai P H, Norby, T, Tan, T T, Donelson, R, Chen, Z D, Li, S. Correlation of oxygen vacancy concentration and thermoelectric properties in $Na_{0.73}CoO_{2-\delta}$, Appl. Phys. Lett. 2010, 96(14), 141905.

[22]  Krasutskaya N S, Klyndyuk, A I, Evseeva, L E, Tanaeva, S A. Synthesis and properties of $Na_xCoO_2$ (x=0.55, 0.89) oxide thermoelectrics, Inorg. Mater+ 2016, 52(4), 393–399.

[23]  Ito M, Nagira, T, Oda, Y, Katsuyama, S, Majima, K, Nagai, H. Effect of partial substitution of 3d transition metals for Co on the thermoelectric properties of $Na_xCo_2O_4$, Mater. Trans. 2002, 43(3), 601–607.

[24]  Nagira T, Ito, M, Katsuyama, S, Majima, K, Nagai, H. Thermoelectric properties of $(Na_{1-y}M_y)_xCo_2O_4$ (M=K, Sr, Y, Nd, Sm and Yb; y=0.01 similar to 0.35), J. Alloy. Compd. 2003, 348(1–2), 263–269.

[25]  Ito M, Furumoto, D. Effects of mechanical milling and Ag addition on thermoelectric properties of $Na_xCo_2O_4$, Scripta. Mater. 2006, 55(6), 533–536.

[26]  Ito M, Furumoto, D. Microstructure and thermoelectric properties of $Na_xCo_2O_4$/Ag composite synthesized by the polymerized complex method, J. Alloy. Compd. 2008, 450(1–2), 517–520.

[27]  Ito M, Furumoto, D. Effects of noble metal addition on microstructure and thermoelectric properties of $Na_xCo_2O_4$, J. Alloy. Compd. 2008, 450(1–2), 494–498.

[28]  Mallick M M, Vitta, S. Enhancing Thermoelectric Figure-of-Merit of Polycrystalline NayCoO2 by a Combination of Non-stoichiometry and Co-substitution, J. Electron. Mater. 2018, 47(6), 3230–3237.

[29]  Masset A C, Michel, C, Maignan, A, Hervieu, M, Toulemonde, O, Studer, F. et al. Misfit-layered cobaltite with an anisotropic giant magnetoresistance: Ca3Co4O9, Phys. Rev. B. 2000, 62(1), 166–175.

[30]  Muguerra H, Grebille, D, Bouree, F. Disordered misfit [Ca2CoO3][CoO2](1.62) structure revisited via a new intrinsic modulation, Acta Crystallographica Sec. B-Struct. Sci. 2008, 64, 144–153.

[31]  Takeuchi T, Kondo, T, Takami, T, Takahashi, H, Ikuta, H, Mizutani, U. et al. Contribution of electronic structure to the large thermoelectric power in layered cobalt oxides, Phys. Rev. B. 2004, 69(12), 125410.

[32]  Limelette P, Hardy, V, Auban-Senzier, P, Jerome, D, Flahaut, D, Hebert, S. et al. Strongly correlated properties of the thermoelectric cobalt oxide Ca3Co4O9, Phys. Rev. B. 2005, 71(23), 233108.

[33]  Shikano M, Funahashi, R. Electrical and thermal properties of single-crystalline (Ca2CoO3) (0.7)CoO2 with a Ca3Co4O9 structure, Appl. Phys. Lett. 2003, 82(12), 1851–1853.

[34]  Pinitsoontorn S, Lerssongkram, N, Keawprak, N, Amornkitbamrung, V. Thermoelectric properties of transition metals-doped Ca3Co3.8M0.2O9+delta (M = Co, Cr, Fe, Ni, Cu and Zn), J. Mater. Sci. – Mater. Electr. 2012, 23(5), 1050–1056.

[35] Prasoetsopha N, Pinitsoontorn, S, Amornkitbamrung, V. Synthesis and Thermoelectric Properties of Ca3Co4O9 Prepared by a Simple Thermal Hydro-Decomposition Method, Electronic Materials Letters 2012, 8(3), 305–308.

[36] Prasoetsopha N, Pinitsoontorn, S, Kamwanna, T, Amornkitbamrung, V, Kurosaki, K, Ohishi, Y. et al. The effect of Cr substitution on the structure and properties of misfit-layered Ca3Co4-xCrxO9+delta thermoelectric oxides, J. Alloy. Compd. 2014, 588, 199–205.

[37] Katsuyama S, Takiguchi, Y, Ito, M. Synthesis of Ca3Co4O9 ceramics by polymerized complex and hydrothermal hot-pressing processes and the investigation of its thermoelectric properties, J. Mater. Sci. 2008, 43(10), 3553–3559.

[38] Lin Y H, Lan, J, Shen, Z J, Liu, Y H, Nan, C W, Li, J F. High-temperature electrical transport behaviors in textured Ca3Co4O9-based polycrystalline ceramics, Appl. Phys. Lett. 2009, 94(7), 072107.

[39] Lim C H, Seo, W S, Lee, S, Lim, Y S, Kim, J Y, Park, H H. et al. Anisotropy of the thermoelectric figure of merit (ZT) in textured Ca3Co4O9 ceramics prepared by using a spark plasma sintering process, J. Korean Phys. Soc. 2015, 66(5), 794–799.

[40] Kwon O J, Jo, W, Ko, K E, Kim, J Y, Bae, S H, Koo, H. et al. Thermoelectric properties and texture evaluation of Ca3Co4O9 prepared by a cost-effective multisheet cofiring technique, J. Mater. Sci. 2011, 46(9), 2887–2894.

[41] Gunes M, Ozenbas, M. Effect of grain size and porosity on phonon scattering enhancement of Ca3Co4O9, J. Alloy. Compd. 2015, 626, 360–367.

[42] Bittner M, Helmich, L, Nietschke, F, Geppert, B, Oeckler, O, Feldhoff, A. Porous Ca3Co4O9 with enhanced thermoelectric properties derived from Sol-Gel synthesis, J. Eur. Ceram. Soc. 2017, 37(13), 3909–3915.

[43] Rebola A, Klie, R, Zapol, P, Ogut, S. First-principles study of the atomic and electronic structures of misfit-layered calcium cobaltite (Ca2CoO3)(CoO2)(1.62) using rational approximants, Phys. Rev. B. 2012, 85(15), 155132.

[44] Srepusharawoot P, Pinitsoontorn, S, Maensiri, S. Electronic structure of iron-doped misfit-layered calcium cobaltite, Comput. Mater. Sci. 2016, 114, 64–71.

[45] Xu L X, Li, F, Wang, Y. High-temperature transport and thermoelectric properties of Ca3Co4-xTixO9, J. Alloy. Compd. 2010, 501(1), 115–119.

[46] Wang Y, Sui, Y, Wang, X, Sui, W, Liu, X. Enhanced high temperature thermoelectric characteristics of transition metals doped Ca3Co4O9+delta by cold high-pressure fabrication, J. Appl. Phys. 2010, 107, 033708.

[47] Nong N V, Yanagiya, S, Monica, S, Pryds, N, Ohtaki, M. High-Temperature Thermoelectric and Microstructural Characteristics of Cobalt-Based Oxides with Ga Substituted on the Co-site, J. Electron. Mater. 2011, 40(5), 716–722.

[48] Tian R M, Donelson, R, Ling, C D, Blanchard, P E R, Zhang, T S, Chu, D W. et al. Ga Substitution and Oxygen Diffusion Kinetics in Ca3Co4O9+delta-Based Thermoelectric Oxides, J. Phys. Chem. C 2013, 117(26), 13382–13387.

[49] Prasoetsopha N, Pinitsoontorn, S, Bootchanont, A, Kidkhunthod, P, Srepusharawoot, P, Kamwanna, T. et al. Local structure of Fe in Fe-doped misfit-layered calcium cobaltite: An X-ray absorption spectroscopy study, J. Solid State Chem. 2013, 204, 257–265.

[50] Pinitsoontorn S, Prasoetsopha, N, Srepusharawoot, P, Bootchanont, A, Kidkhunthod, P, Kamwanna, T. et al. Local structure determination of substitutional elements in Ca3Co4-xMxO9 (M = Fe, Cr, Ga) using X-ray absorption spectroscopy, Physica Status Solidi a-Appl. Res. 2014, 211(8), 1732–1739.

[51] Zhou Y Q, Matsubara, I, Horii, S, Takeuchi, T, Funahashi, R, Shikano, M. et al. Thermoelectric properties of highly grain-aligned and densified Co-based oxide ceramics, J. Appl. Phys. 2003, 93(5), 2653–2658.

[52] Butt S, Liu, Y C, Lan, J L, Shehzad, K, Zhan, B, Lin, Y. et al. High-temperature thermoelectric properties of La and Fe co-doped Ca-Co-O misfit-layered cobaltites consolidated by spark plasma sintering, J. Alloy. Compd. 2014, 558, 277–283.

[53] Butt S, Xu, W, Farooq, M U, Ren, G K, Mohmed, F, Lin, Y H. et al. Enhancement of Thermoelectric Performance in Hierarchical Mesoscopic Oxide Composites of Ca3Co4O9 and La0.8Sr0.2CoO3, J Am Ceram Soc. 2015, 98(4), 1230–1235.

[54] Xu G J, Funahashi, R, Shikano, M, Matsubara, I, Zhou, Y Q. Thermoelectric properties of the Bi- and Na- substituted Ca3Co4O9 system, Appl. Phys. Lett. 2002, 80(20), 3760–3762.

[55] Lu Q M, Zhang, J X, Zhang, Q Y, Liu, D M Improved Thermoelectric Properties of $Ca_{3-x}Ba_xCo_4O_9$ (x=0-0.4) Bulks by Sol-Gel and SPS Method. 25th International Conference on Thermoelectrics. Vienna, Austria: IEEE; 2006.

[56] Butt S, Xu, W, He, W Q, Tan, Q, Ren, G K, Lin, Y H. et al. Enhancement of thermoelectric performance in Cd-doped Ca3Co4O9 via spin entropy, defect chemistry and phonon scattering, J. Mater. Chem. A 2014, 2(45), 19479–19487.

[57] Wang Y, Sui, Y, Li, F, Xu, L X, Wang, X J, Su, W H. et al. Thermoelectrics in misfit-layered oxides [(Ca,Ln)(2)CoO3](0.62)[CoO2]: From bulk to nano, Nano Energy 2012, 1(3), 456–465.

[58] Wang Y, Sui, Y, Cheng, J G, Wang, X J, Su, W H. Comparison of the high temperature thermoelectric properties for Ag-doped and Ag-added Ca3Co4O9, J. Alloy. Compd. 2009, 477(1–2), 817–821.

[59] Linderoth S, Nong, N V, Pryds, N, Ohtaki, M. Enhancement of the Thermoelectric Performance of p-Type Layered Oxide Ca(3)Co(4)O(9+delta) Through Heavy Doping and Metallic Nanoinclusions, Adv. Mater. Weinheim 2011, 23(21), 2484–2490.

[60] Carvillo P, Chen, Y, Boyle, C, Barnes, P N, Song, X Y. Thermoelectric Performance Enhancement of Calcium Cobaltite through Barium Grain Boundary Segregation, Inorg. Chem. 2015, 54(18), 9027–9032.

[61] Saini S, Yaddanapudi, H S, Tian, K, Yin, Y N, Magginetti, D, Tiwari, A. Terbium Ion Doping in Ca3Co4O9: A Step towards High-Performance Thermoelectric Materials, Sci Rep 2017, 7, 44621.

[62] Banerjee A N, Kundoo, S, Chattopadhyay, K K. Synthesis and characterization of p-type transparent conducting CuAlO2 thin film by DC sputtering, Thin Solid Films 2003, 440(1–2), 5–10.

[63] Banerjee A N, Maity, R, Ghosh, P K, Chattopadhyay, K K. Thermoelectric properties and electrical characteristics of sputter-deposited p-CuAlO2 thin films, Thin Solid Films 2005, 474(1–2), 261–266.

[64] Feng Y N, Elquist, A, Zhang, Y P, Gao, K Z, Ferguson, I, Tzempelikos, A. et al. Temperature dependent thermoelectric properties of cuprous delafossite oxides, Compos. Part B-Eng. 2019, 156, 108–112.

[65] Kuriyama H, Nohara, M, Sasagawa, T, Takubo, K, Mizokawa, T, Kimura, K, et al. High-temperature thermoelectric properties of Delafossite oxide $CuRh_{1-x}Mg_xO_2$. 25th International conference on Thermoelectrics. Vienna, Austria: IEEE; 2006.

[66] Hayashi K, Sato, K, Nozaki, T, Kajitani, T. Effect of doping on thermoelectric properties of delafossite-type oxide CuCrO2, Jpn. J. Appl. Phys. 2008, 47(1), 59–63.

[67] Ono Y, Satoh, K, Nozaki, T, Kajitani, T. Structural, magnetic and thermoelectric properties of delafossite-type oxide, CuCr1-xMgxO2 (0 ≤ x ≤ 0.05), Jpn. J. Appl. Phys. Part 1-Regular Papers Brief Commun. & Rev. Papers 2007, 46(3a), 1071–1075.

[68] Hayashi K, Nozaki, T, Kajitani, T. Structure and high temperature thermoelectric properties of delafossite-type oxide CuFe1-xNixO2 (0 ≤ x ≤ 0.05), Jpn. J. Appl. Phys. Part 1-Regular Papers Brief Commun. & Rev. Papers 2007, 46(8a), 5226–5229.

[69] Nozaki T, Hayashi, K, Kajitani, T. Electronic Structure and Thermoelectric Properties of the Delafossite-Type Oxides CuFe1-x Ni (x) O-2, J. Electron. Mater. 2009, 38(7), 1282–1286.

[70] Ruttanapun C, Wichainchai, A, Prachamon, W, Yangthaisong, A, Charoenphakdee, A, Seetawan, T. Thermoelectric properties of Cu1-xPtxFeO2 (0.0 ≤ x ≤ 0.05) delafossite-type transition oxide, J. Alloy. Compd. 2011, 509(13), 4588–4594.

[71] Ruttanapun C. Effects of Pd substitution on the thermoelectric and electronic properties of delafossite Cu1-xPdxFeO2 (x=0.01, 0.03 and 0.05), J. Solid State Chem. 2014, 215, 43–49.

[72] Ruttanapun C, Boonchom, B, Vittayakorn, N, Harnwunggmoung, A, Charoenphakdee, A. Synthesis and Thermoelectric Properties of Cu0.95Pt0.05Fe0.97Sn0.03O2 Delafossite-Oxide, Ferroelectrics 2013, 453(1), 75–83.

[73] Zhao L D, Berardan, D, Pei, Y L, Byl, C, Pinsard-Gaudart, L, Dragoe, N. Bi1-xSrxCuSeO oxyselenides as promising thermoelectric materials, Appl. Phys. Lett. 2010, 97(9), 092118.

[74] Barreteau C, Pan, L, Amzallag, E, Zhao, L D, Berardan, D, Dragoe, N. Layered oxychalcogenide in the Bi-Cu-O-Se system as good thermoelectric materials, Semicond Sci Tech 2014, 29(6), 064001.

[75] Zhao L D, He, J Q, Berardan, D, Lin, Y H, Li, J F, Nan, C W. et al. BiCuSeO oxyselenides: new promising thermoelectric materials, Energy Environ. Sci. 2014, 7(9), 2900–2924.

[76] Zhang X X, Chang, C, Zhou, Y M, Zhao, L D. BiCuSeO Thermoelectrics: An Update on Recent Progress and Perspective, Materials 2017, 10(2), 198.

[77] Liu Y, Zhao, L D, Zhu, Y, Liu, Y, Li, F, Yu, M. et al. Synergistically Optimizing Electrical and Thermal Transport Properties of BiCuSeO via a Dual-Doping Approach, Adv Energy Mater 2016, 6(9), 1502423.

[78] Pan L, Lang, Y D, Zhao, L, Berardan, D, Amzallag, E, Xu, C. et al. Realization of n-type and enhanced thermoelectric performance of p-type BiCuSeO by controlled iron incorporation, J. Mater. Chem. A 2018, 6(27), 13340–13349.

[79] Barreteau C, Pan, L, Pei, Y L, Zhao, L D, Berardan, D, Dragoe, N. Oxychalcogenides as new efficient p-type thermoelectric materials, Funct. Mater. Lett. 2013, 6(5), 134007.

[80] Liu Y, Zhao, L D, Liu, Y C, Lan, J L, Xu, W, Li, F. et al. Remarkable Enhancement in Thermoelectric Performance of BiCuSeO by Cu Deficiencies, J. Am. Chem. Soc. 2011, 133(50), 20112–20115.

[81] Li Z, Xiao, C, Fan, S J, Deng, Y, Zhang, W S, Ye, B J. et al. Dual Vacancies: An Effective Strategy Realizing Synergistic Optimization of Thermoelectric Property in BiCuSeO, J. Am. Chem. Soc. 2015, 137(20), 6587–6593.

[82] Li F, Li, J F, Zhao, L D, Xiang, K, Liu, Y, Zhang, B P. et al. Polycrystalline BiCuSeO oxide as a potential thermoelectric material, Energy Environ. Sci. 2012, 5(5), 7188–7195.

[83] Ishizawa M, Yasuzato, Y, Fujishiro, H, Naito, T, Katsui, H, Goto, T. Oxidation states and thermoelectric properties of BiCuSeO bulks fabricated under Bi or Se deficiencies in the nominal composition, J. Appl. Phys. 2018, 123(24), 245104.

[84] Das S, Chetty, R, Wojciechowski, K, Suwas, S, Mallik, R C. Thermoelectric properties of Sn doped BiCuSeO, Appl. Surf. Sci. 2017, 418, 238–245.

[85] Pei Y L, He, J Q, Li, J F, Li, F, Liu, Q J, Pan, W. et al. High thermoelectric performance of oxyselenides: intrinsically low thermal conductivity of Ca-doped BiCuSeO, Npg Asia Mater. 2013, 5, e47.

[86] Farooq M U, Butt, S, Gao, K W, Zhu, Y C, Sun, X G, Pang, X L. et al. Cd-doping a facile approach for better thermoelectric transport properties of BiCuSeO oxyselenides, RSC Adv. 2016, 6(40), 33789–33797.

[87] Li J, Sui, J, Pei, Y L, Meng, X F, Berardan, D, Dragoe, N. et al. The roles of Na doping in BiCuSeO oxyselenides as a thermoelectric material, J. Mater. Chem. A 2014, 2(14), 4903–4906.

[88] Feng B, Li, G Q, Pan, Z, Hu, X M, Liu, P H, Li, Y W. et al. Enhanced thermoelectric performances in BiCuSeO oxyselenides via Er and 3D modulation doping, Ceram Int 2019, 45 (4), 4493–4498.

[89] Lan J L, Liu, Y C, Zhan, B, Lin, Y H, Zhang, B P, Yuan, X. et al. Enhanced Thermoelectric Properties of Pb-doped BiCuSeO Ceramics, Adv. Mater. Weinheim 2013, 25(36), 5086–5090.

[90] Yang D W, Su, X L, Yan, Y G, Hu, T Z, Xie, H Y, He, J. et al. Manipulating the Combustion Wave during Self-Propagating Synthesis for High Thermoelectric Performance of Layered Oxychalcogenide Bi1-xPbxCuSeO, Chem. Mater. 2016, 28(13), 4628–4640.

[91] Li J, Sui, J H, Pei, Y L, Barreteau, C, Berardan, D, Dragoe, N. et al. A high thermoelectric figure of merit ZT > 1 in Ba heavily doped BiCuSeO oxyselenides, Energy Environ. Sci. 2012, 5(9), 8543–8547.

[92] Sui J H, Li, J, He, J Q, Pei, Y L, Berardan, D, Wu, H J. et al. Texturation boosts the thermoelectric performance of BiCuSeO oxyselenides, Energy Environ. Sci. 2013, 6(10), 2916–2920.

[93] Pei Y L, Wu, H J, Wu, D, Zheng, F S, He, J Q. High Thermoelectric Performance Realized in a BiCuSeO System by Improving Carrier Mobility through 3D Modulation Doping, J. Am. Chem. Soc. 2014, 136(39), 13902–13908.

[94] Feng B, Li, G Q, Pan, Z, Hu, X M, Liu, P H, He, Z. et al. Effect of synthesis processes on the thermoelectric properties of BiCuSeO oxyselenides, J. Alloy. Compd. 2018, 754, 131–138.

[95] Sun Y Z, Zhang, C C, Cao, C M, Fu, J X, Peng, L M. Co-doping for significantly improved thermoelectric figure of merit in p-type Bi1-2xMgxPbxCuSeO oxyselenides, Ceram Int 2017, 43(18), 17186–17193.

[96] Ren G K, Wang, S Y, Zhu, Y C, Ventura, K J, Tan, X, Xu, W. et al. Enhancing thermoelectric performance in hierarchically structured BiCuSeO by increasing bond covalency and weakening carrier-phonon coupling, Energy Environ. Sci. 2017, 10(7), 1590–1599.

[97] Wen Q, Chang, C, Pan, L, Li, X T, Yang, T, Guo, H H. et al. Enhanced thermoelectric performance of BiCuSeO by increasing Seebeck coefficient through magnetic ion incorporation, J. Mater. Chem. A 2017, 5(26), 13392–13399.

[98] Tang J, Xu, R, Zhang, J, Li, D, Zhou, W P, Li, X T. et al. Light element doping and introducing spin entropy: an effective strategy for enhancement of thermoelectric properties in BiCuSeO, ACS Appl. Mater. Inter. 2019, 11(17), 15543–15551.

Yongbin Zhu, Feng Jiang, and Weishu Liu

# 7 Carrier energy filtering effect in thermoelectric materials

**Abstract:** Carrier energy filtering effect boosting the thermoelectric performance by blocking "cold" carriers is here reviewed and summarized from theoretical and experimental perspectives. Among them, theoretical models including phenomenological interpretation, cut-off model and relax-time model are summarized to discuss the potential of carrier energy filtering effect enhancing the thermoelectric performance. Furthermore, experimental observations are also presented in some common thermoelectric materials systems, such as PbTe, $Bi_2Te_3$, Skutterudite, Half-Heusler systems. According to theory model, it still need the further optimization of the carrier concentration, barrier height, and barrier separated distance by experimental operation to be close to theoretical predication of the enhancement of power factor.

**Keywords:** Carrier energy filtering, Phenomenological interpretation, Cut-off energy model, Relax time model, Experimental observation

## 7.1 Introduction

Carrier energy filtering (CEF) effect could trace back to the earlier efforts to investigate the effect of potential barrier on the thermopower systems [1–3]. The usage of the term "energy filtering" was earlier found in L.W. Whitlow and T. Hirano's work [3]. Simply, it could be interpreted as a new strategy to boost the thermoelectric performance by blocking "cold" carriers [1] or lower energy carriers [3]. Ravich described the favorable condition for the elective scattering effect as the barrier separation $l$ was much greater than the momentum mean free path $l_p$, and adequately smaller than the energy mean free path $l_e$ as follows [1]:

$$l_p < l < l_e \qquad (7.1)$$

In the past 20 years, there were many following up theoretical investigations and experimental verifications. In an excellent perspective written by M. S. Dresselhaus et al. [4], the CEF effect is an important strategy for developing the next generation of thermoelectric materials. Here, we would like to summarize the new advances of CEF effect together with our interpretation. Before moving to the details, we should point out that CEF effect is a kind of selective carrier scattering mechanism according to

**Yongbin Zhu, Feng Jiang, Weishu Liu,** Department of Materials Science and Engineering, Southern University of Science and Technology, ShenZhen, China

https://doi.org/10.1515/9783110596526-007

Ravich's interpretation. Besides the potential barrier at a boundary or an interface, the resonant scattering relative to an atomic point defect is another strategy to induce selective CEF effect, such as doping In and Tl in PbTe [5] or PbSe [6], respectively. Since Joseph P. Heremans has written an excellent book chapter in the CRC thermoelectric handbook [7] with deep interpretations, we would like to only focus on the discussion of the selective scattering of potential barrier or the CEF effect in this chapter.

# 7.2 Theoretical model of CEF effect

In this section, the discussion of the CEF effect will go through from the phenomenological interpretation to the cut-off energy model and the relax time model that theoretically explains how CEF effect enhances Seebeck coefficient through changing energy-dependent carrier scattering process. These theoretical considerations suggest that the CEF effect could achieve higher power factor through partly decoupling the conflicting relationship of electrical conductivity and Seebeck coefficient.

## 7.2.1 Phenomenological interpretation

Thermoelectric phenomenon is associated with the coupling electric current ($j$) and heat flux ($q$) driven by the electric field ($E$) and temperature gradient ($\nabla T$). As both the electron and phonon systems are only slightly deviated from the equilibrium state, the linear relationship between the forces and fluxes could be written as follows [8]:

$$j = \sigma \cdot E - S\sigma \cdot \nabla T, \tag{7.2}$$

$$q = S\sigma T \cdot E - \bar{\kappa} \cdot \nabla T, \tag{7.3}$$

where $T$, $\sigma$, and $S$ are the temperature, electrical conductivity, and Seebeck coefficient, respectively. Here the $\bar{\kappa}$ is the thermal conductivity for $j \neq 0$, that is, $\bar{\kappa} = \kappa_e + S^2\sigma T$. Consider a degenerate free electron gas with Kelvin relations and an isothermal condition (i.e., $\nabla(T) = 0$), the term of electrical field $E$ could be eliminated and the following relationship is obtained:

$$S = \frac{q/T}{j} = \frac{S_{en}}{j}, \tag{7.4}$$

where $S_{en}$ is the entropy current density. The Seebeck coefficient is, therefore, relative to the average electronic entropy. It is worthy to point out that more strict derivation could be followed by the way of Rockwood [9].

For the system with periodic carrier filters, the "cold" carriers with energy lower than the potential barrier ($E_b$) would be totally blocked and only the "hot" carriers could flow from the hot side to the cold side (Figure 7.1). As a result, the system with CEF effect would have larger entropy current density and hence a larger Seebeck coefficient.

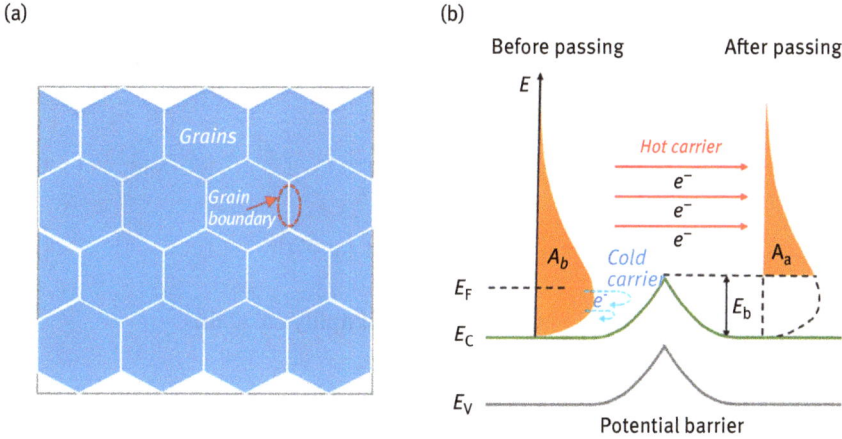

**Figure 7.1:** Schematic illustration of potential barrier scattering at periodic boundary with a barrier height $E_b$. Carriers having energy lower than $E_b$ are strongly scattered and excluded from the transport process. Areas $A_b$ and $A_a$ represent the carrier energy distributions before and after passing through the potential barrier, respectively.

## 7.2.2 Cut-off energy model

In Section 7.2.1, the phenomenological model interpretation gives a simple picture of CEF boosting the Seebeck coefficient. However, the thermoelectric performance is not only determined by the Seebeck coefficient, but also the electrical conductivity. The potential barrier blocks the "cold" carriers, and also reduces the total number of the charge carriers, resulting in a negative impact on the power factor. Kishimoto et al. [10] presented a simple cut-off energy model on the basis of the Boltzmann transport theory with relaxation time approximation. Assuming the energy of conduction band minimum is zero, the cut-off energy is just equal to $E_b$ in number according to Figure 7.1.

In the condition of quasiequilibrium and carrier diffusion, the electrical conductivity and Seebeck coefficient could be expressed by considering the cut-off effect of the potential barrier as follows:

$$\sigma = \frac{2e^2}{3m^*} \int_{E_b}^{\infty} \tau(E)g(E)E\left(-\frac{\partial f(E)}{\partial E}\right) dE, \tag{7.5}$$

$$S = \frac{1}{eT} \left[ \frac{\int_{E_b}^{\infty} \tau(E)g(E)E^2 \left( -\frac{\partial f(E)}{\partial E} \right) dE}{\int_{E_b}^{\infty} \tau(E)g(E)E \left( -\frac{\partial f(E)}{\partial E} \right) dE} - E_F \right], \tag{7.6}$$

where $e$, $m^*$, $\tau(E)$, $g(E)$, and $f(E)$ are the electronic charge, carrier effective mass, relax time, density of state, and Fermi distribution function, respectively. By further including the expression of $\tau(E)$ and $g(E)$, the power factor of $S^2\sigma$ at a given Fermi energy could be simplified as a function of the energy difference parameter $\eta^*$, that is, $\eta^* = (E_b - E_F)/k_B T$. By considering degeneration condition of $E_F/k_B T \gg 1$, eqs. (7.5) and (7.6) could be simplified as follows:

$$\sigma_b = \frac{1}{(1 + e^{\eta^*})} \sigma, \tag{7.7}$$

$$S_b = \frac{k_B}{e} \left[ \eta^* + \left( 1 + e^{\eta^*} \right) \ln\left( 1 + e^{-\eta^*} \right) \right]. \tag{7.8}$$

Bahk et al. [11] also defined a parameter $g(E_b)$ to simplify eq. (7.6):

$$S = \frac{1}{eT} [g(E_b) - E_F]. \tag{7.9}$$

Furthermore, they also proposed the optimized cut-off energy $E_b$ to maximize the power factor as follows:

$$E_b = \frac{[g(E_b) + E_F]}{2}. \tag{7.10}$$

By applying Bahk et al.'s theory, a 6–7 times higher enhancement in the power factor was predicated when the Fermi energy is around 0.3 eV. For the n-type PbTe with a power factor of 35 $\mu$W cm$^{-1}$ K$^{-2}$ at a carrier concentration of $4 \times 10^{18}$ cm$^{-3}$, the power factor could be boosted higher than 80 $\mu$W cm$^{-1}$ K$^{-2}$ at a carrier concentration of $1 \times 10^{20}$ cm$^{-3}$ by considering the optimized cut-off energy $E_b$. However, it is very challenging to obtain such a high carrier concentration.

## 7.2.3 Relax time model

The cut-off model provides a direct and simple picture to interpret the benefits of CEF. However, the quantitative enhancement in power factor is too optimistic since it completely neglects the scattering effect of "hot" carriers. Alternatively, a relax time model is more reasonable by considering the potential barrier as a new scattering center, which could be introduced into the electrical conductivity and the Seebeck coefficient through the Mathiessen's rule:

$$\frac{1}{\tau(E)} = \sum_i \frac{1}{\tau_i(E)}. \tag{7.11}$$

where the $\tau_i(E)$ is the relax time of different scattering mechanisms. Popescu et al. [12] deduced the relax time relative to the CEF by multiplying the transmission probability $T(E)$ after scattering from the $N$ barriers, as shown in eq. (7.12). It used an assumption of a rectangular potential with an average height of $E_b$, width of $w$, and distance of $L$ between them, and a standard quantum-mechanical solution of $T(E)$ by considering the plane-wave-like behavior of the incident carrier:

$$\tau_b(E) = \begin{cases} L\sqrt{\dfrac{m^*}{2E}}\left[1 + \dfrac{4\dfrac{E}{E_b}\left(1 - \dfrac{E}{E_b}\right)}{\sinh^2\left[\sqrt{\dfrac{8\pi^2 m^* E_b w^2}{h^2}\left(1 - \dfrac{E}{E_b}\right)}\right]}\right], & E < E_b \\[4ex] L\sqrt{\dfrac{m^*}{2E}}\left[1 + \dfrac{4\dfrac{E}{E_b}\left(\dfrac{E}{E_b} - 1\right)}{\sin^2\left[\sqrt{\dfrac{8\pi^2 m^* E_b w^2}{h^2}\left(\dfrac{E}{E_b} - 1\right)}\right]}\right], & E > E_b \end{cases} \tag{7.12}$$

Furthermore, Popescu et al. obtained an excellent agreement between the experimental results of Ag-doped PbTe and the theoretical calculation by including the $\tau_b(E)$ of eq. (7.12) with parameter of $E_b = 60$ meV, $w = 50$ nm, and $L = 300$ nm, together with acoustic phonon scattering ($\tau_{a\text{-ph}}(E)$), optical phonon scattering ($\tau_{o\text{-ph}}(E)$), and impurity phonon scattering ($\tau_{imp}(E)$), as shown in Figure 7.2. They also theoretically showed that the CEF effect would increase $S$ in the system with higher barrier (larger $E_b$), higher frequency (smaller $L$), and smaller transmission (larger $w$), but with the cost of decreased $\sigma$. An enhanced power factor (5%) was predicated by carefully tuning the potential barrier parameters.

It is noted that the enhanced power factor is not only impacted by the potential barrier parameters, but also the Fermi energy. Bachmann et al. [13] suggested that higher carrier concentration was necessary for a reasonable enhancement of power factor from the CEF effect. In the case of low doping concentration, the Seebeck coefficient indeed increased, whereas the electrical conductivity strongly decreased, resulting in an even lower power factor. Furthermore, they also suggested that the grain boundary fitted with the double Schottky barrier model would have significant impact on the thermoelectric parameters. It seems that the grain boundary with additional disorder structure would be more positive to the enhanced power factor.

Faleev and Léonard proposed a relax time $\tau_i$ for the scattering at metal/semiconductor interface in nanocomposites [14]. The potential barrier is raised from the band bending away from the interface as shown in Figure 7.3. The transport cross section for scattering on the spherically symmetric potential was calculated by

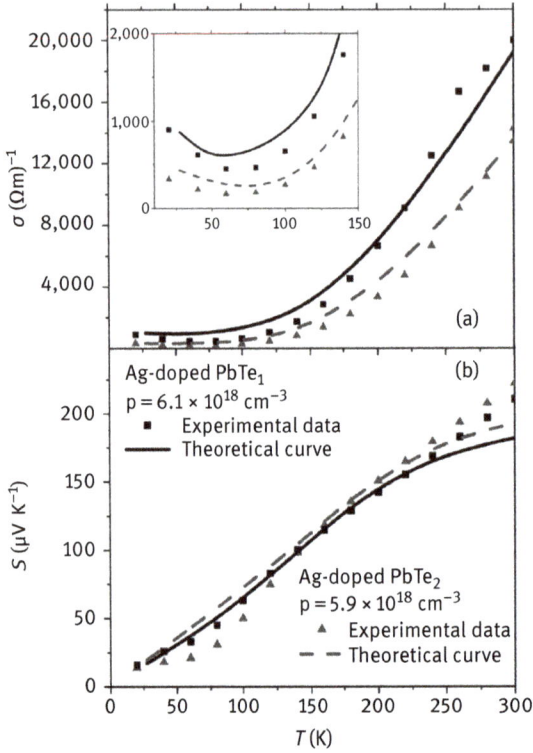

**Figure 7.2:** Temperature-dependent thermoelectric properties of Ag-doped PbTe: (a) electrical conductivity $\sigma$ and (b) Seebeck coefficient $S$ [12].

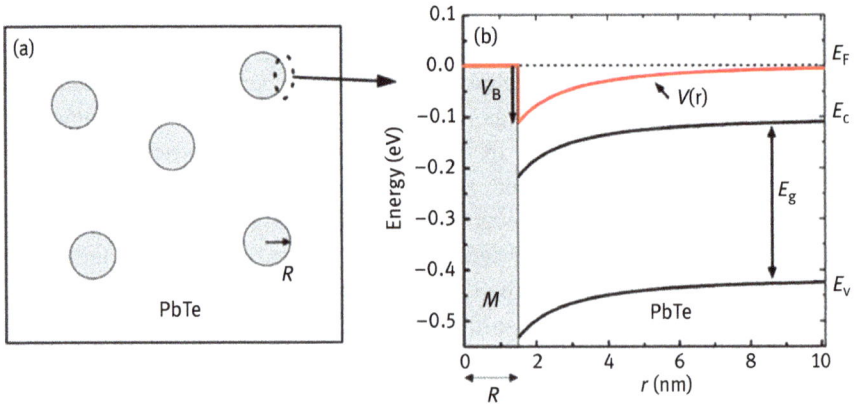

**Figure 7.3:** (a) Schematic of the semiconductor host with metallic nanoinclusions, (b) the calculated potential $V(r)$ and the energy diagram for PbTe at $T = 300$ K, $n = 2.5 \times 10^{19}$ cm$^{-3}$, $V_B = -0.11$ eV, and $R = 1.5$ nm [14].

solving the Schrödinger equation. An analytical description of $\tau_i$ was obtained from the Born approximation by using Fermi's golden rule as follows:

$$\tau_i(E) = E^{3/2} \frac{(1 + E/E_g)^{3/2}}{(1 + 2E/E_g)} \frac{R}{x} \frac{4\sqrt{2m_{dl}^*}}{3\alpha(E, R)},$$
(7.13)

where

$$\alpha(E, R) = \int_0^{2kR} \left| \int_1^\infty \sin(yt) V(yR) y \, dy \right|^2 t \, dt.$$
(7.14)

It is noted that the function $\alpha(E,R)$ varies slowly with both $E$ and $R$, so the inverse relaxation time due to electron scattering by inclusions can be approximated as $\tau_i^{-1} \approx CE^{-3/2}$, where the constant $C$ depends on potential $V_B$, volume fraction $x$, and radius $R$, and mostly through the combination $C \propto V_B^2 x/R$. Figure 7.4 shows the calculated Seebeck coefficient and electrical conductivity, and hence the power factor for PbTe as a function of the interface potential $V_B$. In general, the power factor increases when compared to that without potential barrier, in a range of $-0.15$ eV $< V_B < 0.15$ eV, where the optimal interface potential $V_B \approx \pm 0.07$ eV. Furthermore, eq. (7.13) provides a new dimension to boost the thermoelectric performance by fixing the $V_b$ and changing the volume fraction and radius, which is equivalent to tuning the parameter of $L$ in the case of Popescu's derivation.

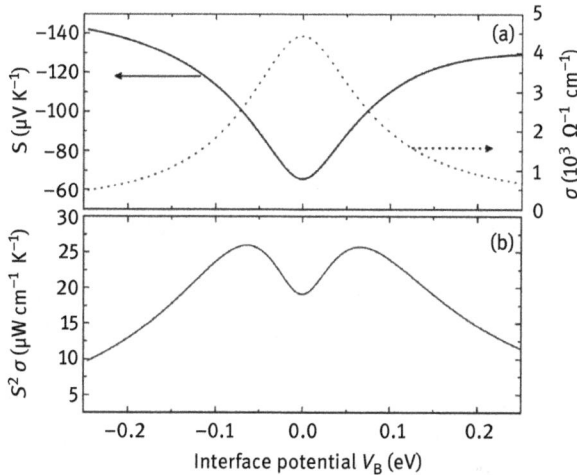

**Figure 7.4:** Theoretical thermoelectric properties of PbTe as a function of the interface potential $V_B$: (a) Seebeck coefficient and electrical conductivity, and (b) power factor. Parameters are $R = 1.5$ nm, $T = 300$ K, $x = 5\%$, and $n = 2.5 \times 10^{19}$ cm$^{-3}$ [14].

In summary, both the relax time mode for grain boundary and nanoinclusion suggest that the CEF effect could be more promising to enhance the power factor and hence ZT for the case with higher carrier concentration, which is consistent with the simple picture of the cut-off model.

# 7.3 Experimental observation of CEF effect

The CEF effect has also been investigated experimentally in a variety of thermoelectric material systems. These experimental results verified the feasibility of the CEF effect on enhancing the Seebeck coefficient and power factor. Here, we give a simple review of several typical thermoelectric materials to explain the effectiveness of the CEF effect.

## 7.3.1 PbTe system

The earliest experimental observation of CEF effect could trace back to the work of Ioffe Technical Physical Institute in PbTe film in 1970s [4]. Kishimoto et al. [10] reported significantly enhanced Seebeck coefficient in the PbTe thin film by sputtering as compared with the bulk counterpart as shown in Figure 7.5. As the carrier concentration is higher than $6 \times 10^{19}$ cm$^{-3}$, the power factor of the thin film is remarkably higher than the bulk materials. The estimated potential barrier is around 50–250 meV, which is significantly dependent on the substrate temperature. It was believed that the defects (probably Te vacancies) at grain boundary would be the origin of the potential barrier.

The CEF effect was also observed in the bulk nanocrystalline PbTe or PbTe with nanoinclusions. Martin et al. reported a p-type Ag-doped PbTe nanocrystals (with average grains of 100 nm) showing a 20–30% higher Seebeck coefficient than the normal PbTe bulk materials and hence a significant enhancement in the power factor [15]. Furthermore, the potential barrier was estimated to be around 60 meV according to the unique temperature-dependent carrier mobility. The chemisorption of oxygen was believed to serve as a trapping center for the electrons at the grain boundary, that is, a filtering of the low-energy electrons. A similar CEF effect was also observed in a PbTe with Ag-rich nanoinclusions [16]. However, only a decreased power factor was observed because the strong reduction in carrier mobility from 821 to 185 cm$^2$ V$^{-1}$ s$^{-1}$ canceled out the benefits from the increased Seebeck coefficient from 260 to 328 μV K$^{-1}$. A notable CEF effect was also observed in the n-type PbTe with Pb precipitates of 30–40 nm, which was confirmed by the relationship between the Seebeck coefficient and carrier concentration and a high scattering parameter of $s$ larger than 2 [17].

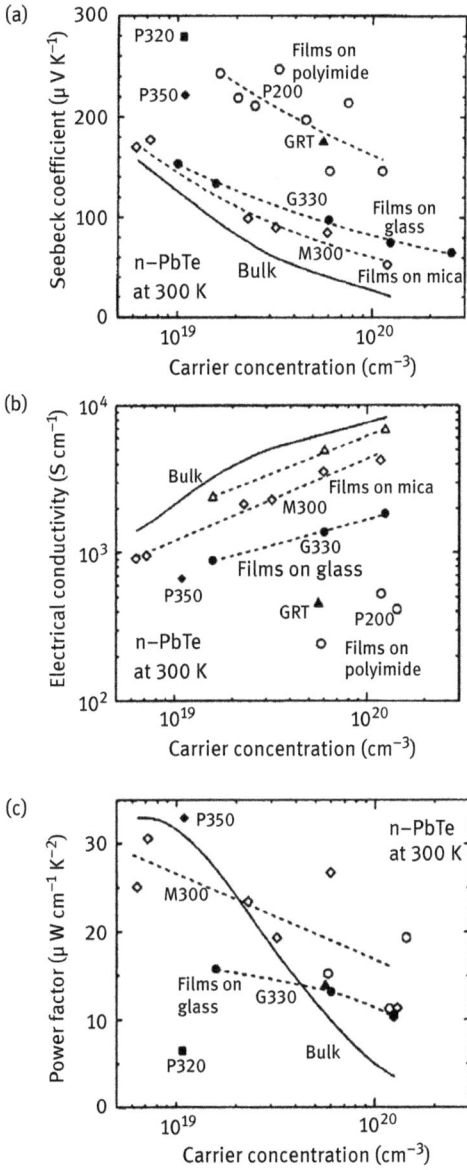

**Figure 7.5:** Thermoelectric properties of n-type PbTe as a function of carrier concentration for the films prepared on heated glass substrates (•) in comparison with that for bulk samples (solid line) at room temperature: (a) Seebeck coefficient, (b) electrical conductivity, and (c) power factor [10].

## 7.3.2 Bi$_2$Te$_3$ system

The CEF effect is also widely observed in the Bi$_2$Te$_3$ family, which is widely used in solid-state refrigeration and power generation at near room temperatures. Sumithra et al. [18] reported a 30% increase in power factor for the n-type Bi$_2$Te$_3$ nanocomposite with 7% extra Bi because of a near doubling of the electrical conductivity while only a 25% reduction of the Seebeck coefficient, as shown in Figure 7.6(a). In contrast, Li et al. [19] reported a CEF effect in p-type Bi$_{0.3}$Sb$_{1.7}$Te$_3$ by including the 4 vol.% SiC nanoparticle with a simultaneously increased carrier concentration from 1.81 to 3.39 × 10$^{19}$ cm$^{-3}$ and Seebeck coefficient from 188 to 196 μV K$^{-1}$. It provided a new way to use semiconductor–semiconductor interface to tune the potential barrier, as shown in Figure 7.6(b). It is believed that the coherent interfaces with Bi$_{0.3}$Sb$_{1.7}$Te$_3$ matrix and added SiC nanoparticles could weaken the obstructing effect of nanoinclusions against carrier transport. Furthermore, the CEF effect has also been reported in p-type Sb$_2$Te$_3$ by embedding Pt nanocrystals as shown in Figure 7.6(c) [20]. Similarly, bringing in foreign nanoparticles to induce heterogeneous interface barrier can also achieve the same effect as that with nanoinclusion, for example, (Bi$_2$Te$_3$)$_{0.2}$(Sb$_2$Te$_3$)$_{0.8}$ embedded with amorphous SiO$_2$ nanoparticles [21].

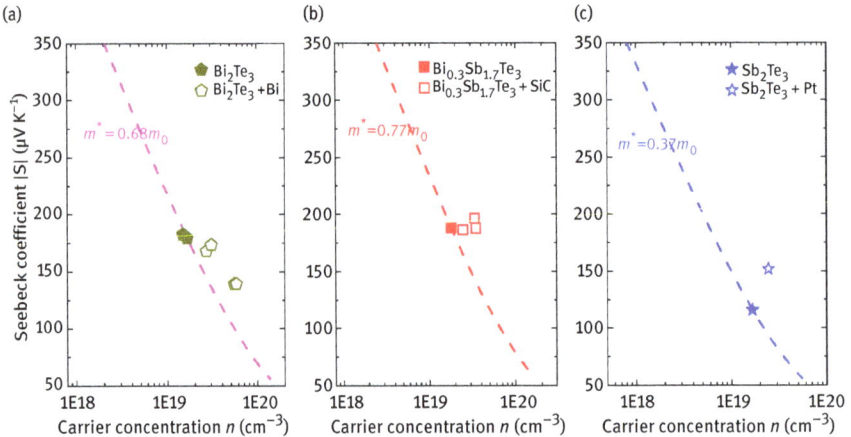

**Figure 7.6:** Seebeck coefficient as a function of carrier concentration for different Bi$_2$Te$_3$-based samples. (a) Bi$_2$Te$_3$ + Bi [18], (b) Bi$_{0.3}$Sb$_{1.7}$Te$_3$ + SiC [19], and (c) Sb$_2$Te$_3$ + Pt [20]. The dashed line refers to the Pisarenko curves.

Moreover, CEF effect has also been observed in case of homogeneous interface, that is, grain boundary. An almost doubling Seebeck coefficient (from 125 to 260 μV K$^{-1}$ at 300 K) has been obtained in polycrystalline Bi$_2$Te$_{2.7}$Se$_{0.3}$ made by ball milling and spark plasma sintering route [22]. Utilizing the interface potential caused by band gap difference between Bi$_2$Te$_3$ and Bi$_2$Se$_3$, about 50% enhancement of Seebeck coefficient has been obtained throughout the measuring temperature range compared to $Bi_2Te_3$ matrix for $Bi_2Te_3/Bi_2Se_3$ nanoflake composites [23].

## 7.3.3 Skutterudite system

Skutterudite is a medium-temperature thermoelectric material. CEF effect has been reported in both homogenous grain boundaries [24] and heterogeneous interfaces [25]. Xiong et al. reported a significantly enhanced power factor in $Ba_{0.22}Co_4Sb_{12}$ skutterudites with the addition of nanosized $TiO_2$ particles. As the content of $TiO_2$ was less than 0.8%, the Seebeck coefficient increased with a slight deterioration of electrical conductivity, and hence obtaining a relative high power factor. As the content of $TiO_2$ reached to 1.8%, a significant reduction in the electrical conductivity was observed, as shown in Figure 7.7. It was believed that nanosized $TiO_2$ had a large band

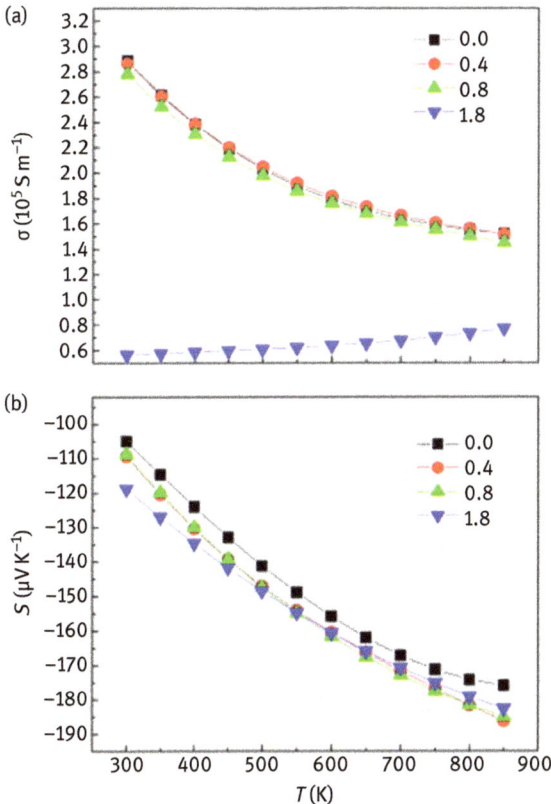

**Figure 7.7:** Thermoelectric properties of $Ba_{0.22}Co_4Sb_{12}$ with $x$% nano-$TiO_2$: (a) electrical conductivity and (b) the Seebeck coefficient [25].

gap; hence, it served as the CEF effect at the grain boundaries. Similar strategies by introducing nanoinclusions at boundary as energy barriers with slightly deteriorating electrical conductivity to largely improve the Seebeck coefficient are also investigated by Ag-nanoparticle inclusions in Ba-filled skutterudite [26], and nanosized $CeO_2$ particles in $Co_{0.97}Pd_{0.03}Sb_3$ materials [27].

Furthermore, a modified grain boundary reported an effective CEF as shown in Figure 7.8(a–b). Meng and coworkers [28] observed dislocation arrays in $Yb_{0.3}Co_4Sb_{12+x}$ with 20% Sb (Sb–MS), as compared with the one without additional Sb (MS). Furthermore, they observed a significantly enhanced Seebeck coefficient in

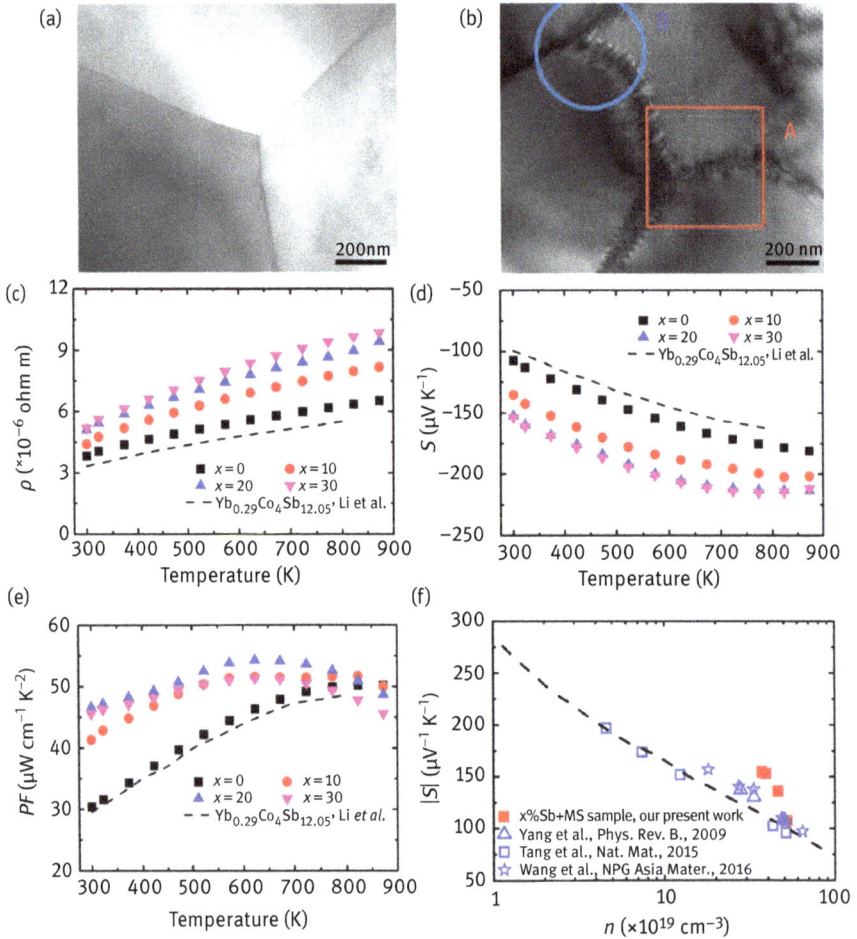

**Figure 7.8:** Structures and thermoelectric properties of $Yb_{0.3}Co_4Sb_{12}+x\%Sb$ ($x = 0$, 10, 20, and 30). Low-magnification TEM image of (a) Sb-free sample and (b) 20% Sb sample, (c) electrical resistivity, (d) Seebeck coefficient, and (e) power factor [28].

$Yb_{0.3}Co_4Sb_{12+x}$ with 20% additional Sb. Although the electrical conductivity got a decrease as the content of Sb increased, the Seebeck coefficient of Sb samples was still much higher than the reported materials at the same carrier concentration, as shown in Figure 7.8(c–e). Figure 7.8(f) compares the Seebeck coefficient and carrier concentration of the Sb-MS samples with Pisarico curve of regularly filled skutterudites, which highly supports the existence of the CEF effect.

## 7.3.4 Half-Heusler system

Half-Heusler alloy is another widely investigated thermoelectric material. Chen et al. [29] reported a significant increase in Seebeck coefficient with little deterioration to electrical conductivity in $Zr_{0.5}Hf_{0.5}Ni_{0.8}Pd_{0.2}Sn_{0.99}Sb_{0.01}/ZrO_2$ nanocomposites, as shown in Figure 7.9. The optimized power factor was obtained with 6 vol.% $ZrO_2$ nanoparticles. It was found that a higher fraction of $ZrO_2$ nanoparticles scattered the electrons and cancelled the benefits of the enhanced Seebeck coefficient. A remarkable enhancement in power factor was also observed in (Ti, Zr, Hf)(Co, Ni)Sb alloy by

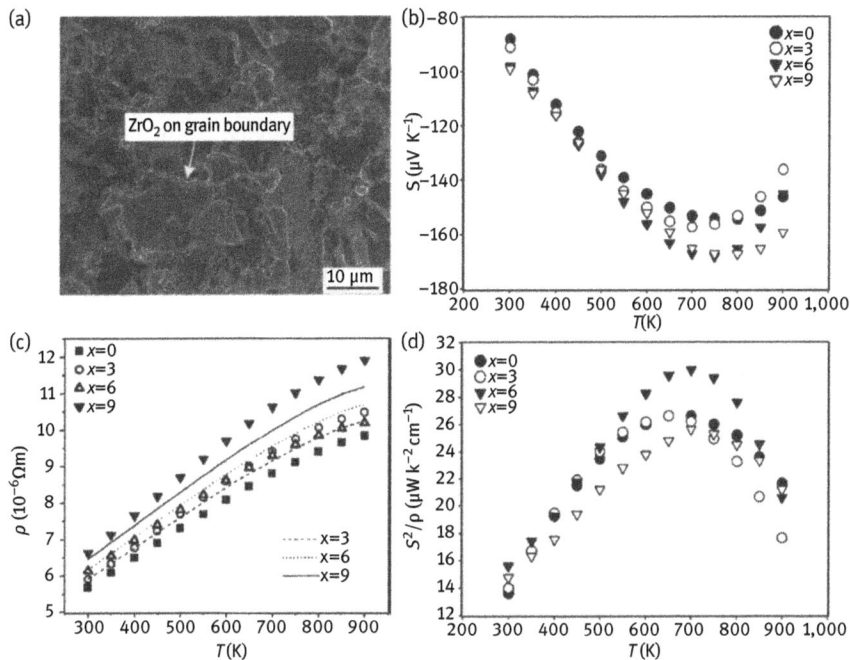

**Figure 7.9:** Structures and thermoelectric properties of $Zr_{0.5}Hf_{0.5}Ni_{0.8}Pd_{0.2}Sn_{0.99}Sb_{0.01}$ dispersed with x vol.% $ZrO_2$ nanoparticles. (a) SEM micrograph of the nanocomposite containing 3 vol.% $ZrO_2$, (b) Seebeck coefficient, (c) electrical resistivity, and (d) power factor [29].

forming InSb nanoinclusion. It is believed that the CEF effects simultaneously increased the effective concentration and the Seebeck coefficient [30]. A similar CEF effect was also observed by adding InSb particles in the Ti(Co, Fe)Sb material [31].

Additionally, the CEF effect did not only shift the carrier concentration-dependent Seebeck coefficient from a regular Pisarenko curve, and was also reported to suppress the bipolar thermal conductivity if the potential barrier could selectively scatter the minor carrier rather than the major carrier [32, 33].

## 7.4 Conclusion

In this chapter, we have reviewed the strategy of CEF effect to enhance the thermoelectric performance. The theoretical models including phenomenological interpretation to the cut-off energy model and relax time model have been discussed, suggesting a very promising enhancement in thermoelectric power factor for the system having a high carrier concentration. In experimental observation, many researchers have claimed an enhanced Seebeck coefficient relative to the CEF effect. However, the enhancement of power factor is still humble when compared with the theoretical predication. Although many experimental works have confirmed the existence of the CEF effect based on the shift of carrier concentration-dependent Seebeck coefficient from a regular Pisarenko curve, it still lacks further optimization relative to the carrier concentration, barrier height, and barrier separated distance. The suppression of bipolar thermal conductivity provides a different direction to use the CEF effect. The theoretical consideration for both the major carrier and minor carrier would be desired in the future.

## References

[1] Ravich Y I, Rowe, D M. CRC Handbook of thermoelectrics. Chapter 7: Selective carrier scattering in thermoelectric materials, New York, CRC Press, 1995, 67–73.
[2] Moizhes B Y, Nemchinsky, V A. 11th International Conferences Thermoelectrics, Texas, (University of Texas Press, Texas, 1993), 232, 1993.
[3] Whitlow L W, Hirano, T. Superlattice applications to thermoelectricity, J. Appl. Phys. 1995, 78, 5460–5466.
[4] Minnich A, Dresselhaus, M S, Ren, Z F et al. Bulk nanostructured thermoelectric materials: Current research and future prospects, Energy Environ. Sci. 2009, 2, 466–479.
[5] Heremans J P, Jovovic, V, Toberer, E S et al. Enhancement of thermoelectric efficiency in PbTe by distortion of the electronic density of states, Science 2008, 321, 554–557.
[6] Androulakis J, Lee, Y, Todorov, I et al. High-temperature thermoelectric properties of n-type PbSe doped with Ga, In, and Pb, Phys. Rev. B 2011, 83, 195209.
[7] Rowe D M. Thermoelectrics and its energy harvesting, 2-Volume Set. Chapter 12: The Effect of Resonant Energy Levels on the Thermoelectric Power and Thermoelectric Power Factor, CRC Press, 2018.

[8]   Zlatic V, René, M. Modern theory of thermoelectricity Chpater 4: Physical interpretation, OUP Oxford, 2014.

[9]   Rockwood A L. Relationship of thermoelectricity to electronic entropy, Phys. Rev. A 1984, 30, 2843.

[10]  Kishimoto K, Tsukamoto, M, Koyanagi, T. Temperature dependence of the Seebeck coefficient and the potential barrier scattering of n-type PbTe films prepared on heated glass substrates by RF sputtering, J. Appl. Phys. 2002, 92, 5331–5339.

[11]  Bahk J H, Bian, Z X, Shakouri, A. Electron energy filtering by a nonplanar potential to enhance the thermoelectric power factor in bulk materials, Phys. Rev. B 2013, 87, 075204.

[12]  Popescu A, Woods, L M, Martin, J et al. Model of transport properties of thermoelectric nanocomposite materials, Phys. Rev. B. 2009, 79, 205302.

[13]  Ravich Y I, Efimova, B A, Smirnov, I A. Semiconducting lead chalcogenides, New York, Plenum Press, 1970.

[14]  Faleev S V, François, L. Theory of enhancement of thermoelectric properties of materials with nanoinclusions, Phys. Rev. B 2008, 77, 214304.

[15]  Martin J, Wang, L, Chen, L D et al. Enhanced Seebeck coefficient through energy-barrier scattering in PbTe nanocomposites, Phys. Rev. B 2009, 79, 115311.

[16]  Paul B, Kumar, A, Banerji, P. Embedded Ag-rich nanodots in PbTe: enhancement of thermoelectric properties through energy filtering of the carriers, J. Appl. Phys. 2010, 108, 064322.

[17]  Heremans J P, Thrush, C M, Morelli, D T. Thermopower enhancement in PbTe with Pb precipitates, J. Appl. Phys. 2005, 98, 063703.

[18]  Sumithra S, Takas, N J, Misra, D K et al. Enhancement in thermoelectric figure of merit in nanostructured $Bi_2Te_3$ with semimetal nanoinclusions, Adv. Energy Mater. 2011, 1, 1141–1147.

[19]  Li J H, Tan, Q, Li, J F et al. BiSbTe-based nanocomposites with high $ZT$: the effect of SiC nanodispersion on thermoelectric properties, Adv. Funct. Mater. 2013, 23, 4317–4323.

[20]  Ko D K, Kang, Y J, Murray, C B. Enhanced thermopower via carrier energy filtering in solution-processable $Pt$-$Sb_2Te_3$ nanocomposites, Nano Lett. 2011, 11, 2841–2844.

[21]  Dou Y C, Qin, X Y, Li, D et al. Enhanced thermopower and thermoelectric performance through energy filtering of carriers in $(Bi_2Te_3)_{0.2}(Sb_2Te_3)_{0.8}$ bulk alloy embedded with amorphous $SiO_2$ nanoparticles, J. Appl. Phys. 2013, 114, 044906.

[22]  Soni A, Shen, Y Q, Yin, M et al. Interface driven energy filtering of thermoelectric power in spark plasma sintered $Bi_2Te_{2.7}Se_{0.3}$ nanoplatelet composites, Nano Lett. 2012, 12, 4305–4310.

[23]  Min Y H, Roh, J W, Yang, H et al. Surfactant-free scalable synthesis of $Bi_2Te_3$ and $Bi_2Se_3$ nanoflakes and enhanced thermoelectric properties of their nanocomposites, Adv. Mater. 2013, 25, 1425–1429.

[24]  Chen S, Ren, Z F. Recent progress of half-Heusler for moderate temperature thermoelectric applications, Mater. Today 2013, 16, 387.

[25]  Xiong Z, Chen, X H, Zhao, X Y et al. Effects of nano-$TiO_2$ dispersion on the thermoelectric properties of filled-skutterudite $Ba_{0.22}Co_4Sb_{12}$, Solid State Sci. 2009, 11, 1612–1616.

[26]  Zhou X Y, Wang, G Y, Zhang, L et al. Enhanced thermoelectric properties of Ba-filled skutterudites by grain size reduction and Ag nanoparticle inclusion, J. Mater. Chem. 2012, 22, 2958–2964.

[27]  Alleno E, Gaborit, M, Lenoir, B et al. Improved thermoelectric performances in nano-Co0.97Pd0.3Sb3, AIP conference proceedings, 2012, 1, 271–274.

[28]  Meng X F, Liu, Z H, Cui, B et al. Grain boundary engineering for achieving high thermoelectric performance in n-type skutterudites, Adv. Energy Mater. 2017, 7, 1602582.

[29]  Chen L D, Huang, X Y, Zhou, M et al. The high temperature thermoelectric performances of $Zr_{0.5}Hf_{0.5}Ni_{0.8}Pb_{0.2}Sn_{0.99}Sb_{0.01}$ alloy with nanophase inclusions, J. Appl. Phys. 2006, 99, 064305.

[30]  Xie W J, He, J, Zhu, S et al., Simultaneously optimizing the independent thermoelectric properties in (Ti,Zr,Hf)(Co,Ni)Sb alloy by in situ forming InSb nanoinclusions, Acta Mater. 2010, 58, 4705–4713.

[31]  Xie W J, Yan, Y G, Zhu, S et al. Significant *ZT* enhancement in p-type Ti(Co,Fe)Sb–InSb nanocomposites via a synergistic high-mobility electron injection, energy-filtering and boundary-scattering approach, Acta Mater. 2013, 61, 2087–2094.

[32]  Poudel B, Hao, Q, Ma, Y et al., High-thermoelectric performance of nanostructured bismuth antimony telluride bulk alloys, Science 2008, 320, 634–638.

[33]  Liu W S, Ren, Z F, Chen, G. Thermoelectric nanomaterials, Springer Series in Materials Science,2013, 11.

Naiming Liu, Jerrold A. Floro, and Mona Zebarjadi

# 8 Charge transfer and interface design in thermoelectric materials

**Abstract:** Thermoelectric modules have applications in power generation, refrigeration, active cooling, and thermal switching. In all these applications, the key is to have a thermoelectric material with large power factor (PF). Simultaneously, large electrical conductivity and Seebeck coefficient values are needed to achieve large thermoelectric PF. Heavily doped semiconductors are by far the most studied class of thermoelectric materials. At low temperatures, the carrier mobility of heavily doped semiconductors is limited by the ionized impurity scattering. This chapter is an overview of the strategies that are proposed to enhance the charge carrier mobility of heavily doped semiconductors in their nanostructured form, including: selective doping, modulation doping, surface doping, and percolation. These strategies are demonstrated in multiphase samples in which the charge transfer may occur at the vicinity of the heterointerfaces. Therefore, interfaces are unavoidable and their impact on the thermoelectric transport is crucial in determining the performance of the thermoelectric materials.

**Keywords:** Thermoelectric, Interfaces, Carrier mobility, Charge transfer

## 8.1 Applications and different operational modes of thermoelectric modules

A thermoelectric module can operate in different operational modes as shown in Figure 8.1. It can serve as a power generator wherein thermal energy is converted into electricity. In this operational mode, thermoelectric modules have been used for half a century in space applications, where reliability and low maintenance are the key required features. This mode is also useful for waste heat recovery, wearable electronics, and solar thermal applications.

Thermoelectric modules (or Peltier modules) can operate in the refrigeration mode, where electrical energy is used to pump heat from the cold side to the hot side for applications such as portable refrigerators and small-scale air conditioners. If the direction of the applied current is reversed, the Peltier modules pump heat from the hot side

**Naiming Liu, Jerrold A. Floro,** Department of Materials Science and Engineering, University of Virginia, Charlottesville, VA, USA
**Mona Zebarjadi,** Department of Materials Science and Engineering, University of Virginia, Charlottesville, VA, USA; Department of Electrical and Computer Engineering, University of Virginia, Charlottesville, VA, USA

https://doi.org/10.1515/9783110596526-008

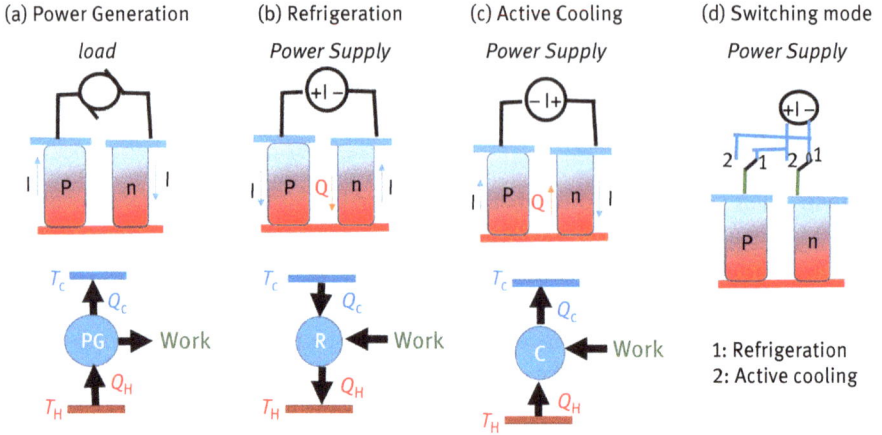

**Figure 8.1:** Different operational modes of thermoelectric modules.

to the cold side similar to a heat spreader. This mode is referred to as active cooling mode and could be used for electronic cooling and heat management in batteries [1, 2].

Finally, in applications where temperature regulation in a finite range is needed, thermoelectric modules can be used as thermal switches, switching between low heat flux mode and large heat flux mode. [3]

In all operational modes, a large thermoelectric power factor (PF), $\sigma S^2$, is needed, where $\sigma$ is the electrical conductivity and $S$ is the Seebeck coefficient. The conversion efficiency of the power generation mode and the coefficient of performance of the refrigeration mode are increasing functions of a dimensionless figure of merit $ZT$ expressed as

$$ZT = \frac{S^2 \sigma}{\kappa} T,$$
(8.1)

where $\kappa$ is the thermal conductivity and $T$ is the temperature. Therefore, in these two modes, materials with low thermal conductivity are needed. However, in the active cooling mode, high thermal conductivity materials are desired.

## 8.2 Thermoelectric power factor

A large PF requires simultaneously large electrical conductivity and large Seebeck coefficient. This is inherently difficult as a large Seebeck coefficient means charge accumulation on the sides of the sample. A simultaneously large electrical conductivity would then shorten the sample and result in the back diffusion of the charge carriers.

In general, transport coefficients, including electrical conductivity and Seebeck coefficients, are defined as

$$L_{\alpha\beta}^{n} = q^2 \int D(\varepsilon)\tau v_\alpha v_\beta (\varepsilon - \mu)^n \frac{\partial f}{\partial \mu} d\varepsilon, \tag{8.2}$$

$$\sigma_{\alpha\beta} = L_{\alpha\beta}^{0}, \tag{8.3}$$

$$S_{\alpha\beta} = -\frac{L_{\alpha\beta}^{1}}{qTL_{\alpha\beta}^{0}}, \tag{8.4}$$

$$\text{PF} = \sigma_{\alpha\beta} S_{\alpha\beta}^{2} = \frac{\left(L_{\alpha\beta}^{1}\right)^2}{q^2 T^2 L_{\alpha\beta}^{0}}, \tag{8.5}$$

$$\kappa_{\alpha\beta} = \frac{1}{q^2 T}\left(L_{\alpha\beta}^{2} - \frac{\left(L_{\alpha\beta}^{1}\right)^2}{L_{\alpha\beta}^{0}}\right), \tag{8.6}$$

where $q$ is the elementary charge (=1.6 × $10^{-19}$ C), $L$ is the transport response matrix, $\alpha$ and $\beta$ refer to different crystallographic directions, $D(\varepsilon)$ is the density of states at given energy $\varepsilon$, $v_\alpha$ is the group velocity in the $\alpha$ direction, $\mu$ is the chemical potential, $\tau$ is the relaxation time, and $f$ is the Fermi Dirac function or the electronic distribution function. The thermal conductivity in eq. (8.6) only accounts for the electronic contribution.

All transport properties are a function of differential conductivity, defined as $\sigma(\varepsilon) = D(\varepsilon)\tau v_\alpha v_\beta$, and the position of the chemical potential. The **electrical conductivity**, $\sigma$, is a function of the carrier concentration of each band ($n/p$), their relaxation time, and their effective mass ($m^\star$) expressed as follows:

$$\sigma = \frac{nq^2\tau}{m^*}. \tag{8.7}$$

The Seebeck coefficient is a result of asymmetry in the differential conductivity at the vicinity of the chemical potential. In metals, the chemical potential is deep within the band; therefore, they usually possess high electrical conductivity due to high carrier density (>$10^{22}$ cm$^{-3}$). However, their Seebeck coefficient values are generally small. Semiconductors usually have a Seebeck coefficient that is several orders of magnitude larger than metals but have considerably smaller electrical conductivity. The large Seebeck coefficient of semiconductors is due to the presence of a bandgap that breaks the symmetry between the electrons and the holes and creates asymmetry in the differential conductivity. Over the years, many other strategies to create asymmetry in the differential conductivity including resonant impurity states [4], magnon–drag [5], electron–hole puddles [6], electron energy filtering [7], and asymmetry in electron–hole masses in the case of semimetals [8, 9] have been explored.

While the presence of a bandgap is not necessary to create asymmetry in the differential conductivity, it is perhaps the most studied case. As the semiconductor doping concentration increases, the electrical conductivity increases but the Seebeck coefficient decreases. Therefore, there is an optimum carrier concentration in which the thermoelectric PF is maximum. The optimized carrier concentration is generally on the order of $10^{19}$~$10^{20}$ cm$^{-3}$. For this reason, most high-performance thermoelectric materials are heavily doped semiconductors.

In this chapter, we focus on heavily doped semiconductors. While the electrical conductivity and the Seebeck coefficient are tunable in this class of materials, they are interrelated as explained before. Since these materials are heavily doped, their carrier mobility is limited by the electron–impurity scatterings especially at low temperatures. We review strategies to improve the carrier mobility in semiconducting thermoelectric materials.

## 8.3 The effect of interfaces on charge transport

Many of the modern thermoelectric materials are multiphase nanocomposites. Nanocomposites have been shown to be effective in reducing thermal conductivity [10, 11]. However, heterointerfaces are usually associated with excess defects and may also exhibit band discontinuities that degrade charge carrier mobility. Therefore, *control of interfaces* (in terms of minimizing charge-trapping defects) and *band alignment* [12, 13] have been exploited in order to minimize the degradation of electrical transport as a result of nanostructuring.

*Control of interfaces* is crucial in multiphase systems aimed at decoupling the electrical and thermal transport. In the case of *coherent interfaces*, the surrounding atomic order is not disrupted compared with incoherent interfaces, in which interfacial dislocations can act as charge traps that are detrimental to the electrical transport, resulting in limited ZT improvement in spite of increased phonon scattering. Coherent interfaces can be present when the constituent phases are chemically and structurally similar, usually by phase precipitation or separation via metallurgical methods. In general, the embedded phase or precipitates are kept very small (<10 nm) to avoid dislocation formation. Meanwhile, due to the lattice mismatch, there is certain degree of strain associated with the interfaces. These so-called endotaxial strained interfaces are expected to facilitate charge transport but strongly block/scatter phonon propagation [14, 15].

Since minimizing grain size is pervasive in nanostructured thermoelectric materials, a high density of grain boundaries is also of great concern regarding the electrical transport. The use of solid state, displacive phase transformations to produce a fine distribution of coherent domain boundaries could provide a new and useful route to interfaces that do not disturb electrons [16]. Yu et al. used a so-called liquid state manipulation method followed by melt spinning, ball mill, and spark plasma sintering to synthesize $Bi_{0.5}Sb_{1.5}Te_3$ alloy with large density of homogeneously distributed twin boundaries [17]. It was found that extensive twin boundaries are effective in reducing the lattice thermal conductivity while maintaining a good electrical conductivity. Moreover, the Seebeck coefficient is enhanced due to an "energy filtering" effect, by which the low-energy charge carriers are preferentially scattered.

The other highlight of utilizing phase transformations for nanocomposite synthesis is that "self-assembled" lamellar/plate structures can be obtained in, for instance, eutectic or eutectoid decomposition. Such ordered structures are not readily obtained by mechanical alloying/mixing and sintering, and less impurities and defects are associated with the "in situ" constructed interfaces. More importantly, in most cases, the lamellar phase may possess specific orientation relationships with the matrix phase. Figure 8.2 shows the $Sb_2Te_3$–PbTe pseudobinary phase diagram, where lamellar growth of $Sb_2Te_3$ and PbTe can be achieved by eutectoid decomposition from metastable $Pb_2Sb_6Te_{11}$ phase [18]. Orientation relationship is illustrated in Figure 8.2(b), where low lattice mismatch between the PbTe (111) and $Sb_2Te_3$ (001) planes leads to coherent interfaces between the lamellae.

Rapid solidification can access small length scales in eutectics, again with preferred orientation relationships being possible (Figure 8.3). Jensen et al. [19] produced Ge-rich lamella structures embedded in $\alpha$-$FeSi_{2+\delta}$ by arc melting (slow cooling), melt spinning, as well as laser melting. They have shown that by increasing solidification rate, the length scales of Ge-rich lamellae are reduced by orders of magnitude – the thickness of the lamellae, and their spacing, goes from submillimeter for arc melting, to microns for melt spinning, and to tens of nanometers for laser melting [19]. Subsequently, the sample is annealed below Fe–Si eutectoid isotherm (~600 °C), giving rise to a microstructure with ~30 nm DC Si laths embedded in $\beta$-$FeSi_2$ by eutectoid decomposition.

Besides avoiding excess atomic disruption on interfaces, the conduction (or valance) band edge of the matrix and precipitate phase should also match in energies (minimal band offset or energy barrier) to facilitate electron/hole transport. This concept has been experimentally demonstrated in the p-type PbTe–SrTe system, where SrTe nanocrystals are embedded in PbTe matrix [15], as well as in the $CdS_{1-x}Se_x$/$ZnS_{1-x}Se$ system [20], where the valence band edge of each phase can be tuned by the alloy composition. In both of the studies, hole mobility was significantly enhanced due to valence band alignment, which result in ZT enhancement. Moreover, if the two phases are both degenerately doped, in which case the Fermi level lies within the energy bands, the energy barrier will essentially be zero regardless of the band offset. However, it is challenging to deliberately dope the nanoinclusion phase.

(a)

(b)

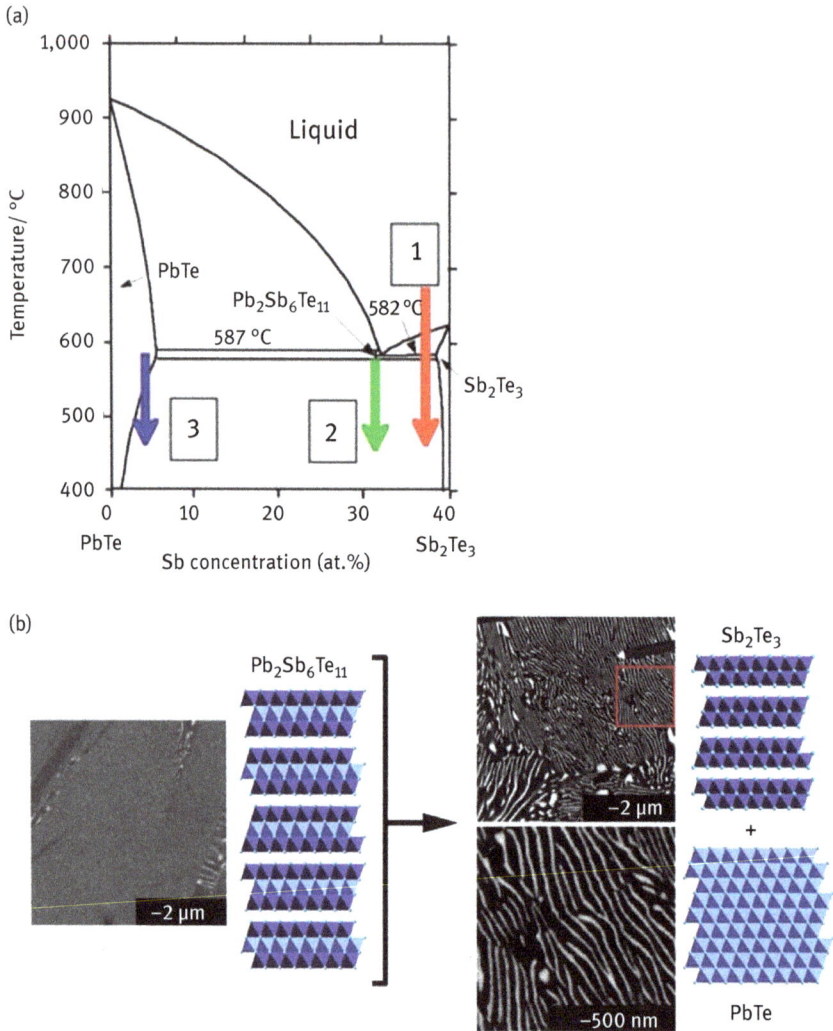

**Figure 8.2:** (a) Sb$_2$Te$_3$–PbTe pseudobinary phase diagram; (b) backscattering electron images showing lamellar growth of Sb$_2$Te$_3$ and PbTe from metastable Pb$_2$Sb$_6$Te$_{11}$ by eutectic decomposition, with schematic illustrating the interface orientation relationship. Reproduced from ref. [18].

## 8.4 Novel doping schemes

### 8.4.1 Selective doping

A minimum energy barrier across the inclusion/matrix heterointerface favors facile charge carrier transport. If the inclusion and matrix phases are both degenerately doped, the heterointerface will be similar to the so-called ohmic contact. In this

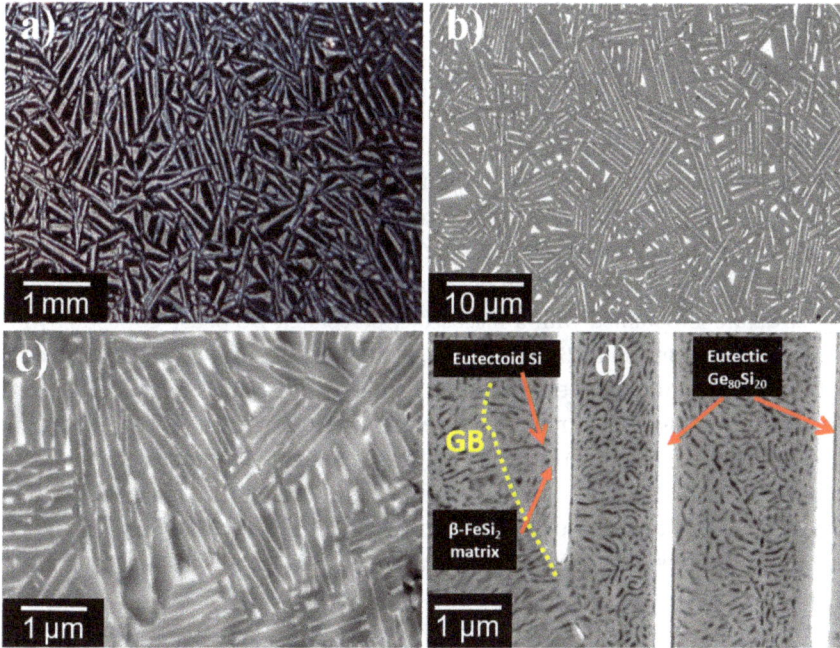

**Figure 8.3:** (a) Optical micrograph of the coarse, slow cooling (arc melting) eutectic microstructure; (b) and (c) backscattered electron image showing the rapid cooling (melt spinning) and ultrafast cooling (laser melting) eutectic microstructure. Reproduced from ref. [19].

case, charge transfer is minimum. Ignoring charge transfer, Bergman mathematically showed that the average composite will always have a thermoelectric figure of merit smaller than its constituent elements [21]. Bergman also ignored thermal conductivity reduction of the interfaces. Therefore, in structures wherein both phases are heavily doped, any enhancement of the figure of merit beyond those of the constituent phases should come from interfacial thermal conductivity reduction.

Seemingly one can easily achieve degenerate multiphase materials by mechanically mixing two heavily doped materials and sintering into a composite. However, the excess defects at the interface will severely trap or scatter the charge carriers irrespective of the energy-level alignment, resulting in lower mobility than expected. Thus, it is more favorable to obtain a self-assembled nanocomposite, with degenerate or close to degenerate doping levels in both phases, using solid-state phase transformations. The challenge associated with this route is to selectively incorporate the impurity ions into constituent phases, especially into the nanoinclusions that are hardly accessible. This becomes even more challenging when the constituent phases have disparate chemistry, in which case different dopant elements are required for each phase. We anticipate selective doping can potentially be achieved by precisely controlling the thermodynamic and kinetic parameters of the doping elements incorporated to their respective parent phase, perhaps assisted by nonequilibrium processing, such as current/field-assisted sintering techniques (or SPS).

In our recent work, intrinsic $\beta$-FeSi$_2$–Si$_{1-y}$Ge$_y$ (($Fe_{29.4}Si_{70.6})_{1-x}Ge_x$) nanocomposites with $x$ = 0–0.1 were synthesized solely by solid-state reaction/transformation without mechanical alloying and post-annealing, with a final microstructure of 100–500 nm SiGe grains/clusters embedded among 200 nm $\beta$-FeSi$_2$ grains [22]. Co and P were introduced as n-type dopants that selectively incorporated in $\beta$-FeSi$_2$ and DC phases. Sb was added to facilitate liquid-phase sintering, which essentially formed a wetting liquid (DC phase) along the $\beta$-FeSi$_2$ grain boundaries during sintering. As a result, a hierarchical structure is obtained with a mesoscale, percolated DC phase coexisting with nanoscale DC inclusions in the $\beta$-FeSi$_2$ matrix, along with selective doping of both constituent phases that to a great extent decouples the thermal and electrical transport. An overall electron mobility at room temperature as high as 12 cm$^2$ V$^{-1} \cdot$ s$^{-1}$, which is much higher than usually observed in $\beta$-FeSi$_2$-based system for thermoelectric materials produced by powder routes, is suggested to be the result of parallel conduction paths in the percolated DC phase [23]. The concept of R/T SPS can be a promising alternative route for synthesizing self-assembled nanocomposite thermoelectric materials.

## 8.4.2 Modulation/transfer doping

The conventional doping strategies for semiconductors all involve introduction of aliovalent ion impurity or defect engineering (i.e., vacancies). The activation process ubiquitously requires solid-state thermal diffusion, which cannot be applied to materials that are heat sensitive. Devices with ultrafine structures, such as nanowires, ultra-thin films, or self-assembled nanofeatures, would also suffer from nonnegligible statistical variation or deactivation of dopants [24]. The other main drawback of conventional doping methods is that they all involve incorporating impurity ions or defects, which create long-range Coulomb potentials that scatter conduction electrons and lowers their mobility. The modulation doping concept has been applied in high mobility transistors, where a heavily doped, wide bandgap material is in contact with an undoped, narrow gap material, for example, AlGaAs/GaAs [25]. In modulation doping (also called remote doping and delta doping), all dopants are concentrated in a thin doping layer. The doping layer is separated from the main transport channel by a spacer. The band alignment between the doped region, the spacer, and the channel results in carrier transport from the doped region to the channel to spatially separate carriers from their parent atoms and therefore to reduce the ionized impurity scattering. The transferred charges form a confined 2D electron gas (2DEG) channel in the vicinity to the heterointerface. An amazing four orders of magnitude enhancement in the mobility has been achieved over 22 years using this strategy [26].

Similar strategy has also been applied in bulk nanostructured thermoelectric materials. There are several possibilities to add dopants that are spatially separated from the charge carriers. Some of these possibilities are shown in Figure 8.4. Zebarjadi et al. pioneered this approach in nanograined SiGe composites, where the dopants are incorporated only into the minority silicon nanograins [27]. These heavily doped grains are finely dispersed (on the order of a few tens of nanometers) and embedded in the undoped nanograined SiGe host, forming a structure similar to Figure 8.4(a) [28]. Due to the band alignment between the grains, the charge carriers spill over from the nanoparticles into the surrounding matrix, while the ionized dopant atoms remain spatially segregated within the nanoparticles. The charge carriers are trapped within the screening length (about few tens of nanometers) of the interfaces. Therefore, to enable charge transport within the bulk matrix, the interfaces should be closely spaced (about twice the screening length). In the demonstrated case, Zebarjadi et al. have observed a 40% enhancement in the PF compared to uniformly alloyed nanocomposites, shown in Figure 8.5, due to the enhanced mobility. This 3D modulation doping strategy has also been widely investigated in other systems [29–33]. We expect the selective doping approach as discussed earlier could be promising in constructing a heavily doped nanophase embedded in undoped matrix.

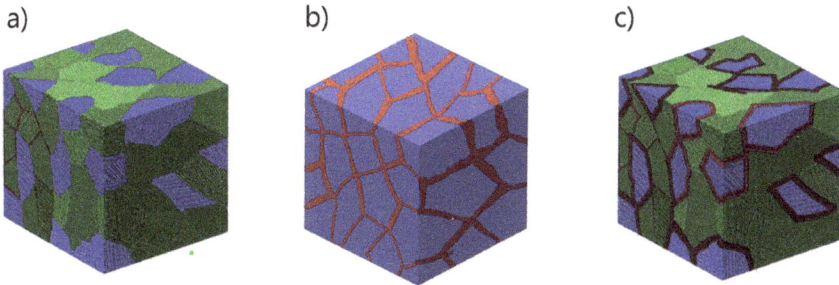

**Figure 8.4:** Different nanostructured modulation doping geometries. (a) Dopants are only included in the blue phase. This phase is mixed with the intrinsic green phase and the mixture is then hot-pressed to form a nanostructured sample. (b) Dopants (or doped phase) are trapped at the red grain boundaries and are donating charge to the undoped host matrix (blue). (c) Doped and updoped regions are separated with a third phase. This third phase acts as spacer to increase the physical separation of charge carriers and their parent atoms.

The 3D modulation doping idea could be combined with band alignment to enhance the properties significantly. Ibanez et al. used silver nanoparticles in PbS matrix and observed that the particles are trapped at the grain boundaries, similar to Figure 8.4(b) [34]. Interestingly, they observed that nanometer size silver particles form a different band alignment when compared to bulk silver with PbS.

As the concentration of silver dopants increases and their spacing decreases, the chemical potential moves closer to the band edge and results in band alignment between silver particles and PbS. This band alignment combined with charge separation

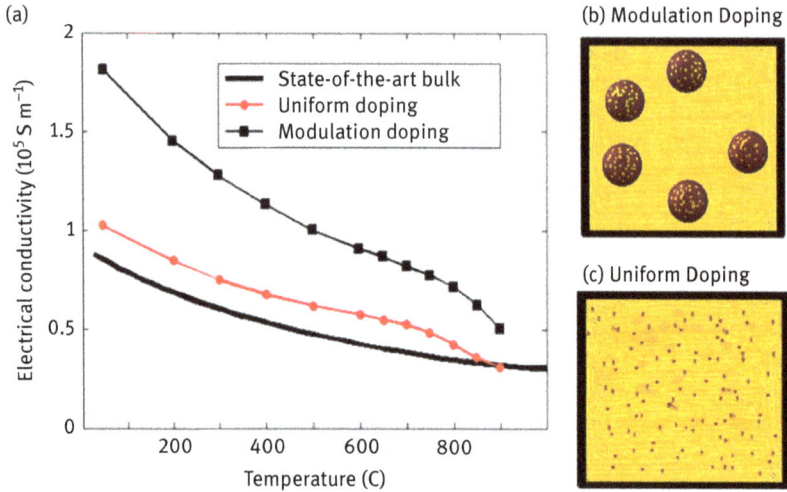

**Figure 8.5:** Schematic and power factor comparison of bulk SiGe with modulation and uniform doping.

results in enhanced carrier mobility and enhanced ZT. Finally, perhaps the best geometry is the one in which spacer layer exists as shown schematically in Figure 8.4(c). The spacer could be a shell around the nanoparticles, that is, core–shell nanoparticles wherein core is heavily doped and shell is undoped could be dispersed in a high mobility intrinsic host matrix. The shell could act as a simple spacer and reduce the coulomb interaction or theoretically, it can be designed to be a cloaking layer to completely screen the nanoparticle scattering. Under the cloaking scenario, electrons can freely move inside the host matrix with much larger mobility values [35–37].

As another unconventional doping approach, polarization-induced doping has been investigated in wide bandgap semiconductors for applications such as laser, light-emitting diode, and high-power electronic devices. One of the technical challenges is to achieve a substantial doping level due to the large thermal activation energy. It is found that the built-in electronic polarization can be utilized to induce charge carriers. In III–VI nitride compounds, for example, Ga(Al)N, the bonds are partially ionic due to different electronegativities of Ga(Al) and N. The ionic nature of the bonds along with the intrinsic asymmetry of the bonding in wurtzite crystal structure gives rise to a large spontaneous polarization. Tensile strain from heteroepitaxy can also result in piezoelectric polarization that adds up to the spontaneous polarization [38]. At *sharp heterojunctions* of AlN/GaN, 2DEG with a high sheet density ~ $5.5 \times 10^{13}$ cm$^{-2}$ (at 77 K) – was formed simply due to the polarization charge-induced electrical field and band bending, similar to the modulation doping scheme, but without deliberately introducing impurity dopants as in GaAs/AlGaAs systems. On the other hand, a very different doping mechanism can be achieved in epitaxially grown *compositionally graded* $Al_xGa_{1-x}N/GaN$ crystals, where the net polarization-induced, unbalanced charges are formed in the graded $Al_xGa_{1-x}N$ unit cells, which essentially form 3D electron slabs [39]. This is because $Al_xGa_{1-x}N$ has higher polarization than that of GaN, and the strength of charge dipoles in each unit cell keeps increasing when $x$ increases. The space charge after compensation in each unit cell would then result in net polarization charges in the bulk, which can effectively dope the material. Polarization direction can be inverted by changing the compositional grading orientation to achieve either p- or n-type doping. The carrier concentration of the 3DEG is insensitive to temperature as it is "field-ionized" and is not subject to thermal freeze-out effect. More importantly, the mobility enhances due to the fact that the ion impurity scattering is absent [39].

Compared with III–VI compounds, the mobility of 2DEG in silicon is couple of orders of magnitude lower. For Si metal–oxide–semiconductor field-effect transistors, the mobility is limited to $\sim 10^4$ cm$^2$ V$^{-1} \cdot$ s$^{-1}$ due to impurity scattering and Si/SiO$_2$ interface roughness [40]. Si/SiGe superlattices with strain-induced heteroband structure modification can generate 2DEG residing in the Si quantum well by modulation doping [41]. A good-quality epitaxy with minimized misfit dislocations and roughness can improve the electron mobility to the order of $\sim 4 \times 10^5$ cm$^2$ V$^{-1} \cdot$ s$^{-1}$ [42]. Nevertheless, the ionized impurity scattering from the supply layer due to the long-range Coulomb potentials would still suppress the mobility from further improvement. Thus, in principle, an undoped Si 2DEG formed by top gating would possess higher mobility than the modulation-doped heterostructures. Mobilities on the order of $\sim 10^6$ cm$^2$ V$^{-1} \cdot$ s$^{-1}$ have been reported [43, 44], which again suggest that electron transport can be greatly enhanced if impurity scattering can be circumvented.

The unconventional doping mechanisms discussed above all involve minimizing ionic impurity scattering for mobility improvement, which is also critical for PF optimization in thermoelectric materials.

While 3D modulation doping strategy has been widely investigated in other bulk systems, studies on 2D modulation doping for thermoelectric device applications are still scarce within literature. Nevertheless, field-induced 2DEG by gating has shown promise for thermoelectric PF improvement. Ohta et al. [45] fabricated an Al$_2$O$_3$/AlGaN/GaN metal oxide semiconductor, high electron mobility transistor, and studied its thermoelectric transport by tuning the gate voltage to modulate the $S$ and $\sigma$ of the 2DEG. It exhibits a high thermoelectric PF of $\approx 9$ mW $\cdot$ m$^{-1}$ K$^{-2}$ at room temperature, which is an order magnitude greater than that of doped GaN bulk, mainly due to mobility enhancement. Seebeck coefficient enhancement was also observed for 2DEG formed in ion-gated transistor with ZnO as the channel material [46]. The quantized DOS of 2DEG can result in a large Seebeck coefficient if the chemical potential is tuned properly. Under a large gate voltage ($\sim 3.5$ V), a sharp band bending at the ZnO surface result in formation of a high-density 2DEG up to $\sim 7 \times 10^{13}$ cm$^{-2}$. Along with the Seebeck coefficient enhancement, the PF was reported to be one order of magnitude larger than that of the 3D-band ZnO materials by impurity doping [46].

In order to make transfer-doped materials practical for power generation, strategies in increasing effective thickness of 2DEG, or even extending 2DEG into bulk 3DEG need to be investigated. 2DEG induced by field effect (gating) consumes electrical power, which is more proper for thermal management applications rather than power generation. In addition, to extend 3D doping with gating seems to be an insurmountable challenge. As for modulation doping or polarization doping schemes, besides the strict epitaxial growth of heterojunctions to minimize the misfit dislocations and roughness, ionized impurity scattering from the remote supply layer would still suppress the mobility, while increasing the spacer thickness would in turn result in lowering the carrier density in the channel material.

In a recent work, we have investigated another doping mechanism utilizing charge transfer at organic–inorganic interface for thermoelectric applications [47, 48]. Physisorption of certain organic molecules on Si can easily be achieved by solution based or physical vapor deposition to initiate charge transfer. Ultimately, it is also feasible to extend this 2D transfer doping to 3D bulk doping by maximizing the surface-to-volume ratio of the inorganic part (e.g., holey Si), so that a 3D network of transferred change carriers can extend through the bulk.

Tetrafluorotetracyanoquinodimethane ($F_4TCNQ$) is a fluorinated TCNQ derivative, and has an exceptionally high electron affinity of $E_a = 5.24$ eV [49]. It is well known as a strong electron acceptor to dope molecules p-type by forming charge transfer organic complex, as well as to enhance hole injection by energy-level alignment at organic–metal interfaces [50, 51]. Therefore, $F_4TCNQ$ has been used for surface doping of inorganic semiconductor surfaces, such as diamond [52], graphene [53], $VO_2$ [54], and Si [55, 56].

The electron affinity of $F_4TCNQ$ is larger than the ionization energy of Si. This favors electron transfer from silicon to $F_4TCNQ$ molecules, and Si is thus p-type doped near the interface. The holes are confined in the direction normal to the interface due to the established electrostatic potential but free to move in the parallel direction.

Wang et al. [48] used first principles calculations to show that physisorbed $F_4TCNQ$ self-assembled monolayers can efficiently dope silicon, achieving hole concentrations as high as $10^{13}$ cm$^{-2}$. Figure 8.6 shows the calculated local density of states of hydrogen terminated Si(100)–TCNQ interface, as well as the charge transfer efficiency using different TCNQ derivatives, including $F_4TCNQ$.

Liu et al. deposited $F_4TCNQ$ molecules on the Si surface. The structure is characterized by transmission electron microscopy (TEM) along Si [110], as shown in Figure 8.7. They showed a factor of 10 enhancement in the electrical conductance of silicon after F4TCNQ deposition and 75% improvement in the thermoelectric PF [47].

(a)

(b)

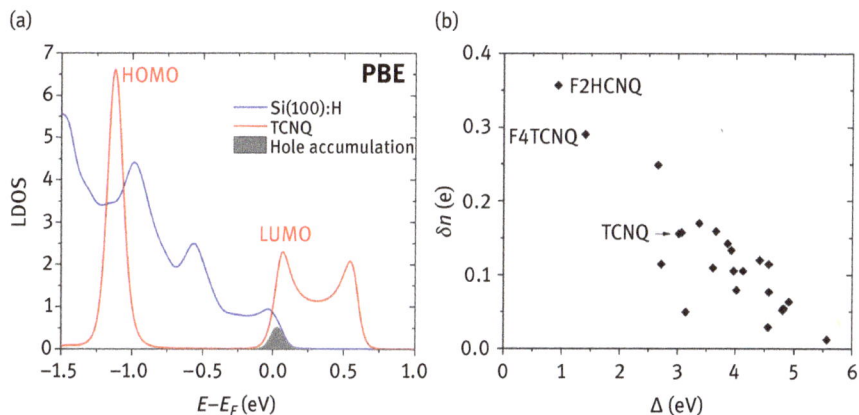

**Figure 8.6:** (a) Local density of states (LDOS) of Si(100):H–TCNQ by DFT (PBE: Perdew–Burke–Ernzerhofb) calculation. (b) Net charge transfer $\delta n$ between the Si(100):H slab calculated by the chemical hardness method. $\Delta$ is the charge transfer energy.

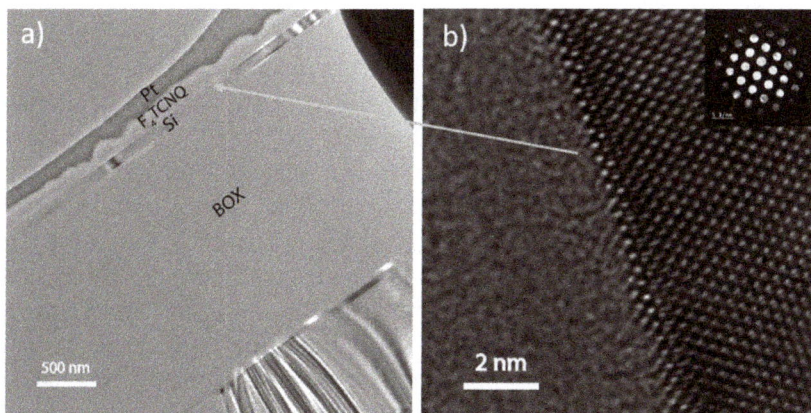

**Figure 8.7:** (a) XTEM of the device by FIB lift-out, with each layer annotated; (b) high-resolution TEM image of F4TCNQ–Si interface along Si [110]. Reproduced after ref. [47].

## 8.4.3 Percolation

Percolation approaches for thermoelectric nanocomposites have mostly been utilized in disordered organic semiconductors that exhibit hopping conduction [57–59]. In inorganic bulk systems, incorporation of a metallic phase has been studied to promote electrical transport by short-range percolation [60]. The volume fraction of the metallic phases has to be kept small, as a complete percolation of metallic phase would deleteriously enhance thermal transport and degrade the Seebeck coefficient. There are also other studies wherein coated powders were produced and subsequently consolidated into bulk, polycrystalline thermoelectric materials, in which the grain boundaries consist of a second phase of nanometer-scale thickness, for instance, $Pb_{0.75}Sn_{0.25}Te$ coated with $Pb_{0.75}Sn_{0.25}Se$ [61] and $La_{0.067}Sr_{0.9}TiO_3$ with small amount of graphene decorating on the grain boundaries [62]. Only moderate improvement in ZT was observed due to reduction in lattice thermal conductivity by interface scattering. Their limited impact on electrical properties, despite being highly conductive, is mainly due to the limited volume fraction, which is far below the percolation threshold despite the fact that it has favorable morphology as being coating on the grain boundaries.

To facilitate the percolation, liquid phase sintering is the key to the process. Liquid phase sintering involves sintering under conditions where solid grains coexist with a wetting liquid. This technique is more often used in ceramic community, as the "liquid" provides a capillary force that pulls the grains together. At the same time, the high temperature softens the solid, further assisting densification. Large diffusion rates are associated with liquids, giving fast sintering or lower sintering temperatures [63]. For thermoelectric materials processing, a low melting point phase can be introduced as the liquid phase, which is expected to wet the nanograins of the matrix materials. Alternatively, one can use a low melting temperature element to facilitate the liquid-phase sintering, and subsequently alloyed or incorporated in the percolated phase. Moreover, one can carefully design the sintering parameters to utilize the liquidus lines in the eutectic systems, in order to have one phase to form a liquid while sintering. Figure 8.8 shows an example of utilizing $Bi_{0.5}Sb_{1.5}Te_3$–Te pseudobinary eutectic system to form Te-coated $Bi_{0.5}Sb_{1.5}Te_3$ platelets for subsequent consolidation [64]. Although this study is focused on the impact of the dislocations generated on the grain boundaries due to expelled liquid Te during sintering, it still provides the feasibility of achieving a percolated composite by liquid phase sintering via eutectic decomposition.

Liu et al. [23] extended the percolation stratagem to an all-semiconducting system to form a composite with a matrix phase exhibiting high Seebeck coefficient, and a secondary phase possessing high carrier mobility and hierarchical length scales. While smaller length scale nanoparticles would contribute to the phonon scattering, the simultaneous presence of larger length scales could permit phase percolation that retains good electrical conductivity. The percolation approach is especially useful when the constituent phases have drastically different crystal and electronic structures, in which case the aforementioned approaches such as constructing coherent interfaces and band alignment are impossible to implement. An overall electron mobility at room temperature as high as 12 $cm^2$ $V^{-1} \cdot s^{-1}$, which is much higher than usually observed in this system for thermoelectric materials produced by powder routes was observed. This large mobility is suggested to be the result of parallel conduction paths in the percolated DC phase. An improved ZT of ~0.67 at 973 K with an average ZT of ~0.65 at a broad temperature range from 823 to 1,073 K was achieved in the β-$FeSi_2$–SiGe nanocomposite. The evidence of percolation is shown in Figure 8.9.

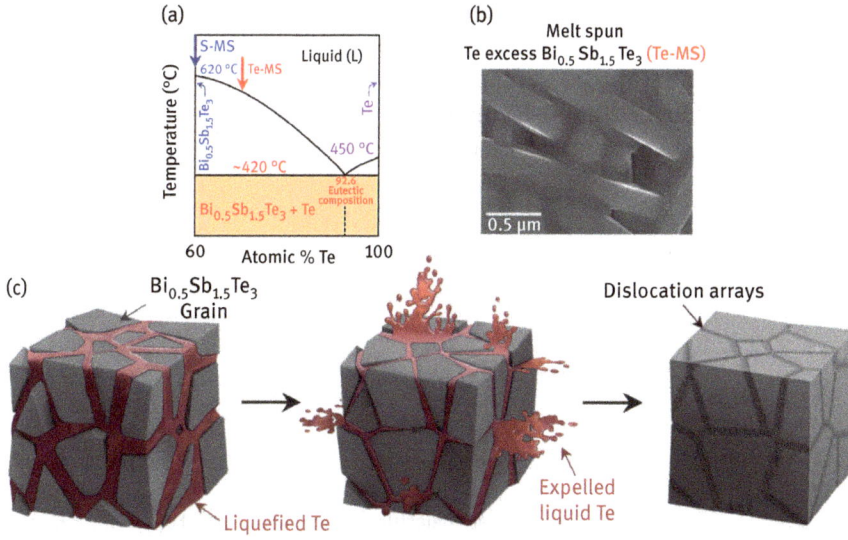

**Figure 8.8:** (a) Phase diagram of $Bi_{0.5}Sb_{1.5}Te_3$–Te system showing a eutectic composition at 92.6 at. % Te; (b) the SEM image of melt-spun ribbon with $Bi_{0.5}Sb_{1.5}Te_3$ platelets coated by Te; (c) Schematic illustration showing the generation of dislocation arrays during the liquid-phase compaction process [64].

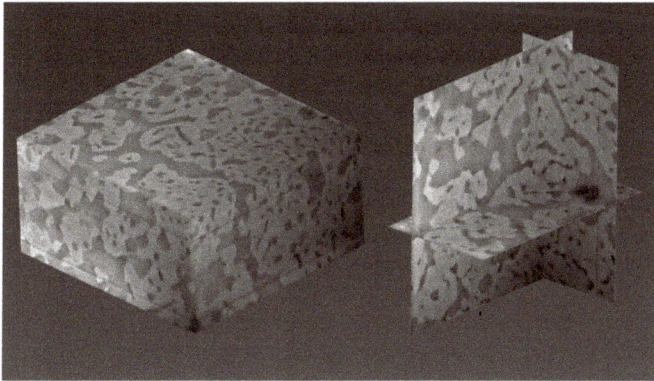

**Figure 8.9:** Three-dimensional reconstruction (12 μm × 12 μm × 5 μm) of sample Sb-1 with percolated DC phase in β-$FeSi_2$ phase [23].

# 8.5 Summary

Heavily doped semiconductors have been the subjects of intense investigation in the thermoelectric field. The carrier mobility of these materials at low temperatures is limited by electron-ionized impurity scattering. Therefore, strategies to separate charge carriers from their parent atoms to reduce the Coulomb interaction are of interest and result in overall improvement in the thermoelectric PF. In this chapter, we summarized some of the latest proposed doping techniques enabling enhanced electrical conductivity in 2D and 3D thermoelectric materials. The complexity and the novelty of many of these strategies require in-depth study and long-term investment in the field.

**Acknowledgments:** This work was supported by the National Science Foundation, Grant number 1653268.

# References

[1]   Adams M J, Verosky M, Zebarjadi M, Heremans J P. Active Peltier coolers based on correlated and magnon-drag metals, Phys. Rev. Appl. 2019.

[2]   Zebarjadi M. Electronic cooling using thermoelectric devices, Appl. Phys. Lett. 2015, 106.

[3]   Adams M J, Verosky M, Zebarjadi M, Heremans J P. High switching ratio variable-temperature solid-state thermal switch based on thermoelectric effects, Int. J. Heat Mass Transf. 2019, 134, 114–118.

[4]   Heremans J P, Wiendlocha B, Chamoire A M. Resonant levels in bulk thermoelectric semiconductors, Energy Environ. Sci. 2012, 5, 5510.

[5]   Watzman S J, Duine R A, Tserkovnyak Y, Boona S R, Jin H, Prakash A, Zheng Y, Heremans J P. Magnon-drag thermopower and Nernst coefficient in Fe, Co, and Ni, Phys. Rev. B 2016, 94, 144407.

[6]   Duan J, Wang X, Lai X, Li G, Watanabe K, Taniguchi T, Zebarjadi M, Andrei, E Y. High thermoelectric power factor in graphene/hBN devices, Proc. Natl. Acad. Sci. U. S. A. 2016, 113, 14272–14276.

[7]   Heremans J P, Jovovic V, Toberer E S, Saramat A, Kurosaki K, Charoenphakdee A, Yamanaka S, Snyder, G J. Enhancement of thermoelectric of the electronic density of states, Science (80-.). 2008, 321, 1457–1461.

[8]   Markov M, Hu X, Liu H-C, Liu N, Poon J, Esfarjani K, Zebarjadi M. Semi-metals as potential thermoelectric materials: case of HgTe, Sci. Rep. 2018, 8.

[9]   Markov M, Rezaei E, Sadeghi S N, Esfarjani K, Zebarjadi M. Thermoelectric properties of semimetals, 2019.

[10]  Poudel B, Hao Q, Ma Y, Lan Y, Minnich A, Yu B, Yan X, Wang D, Muto A, Vashaee D et al. high-thermoelectric performance of nanostructured bismuth antimony telluride bulk alloys, Science 2008, 320, 634–638.

[11]  Biswas K, He J, Blum I D, Wu C-I, Hogan T P, Seidman D N, Dravid V, Kanatzidis M G. High-performance bulk thermoelectrics with all-scale hierarchical architectures, Nature 2012, 489, 414–418.

[12]  Zhao L, He J, Hao S, Wu C, Hogan T P, Wolverton C, Dravid V. P, Kanatzidis M G. Raising the thermoelectric performance of P – Type PbS with endotaxial nanostructuring and valence-band O Ff Set engineering using CdS and ZnS, 2012.

[13]  Shin W H, Roh J W, Ryu B, Chang H J, Kim H S, Lee S, Seo W S, Ahn, K. Enhancing thermoelectric performances of bismuth antimony telluride via synergistic combination of multiscale structuring and band alignment by FeTe2 incorporation, ACS Appl. Mater. Interfaces 2018, 10, 3689–3698.

[14] Tan G, Zhao L D, Kanatzidis M G. Rationally designing high-performance bulk thermoelectric materials, Chem. Rev. 2016, 116, 12123–12149.

[15] Biswas K, He J, Zhang Q, Wang G, Uher C, Dravid V P, Kanatzidis M G. Strained endotaxial nanostructures with high thermoelectric figure of merit, Nat. Chem. 2011, 3, 160–166.

[16] Medlin D L, Snyder G J. Interfaces in bulk thermoelectric materials. a review for current opinion in colloid and interface science, Curr. Opin. Colloid Interface Sci. 2009, 14, 226–235.

[17] Yu Y, He D, Zhang S, Cojocaru-mirédin O, Schwarz T, Sto A, Wang X, Zheng S, Zhu B, Scheu C et al. Simultaneous optimization of electrical and thermal transport properties of Bi0.5Sb1.5Te3 thermoelectric alloy by twin boundary engineering, 2017, 37, 203–213.

[18] Snyder G J, Toberer E S. Complex thermoelectric materials, Nat. Mater. 2008, 7, 105–114.

[19] Jensen W A, Liu N, Donovan B F, Tomko J A, Hopkins P E, Floro J A. Synthesis and thermal transport of eco-friendly Fe-Si-Ge alloys with eutectic/eutectoid microstructure, Mater. Chem. Phys. 2018, 207, 67–75.

[20] Zhao L D, Hao S, Lo S H, Wu C I, Zhou X, Lee Y, Li H, Biswas K, Hogan T P, Uher C et al. High thermoelectric performance via hierarchical compositionally alloyed nanostructures, J. Am. Chem. Soc. 2013, 135, 7364–7370.

[21] Bergman D J, Levy O. Thermoelectric properties of a composite medium, J. Appl. Phys. 1991, 70, 6821.

[22] Liu N, Jensen W A, Zebarjadi M, Floro J A. Tunable B -FeSi 2 E Si 1-Y Ge Y Nanocomposites by a Novel React / Transform spark plasma sintering approach for thermoelectric applications, 2018, 4, 19–27.

[23] Liu N, Rezaei S E, Jensen W A, Song S, Ren Z, Esfarjanit K, Zebarjadi M, Floro J A. Improved thermoelectric performance of eco-friendly β-FeSi2 – SiGe nanocomposite via synergistic hierarchical structuring, phase percolation, and selective doping, Adv. Funct. Mater. 2019.

[24] Björk M T, Schmid H, Knoch J, Riel H, Riess W. Donor deactivation in silicon nanostructures, Nat. Nanotechnol. 2009, 4, 103–107.

[25] Dingle R, Störmer H L, Gossard A C, Wiegmann W. Electron mobilities in modulation-doped semiconductor heterojunction superlattices, Appl. Phys. Lett. 1978, 33, 665.

[26] Pfeiffer L, West K. The role of MBE in recent quantum Hall effect physics discoveries, Phys. E Low-dimensional Syst. Nanostructures 2003, 20, 57–64.

[27] Zebarjadi M, Joshi G, Zhu G, Yu B, Minnich A, Lan Y, Wang X, Dresselhaus M, Ren Z, Chen G. Power factor enhancement by modulation doping in bulk nanocomposites., Nano Lett. 2011, 11.

[28] Yu B, Zebarjadi M, Wang H, Lukas K, Wang H, Wang D, Opeil C, Dresselhaus M, Chen G, Ren Z. Enhancement of thermoelectric properties by modulation-doping in silicon germanium alloy nanocomposites., Nano Lett. 2012, 12, 2077–82.

[29] Hou Q R, Gu B F, Chen Y B, He Y J, Sun J L. Enhancement of the thermoelectric power factor of MnSi1.7 film by modulation doping of Al and Cu, Appl. Phys. A 2013, 114, 943–949.

[30] Pei Y-L, Wu H, Wu D, Zheng F, He J. High thermoelectric performance realized in a BiCuSeO system by improving carrier mobility through 3d modulation doping, J. Am. Chem. Soc. 2014, 136, 13902–13908.

[31] Lee D, Sayed S Y, Lee S, Kuryak C A, Zhou J, Chen G, Shao-Horn Y. Quantitative analyses of enhanced thermoelectric properties of modulation-doped PEDOT:PSS/undoped Si (001) nanoscale heterostructures, Nanoscale 2016, 8, 19754–19760.

[32] Song E, Li Q, Swartzentruber B, Pan W, Wang G T, Martinez J A. Enhanced thermoelectric transport in modulation-doped GaN/AlGaN Core/shell nanowires, Nanotechnology 2016, 27, 15204.

[33] Wang H, Cao X, Takagiwa Y, Snyder G J. Higher mobility in bulk semiconductors by separating the dopants from the charge-conducting band – a case study of thermoelectric PbSe, Mater. Horiz. 2015, 2, 323–329.

[34] Ibáñez M, Luo Z, Genç A, Piveteau L, Ortega S, Cadavid D, Dobrozhan O, Liu Y, Nachtegaal M, Zebarjadi M et al. High-performance thermoelectric nanocomposites from nanocrystal building blocks, Nat. Commun. 2016, 7, 10766.

[35] Liao B, Zebarjadi M, Esfarjani K, Chen G. Cloaking core-shell nanoparticles from conducting electrons in solids, Phys. Rev. Lett. 2012, 109.

[36] Zebarjadi M, Liao B, Esfarjani K, Dresselhaus M, Chen G. Enhancing the thermoelectric power factor by using invisible dopants, Adv. Mater. 2013, 25.

[37] Liao B, Zebarjadi M, Esfarjani K, Chen G. Isotropic and energy-selective electron cloaks on graphene. Phys. Rev. B – Condens. Matter Mater. Phys. 2013, 88.

[38] Bernardini F, Fiorentini V, Vanderbilt D. Spontaneous polarization and piezoelectric constants of III-V Nitrides, Phys. Rev. B – Condens. Matter Mater. Phys. 1997, 56, R10024–R10027.

[39] Simon J, Protasenko V, Lian C, Xing H G, Jena D. Polarization-induced hole doping in semiconductor heterostructures, Science (80-.). 2010, 327, 60–64.

[40] Xiao M, House M G, Jiang H W. Measurement of the spin relaxation time of single electrons in a silicon metal-oxide-semiconductor-based quantum dot, Phys. Rev. Lett. 2010, 104, 3–6.

[41] Abstreiter G, Brugger H, Wolf T, Jorke H, Herzog H J. Strain-Induced two-dimensional electron gas in selectively doped Si / Si X Ge 1 – X Superlattices, Phys. Rev. Lett. 1985, 54, 2441–2444.

[42] Ismail K, Arafa M, Saenger K L, Chu J O, Meyerson B S. Extremely high electron mobility in Si/SiGe modulation-doped heterostructures, Appl. Phys. Lett. 1995, 66, 1077–1079.

[43] Huang S H, Lu T M, Lu S C, Lee C H, Liu C W, Tsui D C. Mobility enhancement of strained Si by optimized SiGe/Si/SiGe structures, Appl. Phys. Lett. 2012, 101, 99–102.

[44] Lu T M, Tsui D C, Lee C H, Liu C W. Observation of two-dimensional electron gas in a Si quantum well with mobility of 1.6 × 106 cm2 /V S, Appl. Phys. Lett. 2009, 94, 4–7.

[45] Ohta H, Kim S W, Kaneki S, Yamamoto A, Hashizume T. High thermoelectric power factor of high-mobility 2D electron gas, Adv. Sci. 2017, 5, 1700696.

[46] Iwasa Y, Tokura Y, Ono S, Shimizu S, Bahramy M S, Iizuka T, Miwa K. Enhanced thermopower in ZnO two-dimensional electron gas, Proc. Natl. Acad. Sci. 2016, 113, 6438–6443.

[47] Liu N, Peters J, Ramu A, Floro J A, Bowers J E, Zebarjadi M. Thermoelectric transport at F 4 TCNQ–silicon interface, APL Mater. 2019, 7, 21104.

[48] Wang X, Esfarjani K, Zebarjadi M. First-principles calculation of charge transfer at the silicon-organic interface, J. Phys. Chem. C 2017, 121.

[49] Gao W, Kahn A. Electronic structure and current injection in zinc phthalocyanine doped with tetrafluorotetracyanoquinodimethane: interface versus bulk effects, Org. Electron. physics, Mater. Appl. 2002, 3, 53–63.

[50] Koch N, Duhm S, Rabe J P, Vollmer A, Johnson R L. Optimized hole injection with strong electron acceptors at organic-metal interfaces, Phys. Rev. Lett. 2005, 95, 4–7.

[51] Braun S, Salaneck W R, Fahlman M. Energy-level alignment at organic/metal and organic/organic interfaces, Adv. Mater. 2009, 21, 1450–1472.

[52] Qi D, Chen W, Gao X, Wang L, Chen S, Kian P L, Wee A T S. Surface transfer doping of diamond (100) by tetrafluoro-tetracyanoquinodimethane, J. Am. Chem. Soc. 2007, 129, 8084–8085.

[53] Chen W, Chen S, Dong C Q, Xing Y G, Wee A T S. Surface transfer p-type doping of epitaxial graphene, J. Am. Chem. Soc. 2007, 129, 10418–10422.

[54]  Wang K, Zhang W, Liu L, Guo P, Yao Y, Wang C H, Zou, C, Yang Y W, Zhang G, Xu, F. Holes doping effect on the phase transition of VO2film via surface adsorption of F4TCNQ molecules, Appl. Surf. Sci. 2018, 447, 347–354.

[55]  Mukai K, Yoshinobu J. Observation of charge transfer states of F4-TCNQ on the 2-Methylpropene Chemisorbed Si(1 0 0)(2 × 1) surface, J. Electron Spectros. Relat. Phenomena 2009, 174, 55–58.

[56]  Yoshimoto S, Furuhashi M, Koitaya T, Shiozawa Y, Fujimaki K, Harada Y, Mukai K, Yoshinobu J. Quantitative analysis of chemical interaction and doping of the Si(111) native oxide surface with tetrafluorotetracyanoquinodimethane, J. Appl. Phys. 2014, 115.

[57]  Cho C, Stevens B, Hsu J, Bureau R, Hagen D A, Regev O, Yu C, Grunlan J C. Completely organic multilayer thin film with thermoelectric power factor rivaling inorganic tellurides, Adv. Mater. 2015, 27, 2996–3001.

[58]  Lu N, Li L, Liu M. Universal carrier thermoelectric-transport model based on percolation theory in organic semiconductors, Phys. Rev. B 2015, 91, 195205.

[59]  Tessler B N, Preezant Y, Rappaport N, Roichman Y. Charge transport in disordered organic materials and its relevance to thin-film devices : a tutorial review, 2009, 2741–2761.

[60]  Gharleghi A, Liu Y, Zhou M, He J, Tritt T M, Liu C J. Enhancing the thermoelectric performance of nanosized $CoSb_3$: Via short-range percolation of electrically conductive WTe2inclusions, J. Mater. Chem. A 2016, 4, 13874–13880.

[61]  You A, Be M A Y, In I. Controlled two-dimensional coated nanostructures for bulk thermoelectric composites, 2015, 163114, 1–4.

[62]  Lin Y, Norman C, Srivastava D, Azough F, Wang L, Robbins M, Simpson K, Freer R, Kinloch I A. Thermoelectric power generation from lanthanum strontium titanium oxide at room temperature through the addition of graphene, 2015.

[63]  German R M, Suri P, Park S J. Review: Liquid phase sintering, J. Mater. Sci. 2009, 44, 1–39.

[64]  Kim S, Lee K H, Mun H A, Kim H S, Hwang S W, Roh J W, Yang D J, Shin W H, Li X S, Lee Y H et al. Dense dislocation arrays embedded in grain boundaries for high-performance bulk thermoelectrics, Science (80-.). 2015, 348.

Priyanka Jood, Raju Chetty, and Michihiro Ohta

# 9 All-scale phonon scattering

**Abstract:** Thermal conductivity is one of the most important parameters determining the performance of a thermoelectric material. Recently, various strategies for effective phonon scattering have been developed which have boosted the thermoelectric figure of merit, ZT of many materials. All-scale phonon scattering is one such strategy where a combination of several scattering mechanisms come into play to impede the phonon transport. In this Chapter we will review pathways of achieving all-scale phonon scattering in traditional and new thermoelectric materials.

**Keywords:** Nanostructuring, Hierarchical architecturing, Phonon meas free path, Band convergence, Minority carrier blocking, Layered structure, Intercalation, Microstructural ordering, Complex crystal structure, Defect, Disorder, Rattling, Module

## 9.1 Introduction

Large efforts have been made to reduce the lattice thermal conductivity ($\kappa_{lat}$) of thermoelectric materials in the last two decades. $\kappa_{lat}$ is a dominant thermal conduction mechanism in most of the semiconductors and alloys over a wide temperature range and can be written as follows [1, 2]:

$$\kappa_{lat} = \frac{1}{3}C_v v l_p = \frac{1}{3}C_v v^2 \tau_q. \tag{9.1}$$

$C_v$ is the heat capacity at constant volume, $v$ is the mean sound velocity, $l_p$ ($=v\tau_q$) is the phonon mean free path, and $\tau_q$ is the phonon relaxation time. In a crystalline solid, $C_v$ and $v$ can be normally regarded as constant at high temperature (for $T > \theta_D$, Debye temperature) [3]. This means that $\kappa_{lat}$ is mainly governed by the $l_p$ or the $\tau_q$.

In simple terms, there are two kinds of phonon scattering events, non-resistive and resistive [1]. The normal (N) processes conserve momentum and do not directly contribute to any thermal resistance; however, these nonresistive processes do redistribute the momentum within different phonons. Determining the contribution of N-processes in the $\kappa_{lat}$ of solids is well explained by Callaway and the information can be found elsewhere [4]. The Umklapp (U) processes, on the other hand, is an example of a resistive event where momentum is not conserved, leading to the

**Priyanka Jood, Raju Chetty, Michihiro Ohta,** Global Zero Emission Research Center and Research Institute for Energy Conservation, National Institute of Advanced Industrial Science and Technology (AIST), Tsukuba, Ibaraki, Japan

https://doi.org/10.1515/9783110596526-009

finite thermal conductivity we observe. Electrons also play an important role in scattering phonons; however, their contribution varies depending on the material [5]. The overall scattering rate is the sum of the scattering rates of all the resistive scattering processes involved (eq. (9.2)), such as Umklapp phonon–phonon scattering ($\tau_U$), electron–phonon interaction ($\tau_{e-p}$), grain boundary scattering ($\tau_B$), resonance scattering ($\tau_R$), point defect scattering ($\tau_{PD}$), and scattering from nanostructures ($\tau_N$):

$$\frac{1}{\tau_q} = \frac{1}{\tau_U} + \frac{1}{\tau_{e-p}} + \frac{1}{\tau_B} + \frac{1}{\tau_R} + \frac{1}{\tau_{PD}} + \frac{1}{\tau_N} + \cdots. \tag{9.2}$$

Grain boundaries are interfaces in a polycrystalline material that can scatter effectively mid- to long-wavelength phonons depending on the grain sizes [6–8]. Rattling-like behavior of guest atoms in certain crystal structures with voids induce resonant scattering, which is also effective in scattering long wavelength phonons [7, 9]. Short-wavelength phonons can be targeted by point defects, due to the $1/\omega^4$ dependency (Rayleigh scattering) of relaxation time, where $\omega$ is the phonon frequency [10]. Nanostructures, such as nanoinclusions, nanopores, nanocomposites, and nanograins, can influence the transport of mid-wavelength phonons [7, 11]. The combination of several extrinsic scattering mechanisms to impede the transport of a wide spectrum of phonons is commonly termed as "all-scale phonon scattering." The detailed description of individual scattering mechanisms can be found in several books [12, 13] and research articles [6, 7, 14, 15]. This chapter is about reviewing pathways for all-scale phonon scattering, which is a relatively new concept.

## 9.2 Approaches for all-scale phonon scattering

The foundation of all-scale phonon scattering is knowing how the length scales and morphology of nano/microstructures influence the phonon transport and how these effects can be mutually adopted for maximum reduction in $\kappa_{lat}$. Alloying has been a traditional approach in reducing $\kappa_{lat}$, which results in point defect scattering of short mean free path phonons due to atomic-level defects, such as mass fluctuations and local bond strains [11, 13]. The performance of many thermoelectric materials has improved from alloying in the past half century [16–23]. For example, Abrahams *et al.* [17] reported a 70% reduction in room temperature $\kappa_{lat}$ of the InAs–GaAs alloy with the addition of only 10% of either constituent into the respective pure compound. Similarly, SiGe alloy is also an important case study for analyzing the impact of alloying on the thermal and electrical properties [16, 19]. Point defects due to alloying, although effective in scattering short wavelength phonons, have little effect on mid- to long-wavelength phonons.

  This is where "nanoparticles in alloy" approach has shown to further reduce $\kappa_{lat}$, due to the combined effect of point defect scattering of short mean free path phonons and the interfaces between nanostructures and matrix effectively scattering mid-mean free path phonons [24, 25]. For example, $In_{0.53}Ga_{0.47}As$ alloy with randomly distributed 3% ErAs nanoparticles has a $\kappa_{lat}$ ~3 W m$^{-1}$ K$^{-1}$, which is ~40% lower than the parent alloy $In_{0.53}Ga_{0.47}As$ (~5 W m$^{-1}$ K$^{-1}$) and ~90% lower than the individual InAs or GaAs compound (~29–37 W m$^{-1}$ K$^{-1}$), at room temperature [25, 26]. The *ZT* increased to twofold with ErAs nanoinclusions in InGaAs matrix, not only because of reduced $\kappa_{lat}$, but also due to an increase in power factor owing to energy filtering effect.

Mingo and coworkers *et al.* studied 17 different types of silicide nanoparticles in SiGe alloy and predicted an optimal nanoparticle size for reducing $\kappa_{lat}$ [24]. For example, 0.8% volume fraction of silicide in SiGe alloy with 2–10 nm particle size could result in four times smaller minimum thermal conductivity compared to the value for SiGe alloy at room temperature. Their study also pointed out that the nanoparticles should not obstruct the electron mean free path if they are placed fairly apart, which in their case was controlled by 0.8% volume fraction and ≥2 nm size of the nanoparticles. Also, lattice match between the nanoparticles and the matrix is critical to produce a composite system with least dislocations for the interest of carrier transport.

Another good example of materials for which great reduction in $\kappa_{lat}$ has been achieved, primarily by scattering wide spectrum of phonons, is PbTe. The contribution of different mean free path phonons to the $\kappa_{lat}$ of PbTe was calculated using molecular dynamics simulation (Figure 9.1) [27–29]. Phonons with mean free path less than 100 nm contribute to about ~80% of $\kappa_{lat}$ in PbTe, which can be scattered by a combination of point defects and embedding nanoscale precipitates. The remaining ~20% is caused by phonon modes with mean free path of 100–1,000 nm. Effective scattering of these long wavelength phonons by mesoscale grain boundaries attained through pressure-assisted sintering of the fine powder sample has been experimentally achieved, showing a further reduction of $\kappa_{lat}$, compared to nanostructuring alone [28–30]. The PbTe case will be discussed in detail in the next section.

The number of long wavelength phonon modes is small. However, their contribution to $\kappa_{lat}$ is significant because of their large free path lengths [3, 31]. Several techniques of microstructure control and grain boundary engineering to scatter mid- to long-wavelength phonons, in combination with other approaches, are well studied for various materials and have shown considerable decrease in $\kappa_{lat}$ [6, 8, 15]. Recently, a record high *ZT* of ~1.9 at 320 K was achieved for bismuth antimony telluride due to full spectrum phonon scattering, while maintaining the carrier mobility [32]. In this study, periodic dislocations in low-energy grain boundaries achieved through liquid-phase compaction in $Bi_{0.5}Sb_{1.5}Te_3$ effectively scattered mid-wavelength phonons. The short-wavelength phonons were targeted by point defect scattering arising from the Bi–Sb disorder and the long-wavelength phonons were scattered by grain boundaries. Some crystal structures with rattling fillers have shown to scatter long-wavelength phonons quite effectively, such as in skutterudites [33–35], clathrates [36, 37], and tetrahedrites [38–40].

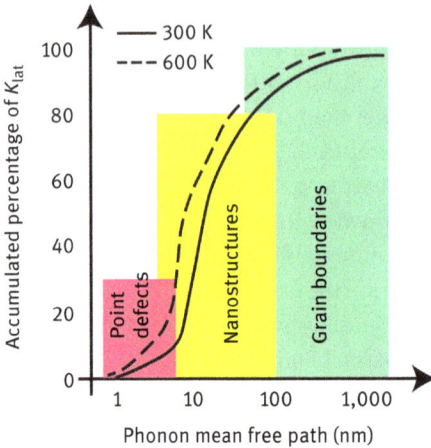

**Figure 9.1:** Contribution of phonons with different mean free paths to the cumulative percentage of $\kappa_{lat}$ for PbTe [27–29].

In this chapter, we will not only review the all-scale phonon scattering approach in PbTe, which is a conventional thermoelectric material, but also revisit some unconventional and emerging thermoelectric sulfides and discuss relevant phonon scattering techniques used.

# 9.3 All-scale phonon scattering in PbTe

## 9.3.1 Nanostructuring and hierarchical architecturing

PbTe is a conventional mid-temperature thermoelectric material, which has been extensively studied since 1950s, leading up to its excellent thermoelectric performance. Last decade has seen a big boost in the $ZT$ of PbTe due to strategies like band convergence to improve effective mass, optimizing carrier concentration to tune Fermi level, and creation of scattering centers at all length scales for strong suppression of phonon propagation commonly termed as hierarchical architecturing [28, 29, 41–54].

One of the most successful approaches for the latter is through nucleation and growth of a secondary phase (or multiple secondary phases) [28, 29, 46, 50, 55–57]. For instance, Figure 9.2 describes the formation of precipitates and other defects pertaining to phase B in the bulk matrix A. The second phase candidate B must be completely soluble in the liquid state but should have no or low solubility in A in the solid state. Therefore, eventually when the melt is quenched or cooled down, B precipitates in A matrix owing to the solubility limit being exceeded. The targeted

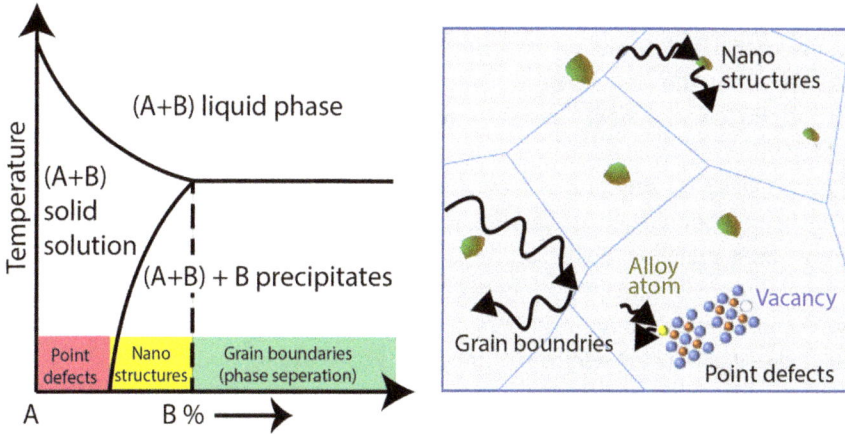

**Figure 9.2:** Phase diagram for the formation of all-scale hierarchical structures along with a schematic describing these hierarchical structures that effectively scatter wide spectrum of phonons [29, 53].

defects, that is, point defects (alloy atoms or/and vacancies) and nanoprecipitates formed with this technique are schematically shown in Figure 9.2. The size and density of defects are sensitive to the processing method, such as annealing temperature and time, and rate of cooling the melt [58]. The short- and mid-mean free path phonons are effectively scattered by the point defects and nanostructures achieved through this approach. Furthermore, the as-obtained ingot can be further crushed into powder, where the crystals are broken down to sizes between 100 nm and 5 µm, followed by the pressure-assisted sintering to form mesoscale grain boundaries where scattering of long mean free path phonons occurs [28] (Figure 9.2). This integrated approach of hierarchical architecturing in one bulk material has proven to reduce the $\kappa_{lat}$ close to the minimum limit ($\kappa_{min}$), both theoretically and experimentally in PbTe.

The comparison of $\kappa_{lat}$ for both p-type and n-type PbTe with values derived from the literature [28, 29, 45, 52, 57, 59–61] is shown in Figure 9.3. Alloy scattering increases when doping content is increased due to an increase in point defects (Na or I) [57, 60]. For example, ~20% reduction in $\kappa_{lat}$ at ~700 K is obtained when Na doping content is increased from ≤2% to 4% in PbTe ingot sample. Nanostructuring using SrTe and CdTe results in a major reduction in $\kappa_{lat}$, ~40% in p-type and ~65% in n-type at room temperature, respectively [59, 60]. Systems such as PbTe–MTe (M = Mg, Ca, Sr, and Ba) where MTe phase precipitates out as nanostructures in PbTe matrix have shown significantly low $\kappa_{lat}$ due to effective scattering of mid-wavelength phonons by the nanostructures and associated strains [56–59]. These nanostructures are coherently or semicoherently embedded in the PbTe matrix with a small band offset between the matrix and nanostructures. This allows smooth transmission of the charge carriers through the nanostructures. The alignment of the valence band energies of PbTe and nanostructures facilitates hole mobility, while the associated strain fields of nanostructures intensify phonon scattering, hence decoupling the electrical and phonon transport. For example, the Hall mobility for both PbTe without SrTe and PbTe with 2% SrTe is similar, $\mu \sim 350$ cm$^2$ V$^{-1}$ s$^{-1}$ at 300 K [59].

Pressure-assisted sintering of a crushed ingot is widely used to promote microstructuring by producing compacts with additional grain boundaries that scatter long mean free path phonons. For instance, sintered sample of Pb$_{0.98}$Na$_{0.02}$Te has up to 28% reduced $\kappa_{lat}$ compared to its ingot equivalent (Figure 9.3) [45]. In case of n-type, higher doping concentration in PbTe$_{0.996}$I$_{0.004}$ sintered sample promotes brittleness, which give rise to a mixed microstructure (grains of two different length scales coexisted) that effectively reduced $\kappa_{lat}$ to as low as $\sim$ 0.4 W m$^{-1}$ K$^{-1}$ at 600 K [61]. Integrated phonon scattering where a broad spectrum of phonons is scattered across multiple length scales has resulted in greatly diminished $\kappa_{lat}$ in p-type PbTe. Sintered Pb$_{0.98}$Na$_{0.02}$Te-4% SrTe combines alloy scattering, scattering

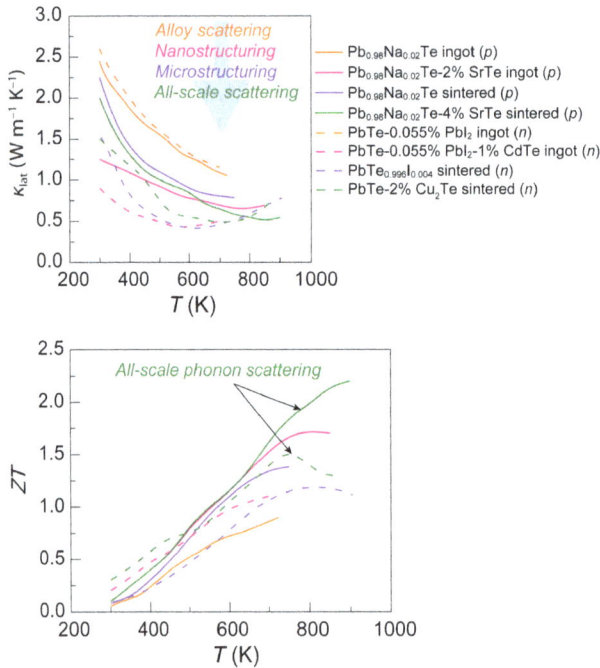

**Figure 9.3:** Temperature dependence of lattice thermal conductivity $\kappa_{lat}$ and thermoelectric figure of merit ZT for the solid solution/alloyed (p-type Pb$_{0.98}$Na$_{0.02}$Te [57] and n-type PbTe-0.055% PbI$_2$ ingot [60]), nanostructured (p-type Pb$_{0.98}$Na$_{0.02}$Te-2% SrTe [59] and n-type PbTe-0.055% PbI$_2$-1% CdTe ingot [60]), microstructured (p-type Pb$_{0.98}$Na$_{0.02}$Te [45] and n-type PbTe$_{0.996}$I$_{0.004}$ [61] sintered sample), and all-scale hierarchical architecture PbTe (p-type Pb$_{0.98}$Na$_{0.02}$Te-4% SrTe [28] and n-type PbTe-2% Cu$_2$Te [52] sintered sample) system.

from nanoprecipitation, and scattering from grain boundaries, resulting in a very low $\kappa_{lat} \sim 0.5$ W m$^{-1}$ K$^{-1}$, while maintaining a high power factor ($S^2/\rho \sim 2{,}400$ µW m$^{-1}$ K$^{-2}$), leading to high $ZT$ value of ~2.2 at 915 K [28]. Compared with the ingot sample of the same composition, sintered sample shows a ~ 30–40% reduction in $\kappa_{lat}$ and a corresponding rise in $ZT$ due to additional scattering of long mean free path phonons at grain boundaries as well as reduction in bipolar contribution to thermal conductivity (discussed in later section).

Recently, high $ZT$ values have been reported for n-type PbTe-2% Cu$_2$Te ($ZT \sim 1.5$ at 723 K) [52] and PbTe-4% InSb ($ZT \sim 1.8$ at 773 K) [51] due to special nanostructures that not only supress $\kappa_{lat}$ but also boost the $S^2/\rho$ (Figure 9.3). Cu atoms fill Pb vacancy sites and hence improve carrier mobility by five times leading to $S^2/\rho \sim 3{,}700$ µW m$^{-1}$ K$^{-2}$ at 423 K. Moreover, nano/microscale Cu-rich precipitates scatter wide spectrum of phonons, leading to a very low $\kappa_{lat}$ (~0.38 W m$^{-1}$ K$^{-1}$) at 700 K in PbTe-5.5% Cu$_2$Te, close to the theoretical minimum value [52]. Addition of InSb in PbTe results in multiphase nanostructures (Pb, In, InSb, and Sb) of sizes between tens and hundreds of nanometers, which offer many phase boundaries and mass fluctuations for effective phonon scattering, as well as energy filtering effect for improved Seebeck coefficient ($S$) [51]. In p-type PbTe, all-scale phonon scattering through multiphase heterogeneities was achieved through minute Ge addition (≤1%), where Ge–Na clusters and Te precipitates with Ge–Na clusters in PbTe were formed in various shapes and sizes, resulting in $\kappa_{lat} \sim 0.5$ W m$^{-1}$ K$^{-1}$ and $ZT \sim 1.9$ at 805 K [50].

## 9.3.2 Band convergence and minority carrier blocking

Alloying in PbTe can result in band convergence, that is, decreasing the offset between light valence band ($L$) and heavy valence band ($\Sigma$), which in turn facilitates charge redistribution between bands [62] (Figure 9.4). Band convergence enhances the hole effective mass and hence $S$ by increasing the valley degeneracy number ($N_v$) [63], $N_v$ is 4 for $L$ band and increases to 12 for the $\Sigma$ band. For example, $S$ increases considerably with an increase in SrTe content in PbTe, from ~60 µV K$^{-1}$ for no SrTe sample to ~90 µV K$^{-1}$ for >6% SrTe added PbTe at 300 K [48].

Band convergence can also suppress bipolar thermal conductivity [46, 48]. Since PbTe is a narrow band gap (~0.30 eV) semiconductor, elevated temperatures easily promote minority carrier jumps across the band gap. The resulting diffusion of electron–hole pairs gives rise to an additional contribution to thermal conductivity, known as the bipolar thermal conductivity ($\kappa_{bi}$). Therefore, the total thermal conductivity ($\kappa_{total}$) is given as $\kappa_{total} = \kappa_{lat} + \kappa_{ele} + \kappa_{bi}$, where $\kappa_{bi}$ becomes significant at high temperatures only. The energy required for bipolar diffusion ($E_g^*$) should be higher than the band gap and can be estimated from $\kappa_{bi} = A \exp(-(E_g^*/2k_B T))$, where $A$ is a constant, $k_B$ is the Boltzmann constant, and $T$ is the temperature [46].

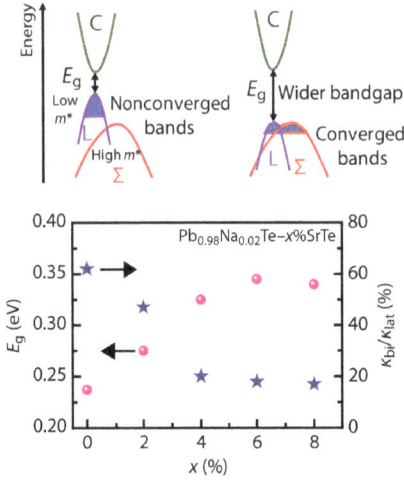

**Figure 9.4:** Schematic showing the relative energies of the valence bands (L and Σ band) and conduction band when no band convergence is present and when the two valence bands converge. Bandgap enlargement of PbTe as a consequence of band convergence due to $x\%$ SrTe alloying, as well as the ratio of bipolar thermal conductivity $\kappa_{bi}$ to lattice thermal conductivity, $\kappa_{lat}$ as a function of SrTe content at 923 K is also shown [48]. It is clear that $\kappa_{bi}$ is greatly suppressed with the enlargement of the band gap.

For suppression of $\kappa_{bi}$, the band gap needs to be enlarged compared to the energy required for bipolar diffusion. An increase in band gap as a consequence of band convergence through alloying is an effective approach to reduce $\kappa_{bi}$, as can be inferred from Figure 9.4, for PbTe with high degree of SrTe alloying [48]. This synergistic approach of alloying leading to nanostructuring and band convergence has resulted in remarkably high $ZT$ values in many alloyed systems, such as PbTe–MgTe [46] ($ZT \sim 2.0$), PbTe–SrTe [48] ($ZT \sim 2.5$), PbTe–PbSe [64] ($ZT \sim 1.8$), PbTe–PbS [65, 66] ($ZT \sim 2.3$), and PbTe–PbSe–SrTe [49] ($ZT \sim 2.3$).

Although band convergence is a powerful approach for reducing $\kappa_{bi}$, it is not always easy to find the right semiconducting bulk material with the desired band structure and the right alloying material that could produce band convergence. Bahk and Shakouri [67] theoretically proposed another approach called minority carrier blocking. Blocking minority carriers by appropriate heterostructure barriers can lead to much needed reduction in $\kappa_{bi}$ at high temperatures. In case of p-type PbTe, the heterostructure barriers must have a large conduction band offset for blocking electrons (minority carriers), but a small valence band offset so that the hole transport (majority carriers) is not affected much by the barriers. Soon after, experimental demonstration of minority carrier blocking in PbTe–Ag$_2$Te nanowire heterostructure-based nanocomposites was made [68]. The band offset between Ag$_2$Te and PbTe provides energy barriers for minority carrier transport, responsible for increase in power factor and $\kappa_{bi}$ reduction, providing $\sim 40\%$ $ZT$ enhancement ($\sim0.66$ at 390 K) compared to Ag$_2$Te. Few other systems have also benefitted from minority carrier blocking [69, 70]; however, this approach is relatively new and remains largely unexplored.

## 9.3.3 Module development

The above-mentioned progress in material development of PbTe-based materials has been timely bridged with the development of high-performance thermoelectric modules, demonstrating record high efficiencies (Table 9.1). First report was made in 2016 of power generation from nanostructured PbTe-based single stage and segmented module [30] (Figure 9.5(a)). The 4% Na-doped PbTe with 2% Mg nanostructuring was used as a p-type leg, having a $ZT$ of ~1.8 at 810 K. The lattice strain fields surrounding the nanostructures provide effective phonon scattering ($\kappa_{lat}$ ~ 0.6 W m$^{-1}$ K$^{-1}$ at 810 K). However, Mg-induced nanostructures form coherent interface with the PbTe matrix, maintaining high $S^2/\rho$ (~2,400 µW m$^{-1}$ K$^{-2}$ at 810 K). PbI$_2$-doped PbTe ($ZT$ ~ 1.4 at 750 K) forms the n-type leg in the module. A high module efficiency ($\eta_{max}$) of ~8.8% for a temperature difference ($\Delta T$) of 570 K was obtained for this single-stage module. The segmented p- and n-type Bi$_2$Te$_3$/nanostructured PbTe legs were made to boost the average $ZT$ (Figure 9.5(b)). A very high $\eta_{max}$ of ~11% was achieved for $\Delta T$ ~ 590 K for the segmented module (Table 9.1). In another study, 0.07% Ge was added to 4% Na-doped PbTe (p-type leg), which resulted in multiphase heterogeneities for all-scale phonon scattering, leading to a $ZT$ of ~1.9 at 805 K [50]. For the Bi$_2$Te$_3$/nano-structured PbTe–Ge cascaded module, the nanostructured PbTe–Ge module was stacked on top of the Bi$_2$Te$_3$ module (Figure 9.5(c)), which achieved an even higher efficiency of $\eta_{max}$ ~ 12% for $\Delta T$ = 590 K [50] (Table 9.1). These works demonstrate the potential of nanostructured materials in thermoelectric application setting.

Table 9.1: Measured maximum power output ($P_{max}$), open-circuit heat flow ($Q_{oc}$), and maximum conversion efficiency ($\eta_{max}$) of the single-stage module of nanostructured PbTe–Mg (p-type) and PbTe (n-type) [30], segmented module of Bi$_2$Te$_3$/nanostructured PbTe–Mg (p-type) and Bi$_2$Te$_3$/PbTe (n-type) [30], and cascaded module of Bi$_2$Te$_3$/nanostructured PbTe–Ge (p-type) and Bi$_2$Te$_3$/PbTe (n-type) [50]. $T_h$ is the hot-side temperature and $T_c$ is the cold-side temperature.

| Module | $T_h$ (K) | $T_c$ (K) | $P_{max}$ (W) | $Q_{oc}$ (W) | $\eta_{max}$ (%) |
|---|---|---|---|---|---|
| Single-stage module of nanostructured Pb$_{0.940}$Mg$_{0.020}$Na$_{0.040}$ (p-type) and PbTe$_{0.9960}$I$_{0.0040}$ (n-type) | 873 | 303 | 3.55 | 30.9 | 8.8 |
| Segmented type module of Bi$_2$Te$_3$/nanostructured Pb$_{0.940}$Mg$_{0.020}$Na$_{0.040}$ (p-type) and Bi$_2$Te$_3$/ PbTe$_{0.9960}$I$_{0.0040}$ (n-type) | 873 | 283 | 2.34 | 15.8 | 11 |
| Cascaded module of Bi$_2$Te$_3$/nanostructured Pb$_{0.953}$Ge$_{0.007}$Na$_{0.040}$Te (p-type) and Bi$_2$Te$_3$/PbTe$_{0.9964}$I$_{0.0036}$ (n-type) | 873 | 283 | 1.77 | 13.2 | 12 |

(a)                        (b)                        (c)

PbTe                              Bi$_2$Te$_3$    PbTe

(d)

Cu$_{26}$Nb$_2$Ge$_6$S$_{32}$

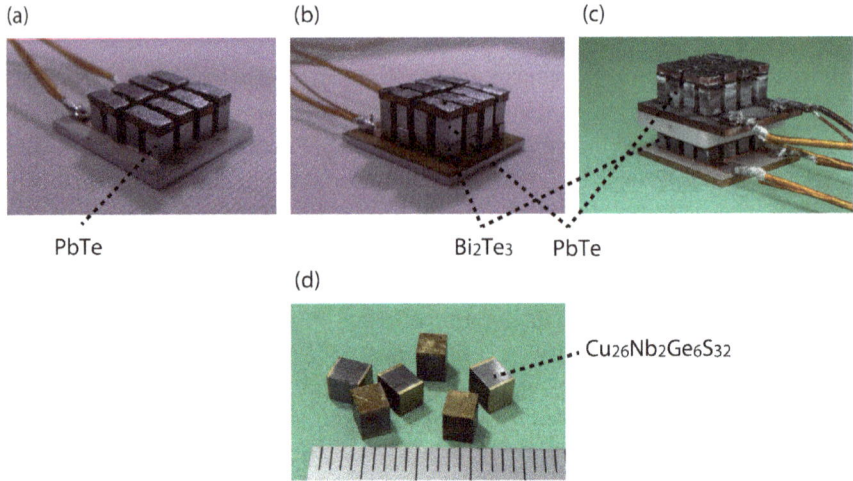

**Figure 9.5:** (a) Single-stage module of nanostructured Pb$_{0.940}$Mg$_{0.020}$Na$_{0.040}$ (p-type) and PbTe$_{0.9960}$I$_{0.0040}$ (n-type) [30], (b) segmented-type module of Bi$_2$Te$_3$/nanostructured Pb$_{0.940}$Mg$_{0.020}$Na$_{0.040}$ (p-type) and Bi$_2$Te$_3$/ PbTe$_{0.9960}$I$_{0.0040}$ (n-type) [30], (c) cascaded module of Bi$_2$Te$_3$/nanostructured Pb$_{0.953}$Ge$_{0.007}$Na$_{0.040}$Te (p-type) and Bi$_2$Te$_3$/PbTe$_{0.9964}$I$_{0.0036}$ (n-type) [50]. (a)–(c) Insulated metal substrate: 18 mm × 15 mm × 1 mm. (d) Cu$_{26}$Nb$_2$Ge$_6$S$_{32}$-based single element with Au diffusion barriers [71]. (a) and (b) Adapted from Ref. [30], (c) adapted from Ref. [50], and (d) adapted from Ref. [71].

Recent work [50] on 24-h stability test performed on an eight-pair module made from nanostructured PbTe shows good repeatability of maximum efficiency and power under various temperature gradients, with no cracks or deformity observed in the nanostructured material. This not only shows that the nanostructures are maintained under high-temperature gradients ($\Delta T$ = 590 K) for 24 h but also confirms good mechanical stability of the module, which exhibits the potential of nanostructured materials in thermoelectric application.

# 9.4 All-scale phonon scattering in thermoelectric sulfides

## 9.4.1 Natural superlattice in layered sulfides

Layered sulfides provide a good opportunity to improve the thermoelectric performance through hierarchical phonon scattering because of their two-dimensional crystal structures (layered structure) [72–75]. Figure 9.6 shows the strategies for $\kappa_{lat}$ reduction through all-scale phonon scattering in the single-layered sulfides (such as TiS$_2$) and misfit-layered sulfides (MS)$_{1+x}$(TS$_2$)$_m$ (M = Sn, Pb, Sb, Bi, and rare-earth

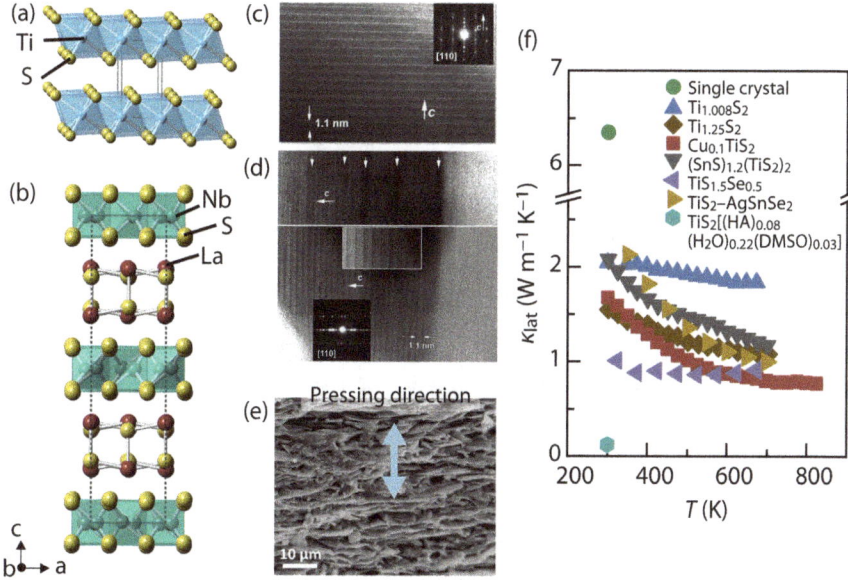

**Figure 9.6:** All-scale phonon scattering for reduction of lattice thermal conductivity ($\kappa_{lat}$) in the single-layered sulfide $TiS_2$ and misfit-layered sulfides $(MS)_{1+x}(TS_2)_m$ (M = Sn, Pb, Sb, Bi, and rare-earth metals; T = Ti, V, Cr, Nb, and Ta; $0.08 < x < 0.28$; and $m = 1, 2, 3$). Crystal structure of (a) $TiS_2$ and (b) $(LaS)_{1.14}NbS_2$ [77]. High-magnification transmission electron microscopy (TEM) images of $(LaS)_{1.14y}NbS_2$ with (c) $y = 1.05$ and (d) $y = 0.95$ [106]. (d) White arrows depict dark contrast bending defect planes resulting from strain. (e) Scanning electron microscopic (SEM) image of $(LaS)_{1.20}CrS_2$ [105]. (f) Temperature dependence of in-plane $\kappa_{lat}$ for the single-crystal sample of near-stoichiometric $TiS_2$ [78] and sintered samples of $Ti_{1.008}S_2$ [79], $Ti_{1.025}S_2$ [90], $Cu_{0.1}TiS_2$ [88], $(SnS)_{1.2}(TiS_2)_2$ [92, 93], $TiS_2$–4% $AgSnSe_2$ [99], and $TiS_2[(HA)_{0.08}(H_2O)_{0.22}(DMSO)_{0.03}]$ (HA: hexylammonium; DMSO: dimethyl sulfoxide) [100]. (c) and (d) Adapted from Ref. [106] and (e) adapted from Ref. [105].

metals; T = Ti, V, Cr, Nb, and Ta; $0.08 < x < 0.28$; and $m = 1, 2, 3$). As shown in Figure 9.6(a), the crystal structure of $TiS_2$ consists of a stacked $CdI_2$-type layer along the crystalline c-axis (space group $P\bar{3}m$). Misfit-layered sulfides are built on an alternate stacking of the distorted NaCl-type MS layers and disulfide $TS_2$ layers, where the crystal structures lack three-dimensional periodicity (Figure 9.6(b)) [76, 77]. In this section, we address the $\kappa_{lat}$ reduction through intercalation of guest atoms/layers into host crystal layers, stacking faults (Figure 9.6(c) and (d)), and microstructural ordering (Figure 9.6(e)). Figure 9.6(f) shows the temperature dependence of in-plane $\kappa_{lat}$ for single- and misfit-layered sulfides. Moreover, the thermoelectric properties of $TiS_2$-based, $CrS_2$-based, and $NbS_2$-based systems are summarized in Table 9.2. While the $TiS_2$-based and $CrS_2$-based systems generally show n-type electrical transport properties, $NbS_2$-based system generally shows p-type properties.

Among the layered sulfides, $TiS_2$-based system is the most studied and shows advanced thermoelectric performance. In $TiS_2$ system, the electrical and thermal

**Table 9.2:** High-temperature Seebeck coefficient ($S$), electrical resistivity ($\rho$), power factor ($S^2/\rho$), lattice thermal conductivity ($\kappa_{lat}$), and thermoelectric figure of merit ($ZT$) of $TiS_2$-based, $CrS_2$-based, and $NbS_2$-based systems.

| Sample | Direction | $T$ (K) | $S$ ($\mu$V K$^{-1}$) | $\rho$ ($\mu\Omega$ m) | $S^2/\rho$ ($\mu$W m$^{-1}$ K$^{-2}$) | $\kappa_{lat}$ (W m$^{-1}$ K$^{-1}$) | $ZT$ | Reference |
|---|---|---|---|---|---|---|---|---|
| Single crystal of near-stoichiometric $TiS_2$ | In-plane | 300 | −251 | 17 | 3,710 | 6.35 | 0.16 | [78] |
| Single crystal of near-stoichiometric $TiS_2$ | Out-of-plane | 300 | – | 13,000 | – | 4.21 | – | [78] |
| Polycrystalline $Ti_{1.008}S_2$ | In-plane | 660 | −148 | 17.4 | 1,270 | 1.8 | 0.34 | [79] |
| Polycrystalline $Ti_{1.008}S_2$ | Out-of-plane | 660 | −152 | 32.0 | 720 | 1.8 | 0.21 | [79] |
| Polycrystalline $Ti_{1.031}S_2$ | In-plane | 660 | −83.6 | 9.20 | 760 | 1.5 | 0.17 | [79] |
| Polycrystalline $Ti_{1.031}S_2$ | Out-of-plane | 660 | −74 | 11.5 | 480 | 1.3 | 0.13 | [79] |
| Polycrystalline $TiS_2$ | In-plane | 700 | −388 | 268 | 560 | 2.36 | 0.16 | [90] |
| Polycrystalline $Ti_{1.025}S_2$ | In-plane | 700 | −192 | 34 | 1,080 | 1.08 | 0.48 | [90] |
| Polycrystalline $TiS_2$ | In-plane | 800 | −276 | 84.2 | 910 | 1.71 | 0.37 | [88] |
| Polycrystalline $Cu_{0.1}TiS_2$ | In-plane | 800 | −142 | 19.1 | 1,060 | 0.77 | 0.47 | [88] |
| Polycrystalline $(BiS)_{1.2}(TiS_2)_2$ | In-plane | 770 | −90 | 10 | 810 | 0.3 | 0.27 | [81, 93] |
| Polycrystalline $(SnS)_{1.2}(TiS_2)_2$ | In-plane | 670 | −140 | 20 | 900 | 0.9 | 0.35 | [92, 93] |
| Polycrystalline $(SnS)_{1.2}(Cu_{0.02}Ti_{0.98}S_2)_2$ | In-plane | 720 | −142 | 30 | 670 | 0.56 | 0.42 | [96] |
| Polycrystalline $Ti_{0.95}Ta_{0.05}S_2$ | In-plane | 700 | −122 | 19.7 | 760 | 1.05 | 0.28 | [97] |
| Polycrystalline $TiS_{1.5}Se_{0.5}$ | In-plane | 700 | −231 | 61.7 | 860 | 1.2 | 0.41 | [98] |

| | | | | | | | | |
|---|---|---|---|---|---|---|---|---|
| Polycrystalline $TiS_2$–4% $AgSnSe_2$ composite | In-plane | 700 | −230 | 34 | 1,550 | 1.0 | 0.8 | [99] |
| Single crystal $TiS_2[(HA)_{0.08}(H_2O)_{0.22}(DMSO)_{0.03}]$ | In-plane | 370 | −90 | 20 | 400 | *0.12 (300 K) | 0.28 | [100] |
| Polycrystalline $(LaS)_{1.20}CrS_2$ | In-plane | 950 | −172 | 171 | 174 | 1.11 | 0.14 | [105] |
| Polycrystalline $(LaS)_{1.20}CrS_2$ | Out-of-plane | 950 | −154 | 278 | 84 | 0.84 | 0.08 | [105] |
| Polycrystalline $(LaS)_{1.14}NbS_2$ | In-plane | 950 | 83 | 16.9 | 405 | 1.86 | 0.12 | [105] |
| Polycrystalline $(LaS)_{1.14}NbS_2$ | Out-of-plane | 950 | 56 | 28.5 | 111 | 0.52 | 0.08 | [105] |
| Polycrystalline $(LaS)_{1.14y}NbS_2$, $y = 0.95$ | In-plane | 950 | 71 | 19 | 260 | 1.2 | 0.11 | [106] |
| Polycrystalline $(LaS)_{1.14y}NbS_2$, $y = 1.05$ | In-plane | 950 | 85 | 16 | 460 | 0.9 | 0.20 | [106] |

transport properties are highly anisotropic because of scattering of both charge carriers and heat-carrying phonons at interface between layers [78–81]. Moreover, the carrier concentration can be tuned through an intercalation of excess Ti or guest atoms, such as Mn, Fe, Co, Ni, and Cu, into host $TiS_6$ octahedral layers [79, 82–91], resulting in the optimized $S^2/\rho$. The $S^2/\rho$ of $TiS_2$ peaks near carrier concentration of $2.8 \times 10^{20}$ $cm^{-3}$ [72]. In the single-crystal sample of near-stoichiometric $TiS_2$ with carrier concentration of $2.8 \times 10^{20}$ $cm^{-3}$, the room-temperature in-plane (direction along $ab$ plane) $\mu$ (~15 $cm^2$ $V^{-1}$ $s^{-1}$) is significantly higher than the out-of-plane ($c$-axis direction) one ($\mu$ ~ 0.017 $cm^2$ $V^{-1}$ $s^{-1}$), leading to $S^2/\rho$ of ~3,710 $\mu W$ $m^{-1}$ $K^{-2}$ for in-plane direction at room temperature (Table 9.2) [78]. This in-plane $S^2/\rho$ is comparable to that of commercially available thermoelectric material, $Bi_2Te_3$. However, the single-crystal sample shows the high $\kappa_{lat}$ for both in-plane and out-plane directions, limiting the $ZT$. The in-plane and out-plane $\kappa_{lat}$ are ~6.35 and ~4.21 W $m^{-1}$ $K^{-1}$, respectively (Table 9.2). An in-plane $ZT$ of ~0.16 at 300 K was reported in near-stoichiometric single-crystal $TiS_2$.

In polycrystals, a pressure-assisted sintering promotes the formation of highly oriented samples for both single- and misfit-layered sulfides. The crystalline in-plane is preferably oriented perpendicular to the pressing direction [79–81]. The higher $\mu$ and $\rho$ were obtained for in-plane direction in sintered samples with highly oriented microstructure, allowing the higher in-plane $S^2/\rho$. For example, an in-plane $S^2/\rho$ of ~1,270 $\mu W$ $m^{-1}$ $K^{-2}$ at 660 K was demonstrated in the sintered sample of $Ti_{1.008}S_2$ (Table 9.2). The carrier concentration of this $Ti_{1.008}S_2$ at 300 K was ~4.5 × $10^{21}$ $cm^{-3}$. In Figure 9.6(e), the in-plane $\kappa_{lat}$ values are compared between the single-crystal sample of near-stoichiometric $TiS_2$ [78] and sintered samples of $Ti_{1.008}S_2$ [79]. The reduced $\kappa_{lat}$ was obtained for sintered samples because of effective phonon scattering at the grain boundary. At 300 K, the in-plane $\kappa_{lat}$ of $Ti_{1.008}S_2$ sintered sample (~2.0 W $m^{-1}$ $K^{-1}$) is lower than that of near-stoichiometric single-crystal sample (~6.35 W $m^{-1}$ $K^{-1}$). A $ZT$ of ~0.34 was obtained at 660 K for the sintered sample of $Ti_{1.008}S_2$.

An intercalation of excess Ti, guest atoms, or guest layers further reduces the $\kappa_{lat}$. Figure 9.6(b) shows the temperature dependence of in-plane $\kappa_{lat}$ for the sintered samples of Ti-rich $Ti_{1.025}S_2$ [90], Cu-intercalated $Cu_{0.1}TiS_2$ [88], BiS-intercalated misfit $(BiS)_{1.2}(TiS_2)_2$ [81, 92, 93], and SnS-intercalated misfit $(SnS)_{1.2}(TiS2)_2$ [92, 93]. At 660 K, $\kappa_{lat}$ for the sintered samples of misfit $(BiS)_{1.2}(TiS_2)_2$ (~0.3 W $m^{-1}$ $K^{-1}$) is lower than that for the sintered samples of $TiS_2$ (~1.8 W $m^{-1}$ $K^{-1}$). Doping effects of alkaline-earth (Mg, Ca, and Sr) and transition metal (V, Cr, Mn, Fe, Co, Ni, Cu, and Zn) elements on the thermoelectric properties of misfit $TiS_2$-based systems were investigated [94–96]. These elements reduce the carrier concentration of the misfit layer sulfides, changing the $S^2/\rho$. A $ZT$ of ~0.42 at 720 K was obtained in Cu-doped system $(SnS)_{1.2}$ $(Cu_{0.02}Ti_{0.98}S_2)_2$ (Table 9.2).

In-plane $\kappa_{lat}$ for the sintered samples of Ta substituted $Ti_{0.95}Ta_{0.05}S_2$ [97], Se substituted $TiS_{1.5}Se_{0.5}$ [98], and composite $TiS_2$–4% $AgSnSe_2$ [99] are listed in Table 9.2. Figure 9.6(f) shows the temperature dependence of in-plane $\kappa_{lat}$ for the sintered samples of $TiS_2$-4% $AgSnSe_2$ [99]. The substituted systems $Ti_{0.95}Ta_{0.05}S_2$ and $TiS_{1.5}Se_{0.5}$ also show reduced $\kappa_{lat}$; for example, the substitution of Ti by Ta slightly reduces $\kappa_{lat}$ to ~1.05 W m$^{-1}$ K$^{-1}$ for $Ti_{0.95}Ta_{0.05}S_2$ at 700 K. Moreover, $TiS_2$-$AgSnSe_2$ composites show the enhanced $S^2/\rho$ and reduced $\kappa_{lat}$, originating from carrier concentration tuning and enhanced phonon scattering, respectively [99]. In $TiS_2$-4% $AgSnSe_2$ composite, a $S^2/\rho$ of ~1,550 μW m$^{-1}$ K$^{-2}$ and reduced $\kappa_{lat}$ of ~ 1.0 W m$^{-1}$ K$^{-1}$ allows a $ZT$ of ~0.8 at 700 K (Table 9.2).

Organic layers, such as $(HA)_x(H_2O)_y(DMSO)_z$ (HA: hexylammonium; DMSO: dimethyl sulfoxide), $(TBA)_x(HA)_y$ (TBA: tetrabutylammonium), and fullerene $C_{60}$, are intercalated into the van der Waals gap between $TiS_2$ layers [100–103]. A significant reduction in the $\kappa_{lat}$ is caused by the organic intercalation, resulting in high $ZT$. For $TiS_2[(HA)_{0.08}(H_2O)_{0.22}(DMSO)_{0.03}]$ the in-plane $\kappa_{lat}$ at 300 K and $ZT$ at 373 K are ~0.12 W m$^{-1}$ K$^{-1}$ and ~0.28, respectively (Table 9.2) [100]. Another important benefit of hybrid inorganic and organic superlattices is that they possess a flexibility. The flexural modulus of $TiS_2[(HA)_{0.08}(H_2O)_{0.22}(DMSO)_{0.03}$ is ~145 MPa at 300 K, which is two orders of magnitude lower than that of a $TiS_2$ single crystal (~14.2 GPa) and lower than a high flexural polymer polytetrafluoroethylene (Teflon) (~455 MPa) [100]. The flexible $TiS_2$-based inorganic and organic systems are a great potential for thermoelectric energy harvesting near room temperature.

The thermoelectric properties of $CrS_2$-based and $NbS_2$-based layered sulfides can be enhanced through the same strategy as that of $TiS_2$-based system. The misfit-layered structures of these sulfides yield low $\kappa_{lat}$ [104, 105]. Moreover, as shown in Figure 9.6(e), a pressure-assisted sintering promotes formation of the highly oriented samples, allowing the high $S^2/\rho$ in the in-plane direction [105]. For $(LaS)_{1.20}CrS_2$, the $\kappa_{lat}$ is less than 1.2 W m$^{-1}$ K$^{-1}$ for both in-plane and out-of-plane directions at 950 K (Table 9.2) [105]. In particular, the out-of-plane $\kappa_{lat}$ is very low (~0.84 W m$^{-1}$ K$^{-1}$) because of the effective phonon scattering at interface between layers. The in-plane $S^2/\rho$ (~174 μW m$^{-1}$ K$^{-2}$) is about twice as large as out-of-plane $S^2/\rho$ (~84 μW m$^{-1}$ K$^{-2}$). A $ZT$ of 0.14 at 950 K was obtained in $(LaS)_{1.20}CrS_2$ for in-plane direction.

$S$ and $\rho$ can be tuned by varying $(MS)/(TS_2)$ compositional ratio [106, 107]. In $(YbS)_{2y}NbS_2$ the room-temperature $S$ and $\rho$ decrease from ~60 μV K$^{-1}$ and ~19 μΩ m for $y = 1.00$ to ~52 μV K$^{-1}$ and ~12 μΩ m for $y = 0.90$, respectively [107]. Moreover, in $(LaS)_{1.14y}NbS_2$, highly oriented texture was formed for (LaS)-rich composition $y = 1.05$ sample [106]. In $(LaS)_{1.14y}NbS_2$, the $ZT$ value increases from 0.12 for $y = 1.00$ sample to 0.20 for $y = 1.05$ sample (Table 9.2) [106].

The mean free path of phonon $(l_p)$ in $(LaS)_{1.14y}NbS_2$ ($y = 0.95$, 1.00, and 1.05) was estimated from $\kappa_{lat}$, heat capacity at constant volume $(C_v)$, and average sound velocity $(v)$ [106] (see eq. (9.1)). Here $C_v$ was approximated as equal to heat capacity at constant pressure $(C_p)$. The both in-plane and out-of-plane $l_p$ values are less than 1.5 nm, reaching the interlayer distance (<1 nm) between LaS and $NbS_2$ layers. This means that the interfaces between the layers effectively scatter the heat-carrying phonons.

High-magnification transmission electron microscopy (TEM) images reveal that stacking faults are induced through varying (MS)/(TS$_2$) compositional ratio. While stacking faults were formed in LaS layers for $y = 0.95$ sample (Figure 9.6(d)), less stacking faults were formed in LaS layers $y = 1.05$ sample (Figure 9.6(c)) [106]. The $ZT$ value will be further boosted through stacking faults tuning, because nanoscale strain-induced stacking faults effectively scatter the heat-carrying phonons.

## 9.4.2 Atomic-scale defects and disorders in colusite

Colusites Cu$_{26}$A$_2$M$_6$S$_{32}$ (A = V, Nb, Ta, Cr, Mo, W; M = Ge, Sn) are among the family of p-type thermoelectric compounds composed of environmentally friendly and cost-effective elements, such as copper and sulfur [108–110]. In 2014, Suekuni et al. found high $ZT$ in Cu$_{26}$V$_2$M$_6$S$_{32}$ (M = Ge, Sn) [111]. In 2016, high $ZT$ was also achieved for vanadium-free Cu$_{26}$A$_2$Sn$_6$S$_{32}$ (A = Nb, Ta) [112]. High $ZT$ in the colusite system arises from both a high $S^2/\rho$ and low $\kappa_{lat}$. Good electrical transport properties originate from the valence band composed mainly of Cu-$3d$ and S-$3p$ hybridized orbitals [111]. In Cu$_{26}$Cr$_2$Ge$_6$S$_{32}$, the smaller size and lower electronegativity of Cr ion among A ions induce less distortion into the Cu–S network, resulting in high $S^2/\rho$ [113]. The $S^2/\rho$ of ~1,940 $\mu$W m$^{-1}$ K$^{-2}$ at 700 K was achieved for Cu$_{26}$Cr$_2$Ge$_6$S$_{32}$ (Table 9.3) [113]. Moreover, a carrier concentration tuning through the changes in chemical composition enhances the $S^2/\rho$, resulting in high $ZT$ [114]. In Sn-poor colusite Cu$_{26}$Ta$_2$Sn$_{5.5}$S$_{32}$, a $ZT$ of ~1.0 at 670 K was obtained, arising from a $S^2/\rho$ of ~800 $\mu$W m$^{-1}$ K$^{-2}$ (Table 9.3) [115]. For the development of a power generator using colusite, Au-based diffusion barrier has been successfully developed [71]. The maximum conversion efficiency of ~3.3% for the hot-side temperature of 570 K and the cold-side temperature of 297 K was demonstrated in the Cu$_{26}$Nb$_2$Ge$_6$S$_{32}$-based single element with Au diffusion barriers. In this section, we address the origin of low $\kappa_{lat}$ and reduction on $\kappa_{lat}$ in the colusite system.

Figure 9.7(a) shows a crystal structure of the colusites. The unit cell belongs to a cubic system with a $P\bar{4}3n$ space group and is composed of 66 atoms. The low $\kappa_{lat}$ originates primarily from the complex crystal structure with a large number of atoms. As shown in Figure 9.7(e), the $\kappa_{lat}$ of colusites is compared with that of sulvanite Cu$_3$VS$_4$ (Cu$_{24}$V$_8$S$_{32}$) [111]. These sulfides have similar chemical compositions and similar cubic crystal structures (sulvanite: space group $P\bar{4}3m$; inset of Figure 9.7(e)), while difference lies in the number of atoms per unit cell. Because the number of atoms per unit cell of colusites is larger than that of sulvanite (eight atoms), the $\kappa_{lat}$ of colusites is substantially lower than that of sulvanite over the temperature range of 300–670 K. For example, at 650 K the $\kappa_{lat}$ values for Cu$_{26-x}$Nb$_2$Sn$_{6+x}$S$_{32}$ with $x = -0.3$ and 1.2 ranges from ~0.47 to ~0.76 W m$^{-1}$ K$^{-1}$ [116] while that for Cu$_3$VS$_4$ is ~1.6 W m$^{-1}$ K$^{-1}$ [108].

**Table 9.3:** High-temperature Seebeck coefficient ($S$), electrical resistivity ($\rho$), power factor ($S^2/\rho$), lattice thermal conductivity ($\kappa_{lat}$), and thermoelectric figure of merit ($ZT$) in colusites.

| Sample | $T$ (K) | $S$ ($\mu$V K$^{-1}$) | $\rho$ ($\mu\Omega$ m) | $S^2/\rho$ ($\mu$W m$^{-1}$ K$^{-2}$) | $\kappa_{lat}$ (W m$^{-1}$ K$^{-1}$) | $ZT$ | Reference |
|---|---|---|---|---|---|---|---|
| $Cu_{26}Cr_2Ge_6S_{32}$ | 700 | 150 | 12 | 1,940 | 0.50 | 0.86 | [113] |
| $Cu_{26}Ta_2Sn_{5.5}S_{32}$ | 670 | 210 | 54 | 790 | 0.35 | 1.0 | [115] |
| $Cu_{26-x}Nb_2Sn_{6+x}S_{32}$ with $x = -0.3$ sintered at 873 K (ordered phase) | 670 | 110 | 21 | 630 | 0.76 | 0.31 | [116] |
| $Cu_{26-x}Nb_2Sn_{6+x}S_{32}$ with $x = 1.2$ sintered at 973 K (disordered phase) | 670 | 340 | 921 | 120 | 0.51 | 0.16 | [116] |
| $Cu_{26-x}Nb_2Sn_{6+x}S_{32}$ with $x = -0.3$ sintered at 973 K (ordered phase with excess Cu in the interstitial sites) | 670 | 210 | 75 | 600 | 0.47 | 0.66 | [116] |

Atomic-scale defects and disorders in colusites are introduced by the loss of sulfur and/or change in atomic ratio of Cu/M [116, 117], allowing effective phonon scattering and a resultant low $\kappa_{lat}$. A small amount of sulfur is lost during sintering at higher temperature, inducing chemical disorder on the cations sublattices (disordered phase) [117]. In the disordered phase, Cu and M ions randomly occupy the ideal positions of the ordered phase. The exsolution texture of ordered and disordered phases were found in natural mineral colusites [118]. The disordered phase was also found by Bourgès et al. in synthetic $Cu_{26}V_2Sn_6S_{32}$ sample sintered at higher temperatures of 1,023 K [116, 117]: this higher-temperature-sintered sample consists of ordered and disordered regions. Figure 9.7(d) shows a high-angle annular dark-field scanning TEM image of ordered and disordered regions found in higher-temperature (1,023 K)-sintered sample of $Cu_{26}V_2Sn_6S_{32}$ [117]. The ordered and disordered regions are coherently intergrown. On the other hand, the lower-temperature (873 K)-sintered sample of $Cu_{26}V_2Sn_6S_{32}$ consists of only ordered region. The volume fraction of the ordered and disordered region varies with the chemical composition change [116]. The $Cu_{26-x}Nb_2Sn_{6+x}S_{32}$ with $x = 1.2$ sample sintered at higher temperature of 973 K consists of only disordered phase.

Moreover, the ordered phase with excess Cu in the interstitial sites called the modified ordered phase (Figure 9.7(c)) was found in Cu-rich colusite $Cu_{26-x}Nb_2Sn_{6+x}S_{32}$ with $x = -0.3$ sample sintered at 973 K [116]. It is also due to the loss of a small amount of sulfur. The excess Cu ions occupy the interstitial sites of the tetrahedral framework.

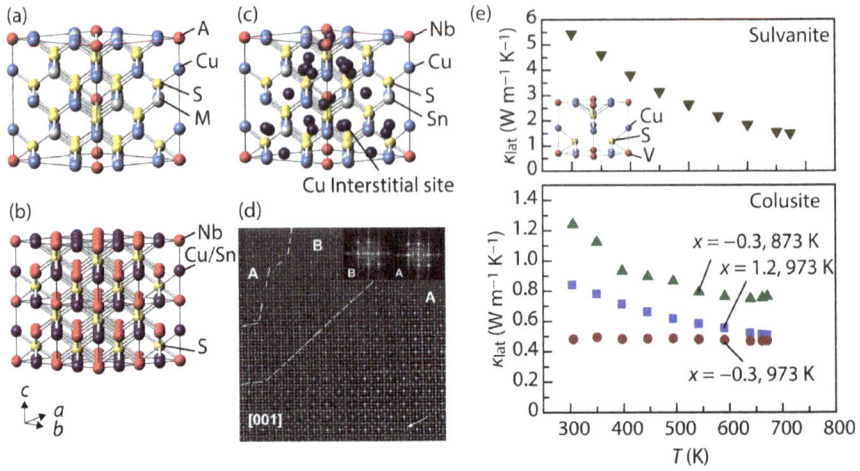

**Figure 9.7:** (a) Crystal structures of colusites $Cu_{26}A_2M_6S_{32}$ (A = V, Nb, Ta, Cr, Mo, W; M = Ge, Sn) (ordered phase), (b) disordered phase found in $Cu_{26-x}Nb_2Sn_{6+x}S_3$ with x = 1.2 sample sintered at 973 K [116], and (c) ordered phase with excess Cu in the interstitial sites found in colusites $Cu_{26-x}Nb_2Sn_{6+x}S_3$ with x = − 0.3 samples sintered at 973 K [116]. (d) High-angle annular dark-field scanning transmission electron microscopy (HAADF-STEM) image of ordered (A) and disordered (B) regions found in higher-temperature (1,023 K)-sintered sample of $Cu_{26}V_2Sn_6S_{32}$ [117]. (e) Temperature (T) dependence of the lattice thermal conductivity ($\kappa_{lat}$) for colusites $Cu_{26-x}Nb_2Sn_{6+x}S_{32}$ [116] and sulvanite $Cu_3VS_4$ for comparison [111]. The $Cu_{26-x}Nb_2Sn_{6+x}S_{32}$ with x = −0.3 sample sintered at 873 K consists of ordered phase. The $Cu_{26-x}Nb_2Sn_{6+x}S_{32}$ with x = −0.3 sample sintered at higher temperature of 973 K consists of ordered phase with excess Cu in the interstitial sites. The $Cu_{26-x}Nb_2Sn_{6+x}S_{32}$ with x = 1.2 sample sintered at 973 K consists of disordered phase. Crystal structure of the sulvanite is given as an inset. (d) Adapted from Ref. [117].

Figure 9.7(e) shows the $\kappa_{lat}$ of $Cu_{26-x}Nb_2Sn_{6+x}S_{32}$ samples sintered at different temperatures of 873 and 973 K and different chemical compositions of x = − 0.3 and x = 1.2. The $\kappa_{lat}$ of higher-temperature-sintered samples is less than that of lower-temperature-sintered samples. The disordered phase in higher-temperature-sintered sample with x = 1.2 and interstitial Cu in higher-temperature-sintered sample with x = − 0.3 allow an effective phonon scattering, reducing the $\kappa_{lat}$. In particular, the extremely low $\kappa_{lat}$ is found in ordered phase with interstitial Cu (x = − 0.3 sample sintered at 973 K). At 670 K, the $\kappa_{lat}$ values of the x = − 0.3 sample sintered at 873 K, x = 1.2 sample sintered at 973 K, and x = − 0.3 sample sintered at 973 K are ~0.76, ~0.51, and ~0.47 W m$^{-1}$ K$^{-1}$, respectively.

As shown in Table 9.3, the atomic-scale defects and disorders in the colusites also affect their electrical transport properties [116, 117]. For $Cu_{26-x}Nb_2Sn_{6+x}S_{32}$, S and ρ in the samples sintered at 973 K are higher than those of the sample sintered at 873 K [116]. This is due to carrier concentration reduction through sulfur loss at higher temperature sintering. Almost the same values of $S^2/\rho$ are obtained for the ordered phase with (~600 μW m$^{-1}$ K$^{-2}$) and without interstitial Cu (~630 μW m$^{-1}$ K$^{-2}$) at

670 K. However, these $S^2/\rho$ values are lower than those of high $S^2/\rho$ system such as $Cu_{26}Cr_2Ge_6S_{32}$ (~1,940 µW m$^{-1}$ K$^{-2}$ at 670 K) [113] and $Cu_{26}Ta_2Sn_{5.5}S_{32}$ (~790 µW m$^{-1}$ K$^{-2}$ at ~670 K) [115]. A $ZT$ value in the colusites will be enhanced through a synergistic combination of $S^2/\rho$ enhancement and $\kappa_{lat}$ reduction.

## 9.4.3 Rattling in tetrahedrites and structural evolution in Chevrel phases

Like colusites, tetrahedrites $Cu_{12}Sb_4S_{13}$ are mainly composed of earth abundant and less toxic elements such as copper and sulfur, owing to which they have gained considerable interest as cost-effective and environmentally friendly thermoelectric materials [38, 39, 108, 119, 120]. These high-temperature thermoelectric properties of tetrahedrites are summarized in Table 9.4. These systems show glass-like low $\kappa_{lat}$ and p-type crystal-like electrical transport properties, resulting in high $ZT$. The $S^2/\rho$ can be increased by tuning the carrier concentration by substituting Mn, Fe, Co, Ni, and Zn for Cu [38, 39, 120, 121, 122]. The $S^2/\rho$ is also enhanced through the substitution of Se for S [123]. The $S^2/\rho$ of ~1,400 µW m$^{-1}$ K$^{-2}$ at 720 K was obtained for Zn-substituted system, $Cu_{11.5}Zn_{0.5}Sb_4S_{13}$ [39].

**Table 9.4:** High-temperature Seebeck coefficient ($S$), electrical resistivity ($\rho$), power factor ($S^2/\rho$), lattice thermal conductivity ($\kappa_{lat}$), and thermoelectric figure of merit ($ZT$) in tetrahedrites.

| Sample | $T$ (K) | $S$ (µV K$^{-1}$) | $\rho$ (µΩ m) | $S^2/\rho$ (µW m$^{-1}$ K$^{-2}$) | $\kappa_{lat}$ (W m$^{-1}$ K$^{-1}$) | $ZT$ | Reference |
|---|---|---|---|---|---|---|---|
| $Cu_{12}Sb_4S_{13}$ | 670 | 120 | 13 | 1,170 | 0.3 | 0.5 | [38] |
| $Cu_{10.5}Ni_{1.5}Sb_4S_{13}$ | 670 | 190 | 40 | 860 | 0.4 | 0.7 | [38] |
| $Cu_{11.5}Zn_{0.5}Sb_4S_{13}$ | 720 | 160 | 20 | 1,400 | 0.1 | 1.0 | [39] |
| $Cu_{10.5}Ni_{1.0}Zn_{0.5}Sb_4S_{13}$ | 723 | 220 | 56 | 850 | 0.4 | 1.0 | [121] |
| $Cu_{11.7}Ge_{0.3}Sb_4S_{13}$ | 665 | 150 | 24 | 990 | 0.6 | 0.65 | [122] |
| $Cu_{11.7}Sn_{0.3}Sb_4S_{13}$ | 665 | 160 | 25 | 960 | 0.6 | 0.65 | [122] |
| $Cu_{13.5}Sb_4S_{13}$ | 723 | 183 | 50 | 720 | 0.25 | 1.0 | [124] |
| $Cu_{13.5}Sb_4S_{12}Se$ | 723 | 170 | 25 | 1,180 | 0.25 | 1.1 | [125] |

As shown in Figure 9.8(a), the crystal structure of tetrahedrites consists of $CuS_4$, $CuS_3$, and $SbS_3$ units and possesses a cubic structure of $I\bar{4}3m$ symmetry. A structural analysis on $Cu_{12}Sb_4S_{13}$ through synchrotron powder X-ray diffraction [38], phonon density-of-states calculation [39], and neutron scattering analyses [40] revealed anisotropic and highly anharmonic vibrations (rattling) of Cu ion perpendicular to the $S_3$

(a)                                        (b)

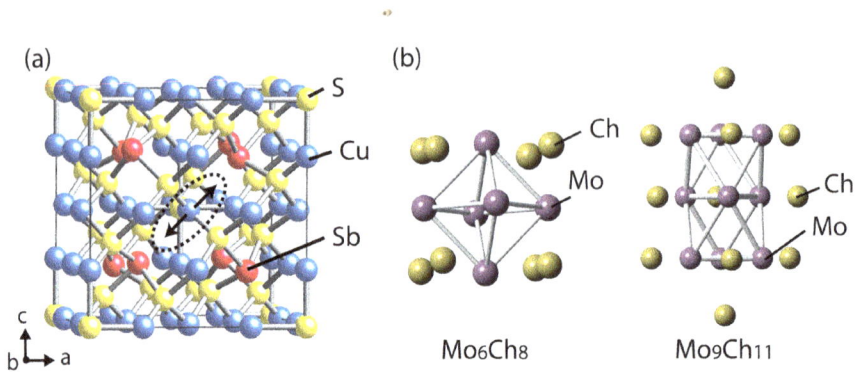

$Mo_6Ch_8$                        $Mo_9Ch_{11}$

**Figure 9.8:** (a) Crystal structures of the tetrahedrites $Cu_{12}Sb_4S_{13}$ and rattling of Cu ion perpendicular to the $S_3$ triangle [38–40]. (b) Host cluster of a $Mo_6Ch_8$ in Chevrel-phase chalcogenides $M_xMo_6Ch_8$ (M = metal; Ch = S, Se, and Te) and structural evolution to $Mo_9Ch_{11}$ [134, 135].

triangle in $CuS_3$ units, which is induced by in-plane chemical pressure of $S_3$ triangle [40]. The rattling of Cu essentially yields low $\kappa_{lat}$. The value of $\kappa_{lat}$ for all samples is low (below 0.6 W m$^{-1}$ K$^{-1}$) over the temperature range of 665–723 K (Table 9.4). The value of $\kappa_{lat}$ of Cu-rich $Cu_{13.5}Sb_4S_{13}$ reduces to ~0.25 W m$^{-1}$ K$^{-1}$ at 723 K through Cu migration [124, 125]. A *ZT* of ~1.1 at 723 K was obtained in Se-substituted and Cu-rich $Cu_{13.5}Sb_4S_{12}Se$ [124]. However, during operation under temperature gradient Cu migration from hot side to cold side leads to degeneration of thermoelectric elements, which reduce the conversion efficiency of the module. Therefore, a different way from liquid-like approach is needed for practical use of tetrahedrite-based thermoelectric module.

The Chevrel-phase chalcogenides $M_xMo_6Ch_8$ (M = metal; Ch = S, Se, and Te) are another family of Cu-based thermoelectric materials and have seen great advances in *ZT* for p-type samples [126–131]. The host structure of a Chevrel phase consists of stacked $Mo_6Ch_8$ clusters (Figure 9.8(b)). Metal M ions fill the space between the $Mo_6Ch_8$ clusters. Chevrel phases generally crystallize in a hexagonal structure (space group $R\bar{3}$), but the metal ion filling results in a distortion of the hexagonal structure to a triclinic symmetry (space group $P\bar{1}$) [132].

Table 9.5 lists the high-temperature thermoelectric properties of Chevrel-phase chalcogenides. The carrier concentration was tuned through the chemical composition change of guest M ions, enhancing the $S^2/\rho$. The $S^2/\rho$ at 950 K was enhanced from ~440 μW m$^{-1}$ K$^{-2}$ for $Cu_{2.5}Mo_6S_8$ to ~810 μW m$^{-1}$ K$^{-2}$ for $Cu_{4.0}Mo_6S_8$ [133]. Moreover, the cage-shape crystal structures act like a rattling structure, leading to low $\kappa_{lat}$. For $Cu_{4.0}Mo_6S_8$, the $\kappa_{lat}$ and *ZT* obtained at 950 K is ~0.9 W m$^{-1}$ K$^{-1}$ and ~0.4, respectively [133].

As shown in Figure 9.8(c), the cage-shape cluster can be modified from $Mo_6Ch_8$ to $Mo_9Ch_{11}$. The compounds of $Ag_{3.6}Mo_9Se_{11}$ with $Mo_9Se_{11}$ cluster [134] and $Ag_3In_2Mo_{15}Se_{19}$ with $Mo_6Se_8$ and $Mo_9Se_{11}$ clusters [135] show crystal-like good electrical transport properties and glass-like thermal transport properties. For $Ag_{3.8}Mo_9Se_{11}$, the $\kappa_{lat}$

**Table 9.5:** High-temperature Seebeck coefficient (S), electrical resistivity ($\rho$), power factor ($S^2/\rho$), lattice thermal conductivity ($\kappa_{lat}$), and thermoelectric figure of merit (ZT) in Chevrel-phase sulfides ($Cu_{2.5}Mo_6S_8$, $Cu_{4.0}Mo_6S_8$, $Cr_{1.3}Mo_6S_8$), $Ag_{3.6}Mo_9Se_{11}$, and $Ag_3In_2Mo_{15}Se_{19}$.

| Sample | T (K) | S ($\mu V\ K^{-1}$) | $\rho$ ($\mu\Omega\ m$) | $S^2/\rho$ ($\mu W\ m^{-1}\ K^{-2}$) | $\kappa_{lat}$ ($W\ m^{-1}\ K^{-1}$) | ZT | Reference |
|---|---|---|---|---|---|---|---|
| $Cu_{2.5}Mo_6S_8$ (space group $R\bar{3}$) | 950 | 76 | 13 | 440 | 1.0 | 0.2 | [133] |
| $Cu_{4.0}Mo_6S_8$ (space group $R\bar{3}$) | 950 | 130 | 22 | 810 | 0.9 | 0.4 | [133] |
| $Cr_{1.3}Mo_6S_8$ (space group $P\bar{1}$) | 973 | 67 | 8 | 570 | 0.2 | 0.17 | [131] |
| $Ag_{3.8}Mo_9Se_{11}$ | 800 | 180 | 50 | 650 | 0.4 | 0.65 | [134] |
| $Ag_3In_2Mo_{15}Se_{19}$ | 1,100 | 110 | 20 | 600 | *0.5 (300 K) | 0.45 | [135] |

and $S^2/\rho$ at 750 K were ~0.4 W m$^{-1}$ K$^{-1}$ and ~650 $\mu$W m$^{-1}$ K$^{-2}$, respectively, leading to ZT of ~0.65 at 800 K [134]. The structural evolution is one way to enhance the thermoelectric performance in Chevrel phase as well as homologous compounds ($PbCh$-$Bi_2Ch_3$) [136–138].

# 9.5 Conclusion

We have clearly entered a new era for thermoelectric research, where focus is not only on advanced material development but also on demonstrating their applicability in modules for power generation and refrigeration. This movement from materials research to module development is because we have materials with sufficiently high ZT owing to the combination of various performance enhancing strategies, good examples of which are PbTe and sulfides. All-scale phonon scattering has provided a big boost to the ZT of various materials in the past decade by reducing the lattice thermal conductivity close to the minimum theoretical limit, as discussed in this chapter. However, in the future, more efforts need to be made for enhancing power factor alongside lattice thermal conductivity reduction to achieve game changing ZT values. Furthermore, there is an urgent need to study the mechanical properties and thermal stability of these advanced materials containing many phonon scattering centers to access their suitability in thermoelectric application setting. Focus should also be made on exploring new materials with environment-friendly elements. The emergence of unconventional materials such as colusites, which are environmentally friendly, and the study of their power generating characteristics is a step in that direction.

**Acknowledgments:** The authors thank Prof. Mercouri G. Kanatzidis of Northwestern University and Argonne National Laboratory, Prof. Shinji Hirai of Muroran Institute of Technology, Dr. Oleg I. Lebedev, Dr. David Berthebaud, and Dr. Emmanuel Guilmeau of Laboratoire CRISMAT, Prof. Toshiro Takabatake of Hiroshima University, Prof. Koichiro Suekuni of Kyushu University, Dr. Yuta Kikuchi, Dr. Yohan Bouyrie, Mr. Atsushi Yamamoto, and Prof. Haruhiko Obara of AIST for plentiful stimulating discussions.

# References

[1]    Yang J H. Theory of thermal conductivity, in: thermal conductivity: theory, properties, and applications, Tritt T M, ed., New York, Kluwer Academic/Plenum Publishers, 2004, 1–17. Doi: 10.1007/b136496.

[2]    Toberer E S, May A F, Snyder G J. Zintl chemistry for designing high efficiency thermoelectric materials, Chem. Mater. 2010, 22, 624–634. Doi: 10.1021/cm901956r.

[3]    Nolas G S, Goldsmid H. Thermal Conductivity of Semiconductors, In: Thermal Conductivity: Theory, Properties, and Applications, Tritt T M, ed., New York, Kluwer Academic/Plenum Publishers, 2004, 105–121. Doi: 10.1007/b136496.

[4]    Callaway J. Model for lattice thermal conductivity at low temperatures, Phys. Rev. 1959, 113, 1046–1051. Doi: 10.1103/PhysRev.113.1046.

[5]    Uher C. Thermal conductivity of metals, In: Thermal Conductivity: Theory, Properties, and Applications, Tritt T M, ed., New York, Kluwer Academic/Plenum Publishers, 2004, 21–91. Doi: 10.1007/b136496.

[6]    Kim W. Strategies for engineering phonon transport in thermoelectrics, J. Mater. Chem. C 2015, 3, 10336–10348. Doi: 10.1039/C5TC01670C.

[7]    Yang J, Xi L L, Qiu W J, Wu L H, Shi X, Chen L D, Yang J, Zhang W Q, Uher C, Singh D J. On the tuning of electrical and thermal transport in thermoelectrics: an integrated theory–experiment perspective, NPJ Comput. Mater 2016, 2, 15015: 1–17. Doi: 10.1038/npjcompumats.2015.15.

[8]    Schrade M, Berland K, Eliassen S N H, Guzik M N, Echevarria-Bonet C, Sørby M H, Jenuš P, Hauback B C, Tofan R, Gunnæs A E, Persson C, Løvvik O M, Finstad T G. The role of grain boundary scattering in reducing the thermal conductivity of polycrystalline XNiSn (X = Hf, Zr, Ti) half-Heusler alloys, Sci. Rep. 2017, 7. 13760: 1–10. Doi: 10.1038/s41598-017-14013-8.

[9]    Pohl R O. Thermal conductivity and phonon resonance scattering, Phys. Rev. Lett. 1962, 8, 481–483. Doi: 10.1103/PhysRevLett.8.481.

[10]   Goldsmid H J. Minimising the Thermal Conductivity, In: Introduction to Thermoelectricity, Hull R, Osgood R M Jr, Parisi J, and Warlimont H, eds, Springer, 63–78. Doi: 10.1007/978-3-642-00716-3.

[11]   Dames C, Chen G. Thermal Conductivity of Nanostructured Thermoelectric Materials, In: Thermoelectrics Handbook: Macro to Nano, Rowe D W, ed., Boca Raton, FL, CRC Press, 2006, 42-1–16. Doi: 10.1201/9781420038903.

[12]   Tritt T M. ed. Thermal Conductivity: Theory, Properties, and Applications, New York, Kluwer Academic/Plenum Publishers, 2004. Doi: 10.1007/b136496.

[13]   Goldsmid H J. Introduction to thermoelectricity, In: Introduction to Thermoelectricity Springer Series in Materials Science., Hull R, Osgood R M Jr, Parisi J, and Warlimont H, eds, Berlin, Heidelberg, Springer,2010, 67–84. Doi: 10.1007/978-3-642-00716-3.

[14]    Liao B L, Qiu B, Zhou J W, Huberman S, Esfarjani K, Chen G. Significant reduction of lattice thermal conductivity by the electron-phonon interaction in silicon with high carrier concentrations: A first-principles study, Phys. Rev. Lett. 2015, 114, 115901:1–6. Doi: 10.1103/PhysRevLett.114.115901.

[15]    Heinz N A, Ikeda T, Pei Y Z, Snyder G J. Applying quantitative microstructure control in advanced functional composites, Adv. Funct. Mater. 2014, 24, 2135–2153. Doi: 10.1002/adfm.201302899.

[16]    Steele M C, Rosi F D. Thermal conductivity and thermoelectric power of Germanium-silicon alloys, J. Appl. Phys. 1958, 29, 1517–1520. Doi: 10.1063/1.1722984.

[17]    Abrahams M S, Braunstein R, Rosi F D. Thermal, electrical and optical properties of (In,Ga)As alloys, J. Phys. Chem. Solids 1959, 10, 204–210. Doi: 10.1016/0022-3697(59)90076-9.

[18]    Smith G E, Wolfe R. Thermoelectric properties of bismuth-antimony alloys, J. Appl. Phys. 1962, 33, 841–846. Doi: 10.1063/1.1777178.

[19]    Dismukes J P, Ekstrom L, Steigmeier E F, Kudman I, Beers D S. Thermal and electrical properties of heavily doped Ge-Si alloys up to 1300°K, J. Appl. Phys. 1964, 35, 2899–2907. Doi: 10.1063/1.1713126.

[20]    Yim W M, Rosi F D. Compound tellurides and their alloys for Peltier cooling – a review, Solid. State. Electron. 1972, 15, 1121–1140. Doi: 10.1016/0038-1101(72)90172-4.

[21]    Poudeu P F P, D'Angelo J, Kong H J, Downey A, Short J L, Pcionek R, Hogan T P, Uher C, Kanatzidis M G. Nanostructures versus solid solutions: low lattice thermal conductivity and enhanced thermoelectric figure of merit in $Pb_{9.6}Sb_{0.2}Te_{10-x}Se_x$ bulk materials, J. Am. Chem. Soc. 2006, 128, 14347–14355. Doi: 10.1021/ja0647811.

[22]    Zaitsev V K, Fedorov M I, Gurieva E A, Eremin I S, Konstantinov P P, Samunin A Y, Vedernikov M V. Highly effective $Mg_2Si_{1-x}Sn_x$ thermoelectrics, Phys. Rev. B 2006, 74, 045207:1–5. Doi: 10.1103/PhysRevB.74.045207.

[23]    LaLonde A D, Pei Y Z, Wang H, Snyder G J. Lead telluride alloy thermoelectrics, Mater. Today 2011, 14, 526–532. Doi: 10.1016/S1369-7021(11)70278-4.

[24]    Mingo N, Hauser D, Kobayashi N P, Plissonnier M, Shakouri A. "Nanoparticle-in-alloy" approach to efficient thermoelectrics: silicides in SiGe, Nano Lett 2009, 9, 711–715. Doi: 10.1021/nl8031982.

[25]    Kim W, Singer S L, Majumdar A, Zide J M O, Klenov D, Gossard A C, Stemmer S. Reducing thermal conductivity of crystalline solids at high temperature using embedded nanostructures, Nano Lett 2008, 8, 2097–2099. Doi: 10.1021/nl080189t.

[26]    Kim W, Zide J, Gossard A, Klenov D, Stemmer S, Shakouri A, Majumdar A. Thermal conductivity reduction and thermoelectric figure of merit increase by embedding nanoparticles in crystalline semiconductors, Phys. Rev. Lett. 2006, 96, 045901:1–4. Doi: 10.1103/PhysRevLett.96.045901.

[27]    Qiu B, Bao H, Zhang G Q, Wu Y, Ruan X L. Molecular dynamics simulations of lattice thermal conductivity and spectral phonon mean free path of PbTe: bulk and nanostructures, Comput. Mater. Sci. 2012, 53, 278–285. Doi: 10.1016/j.commatsci.2011.08.016.

[28]    Biswas K, He J Q, Blum I D, Wu C-I, Hogan T P, Seidman D N, Dravid V P, Kanatzidis M G. High-performance bulk thermoelectrics with all-scale hierarchical architectures, Nature 2012, 489, 414–418. Doi: 10.1038/nature11439. Corrigendum (2012) Nature 490, 570. doi: 10.1038/nature11645.

[29]    Zhao L-D, Dravid V P, Kanatzidis M G. The panoscopic approach to high performance thermoelectrics, Energy Environ. Sci. 2014, 7, 251–268. Doi: 10.1039/C3EE43099E.

[30]    Hu X K, Jood P, Ohta M, Kunii M, Nagase K, Nishiate H, Kanatzidis M G, Yamamoto A. Power generation from nanostructured PbTe-based thermoelectrics: comprehensive development from materials to modules, Energy Environ. Sci. 2016, 9, 517–529. Doi: 10.1039/C5EE02979A.

[31] Goldsmid H J, Penn A W. Boundary scattering of phonons in solid solutions, Phys. Lett. A 1968, 27, 523–524. Doi: 10.1016/0375-9601(68)90898-0.

[32] Kim S I, Lee K H, Mun H A, Kim H S, Hwang S W, Roh J W, Yang D J, Shin W H, Li X S, Lee Y H. et al.. Dense dislocation arrays embedded in grain boundaries for high-performance bulk thermoelectrics, Science 2015, 348, 109–114. Doi: 10.1126/science.aaa4166.

[33] Nolas G S, Cohn J L, Slack G A. Effect of partial void filling on the lattice thermal conductivity of skutterudites, Phys. Rev. B 1998, 58, 164–170. Doi: 10.1103/PhysRevB.58.164.

[34] Sales B C, Mandrus D, Chakoumakos B C, Keppens V, Thompson J R. Filled skutterudite antimonides: electron crystals and phonon glasses, Phys. Rev. B 1997, 56, 15081–15089. Doi: 10.1103/PhysRevB.56.15081.

[35] Yang J, Morelli D T, Meisner G P, Chen W, Dyck J S, Uher C. Effect of Sn substituting for Sb on the low-temperature transport properties of ytterbium-filled skutterudites, Phys. Rev. B 2003, 67, 165207:1–6. Doi: 10.1103/PhysRevB.67.165207.

[36] Avila M A, Suekuni K, Umeo K, Fukuoka H, Yamanaka S, Takabatake T. $Ba_8Ga_{16}Sn_{30}$ with Type-I clathrate structure: drastic suppression of heat conduction, Appl. Phys. Lett. 2008, 92, 041901:1–3. Doi: 10.1063/1.2831926.

[37] Takabatake T, Suekuni K, Nakayama T, Kaneshita E. Phonon-glass electron-crystal thermoelectric clathrates: experiments and theory, Rev. Mod. Phys. 2014, 86, 669–716. Doi: 10.1103/RevModPhys.86.669.

[38] Suekuni K, Tsuruta K, Kunii M, Nishiate H, Nishibori E, Maki S, Ohta M, Yamamoto A, Koyano M. High-performance thermoelectric mineral $Cu_{12-x}Ni_xSb_4S_{13}$ tetrahedrite, J. Appl. Phys. 2013, 113, 043712:1–5. Doi: 10.1063/1.4789389.

[39] Lu X, Morelli D T, Xia Y, Zhou F, Ozolins V, Chi H, Zhou X Y, Uher C. High performance thermoelectricity in earth-abundant compounds based on natural mineral tetrahedrites, Adv. Energy Mater. 2013, 3, 342–348. Doi: 10.1002/aenm.201200650.

[40] Suekuni K, Lee C H, Tanaka H I, Nishibori E, Nakamura A, Kasai H, Mori H, Usui H, Ochi M, Hasegawa T, Nakamura M, Ohira-Kawamura S, Kikuchi T, Kaneko K, Nishiate H, Hashikuni K, Kosaka Y, Kuroki K, Takabatake T. Retreat from stress: rattling in a planar coordination, Adv. Mater. 2018, 30, 1706230:1–6. Doi: 10.1002/adma.201706230.

[41] Sootsman J R, Chung D Y, Kanatzidis M G. New and old concepts in thermoelectric materials, Angew. Chemie – Int. Ed. 2009, 48, 8616–8639. Doi: 10.1002/anie.200900598.

[42] Vineis C J, Shakouri A, Majumdar A, Kanatzidis M G. Nanostructured thermoelectrics: big efficiency gains from small features, Adv. Mater. 2010, 22, 3970–3980. Doi: 10.1002/adma.201000839.

[43] Kanatzidis M G. Nanostructured thermoelectrics: the new paradigm?, Chem. Mater. 2010, 22, 648–659. Doi: 10.1021/cm902195j.

[44] Tan G J, Zhao L-D, Kanatzidis M G. Rationally designing high-performance bulk thermoelectric materials, Chem. Rev. 2016, 116, 12123–12149. Doi: 10.1021/acs.chemrev.6b00255.

[45] Pei Y Z, LaLonde A, Iwanaga S, Snyder G J. High thermoelectric figure of merit in heavy hole dominated PbTe, Energy Environ. Sci. 2011, 4, 2085–2089. Doi: 10.1039/C0EE00456A.

[46] Zhao L D, Wu H J, Hao S Q, Wu C I, Zhou X Y, Biswas K, He J Q, Hogan T P, Uher C, Wolverton C. et al.. All-scale hierarchical thermoelectrics: MgTe in PbTe facilitates valence band convergence and suppresses bipolar thermal transport for high performance, Energy Environ. Sci. 2013, 6, 3346–3355. Doi: 10.1039/C3EE42187B.

[47] Zeier W G, Zevalkink A, Gibbs Z M, Hautier G, Kanatzidis M G, Snyder G J. Thinking like a chemist: intuition in thermoelectric materials, Angew. Chemie Int. Ed. 2016, 55, 6826–6841. Doi: 10.1002/anie.201508381.

[48] Tan G J, Shi F Y, Hao S, Zhao L-D, Chi H, Zhang X M, Uher C, Wolverton C, Dravid V P, Kanatzidis M G. Non-equilibrium processing leads to record high thermoelectric figure of merit in PbTe–SrTe, Nat. Commun. 2016, 7, 12167:1–8. Doi: 10.1038/ncomms12167.

[49]   Pei Y L, Tan G J, Feng D, Zheng L, Tan Q, Xie X B, Gong S K, Chen Y, Li J-F, He J Q. et al..
       Integrating band structure engineering with all-scale hierarchical structuring for high
       thermoelectric performance in PbTe system, Adv. Energy Mater. 2017, 7, 1601450:1–11.
       Doi: 10.1002/aenm.201601450.

[50]   Jood P, Ohta M, Yamamoto A, Kanatzidis M G. Excessively doped PbTe with Ge-induced
       nanostructures enables high-efficiency thermoelectric modules, Joule 2018, 2, 1339–1355.
       Doi: 10.1016/j.joule.2018.04.025.

[51]   Zhang J, Wu D, He D S, Feng D, Yin M J, Qin X Y, He, J Q. Extraordinary thermoelectric
       performance realized in n-type PbTe through multiphase nanostructure engineering,
       Adv. Mater. 2017, 29, 1703148:1–7. Doi: 10.1002/adma.201703148.

[52]   Xiao Y, Wu H J, Li W, Yin M J, Pei Y L, Zhang Y, Fu L W, Chen Y X, Pennycook S J, Huang
       L. et al.. Remarkable roles of Cu To synergistically optimize phonon and carrier transport in
       n-type PbTe-$Cu_2Te$, J. Am. Chem. Soc. 2017, 139, 18732–18738. Doi: 10.1021/jacs.7b11662.

[53]   Xiao Y, Zhao L-D. Charge and phonon transport in PbTe-based thermoelectric materials, npj
       Quantum Mater 2018, 3, 55. Doi: 10.1038/s41535-018-0127-y.

[54]   Tan G J, Ohta M, Kanatzidis M G. Thermoelectric power generation : from new materials to
       devices, Phil. Trans. R. Soc. A 2019, 377 (2152), 20180450. 1–28. Doi: 10.1098/
       rsta.2018.0450.

[55]   Androulakis J, Lin C-H-H, Kong H-J-J, Uher C, Wu C-I-I, Hogan T, Cook B A, Caillat T,
       Paraskevopoulos K M, Kanatzidis M G. Spinodal decomposition and nucleation and growth
       as a means to bulk nanostructured thermoelectrics: enhanced performance in $Pb_{1-x}Sn_xTe$-
       PbS, J. Am. Chem. Soc. 2007, 129, 9780–9788. Doi: 10.1021/ja071875h.

[56]   Biswas K, He J Q, Wang G Y, Lo S-H, Uher C, Dravid V P, Kanatzidis M G. High thermoelectric
       figure of merit in nanostructured p-type PbTe–MTe (M = Ca, Ba), Energy Environ. Sci. 2011, 4,
       4675–4684. Doi: 10.1039/C1EE02297K.

[57]   Ohta M, Biswas K, Lo S-H, He J !, Chung D Y, Dravid V P, Kanatzidis M G. Enhancement
       of thermoelectric figure of merit by the insertion of MgTe nanostructures in p-type PbTe
       doped with $Na_2Te$, Adv. Energy Mater. 2012, 2, 1117–1123. Doi: 10.1002/
       aenm.201100756.

[58]   Wang H, Bahk J-H, Kang C, Hwang J, Kim K, Kim J, Burke P, Bowers J E, Gossard A C,
       Shakouri A. et al.. Right sizes of nano- and microstructures for high-performance and rigid
       bulk thermoelectrics, Proc. Natl. Acad. Sci. U. S. A. 2014, 111, 10949–10954. Doi: 10.1073/
       pnas.1403601111.

[59]   Biswas K, He J Q, Zhang Q, Wang G, Uher C, Dravid V P, Kanatzidis M G. Strained endotaxial
       nanostructures with high thermoelectric figure of merit, Nat Chem 2011, 3, 160–166. Doi:
       10.1038/nchem.955.

[60]   Ahn K, Han M-K, He J Q, Androulakis J, Ballikaya S, Uher C, Dravid V P, Kanatzidis M G. Exploring
       resonance levels and nanostructuring in the PbTe-CdTe system and enhancement of the
       thermoelectric figure of merit, J. Am. Chem. Soc. 2010, 132, 5227–5235. Doi: 10.1021/ja910762q.

[61]   Jood P, Ohta M, Kunii M, Hu X K, Nishiate H, Yamamoto A, Kanatzidis M G. Enhanced average
       thermoelectric figure of merit of n-type $PbTe_{1-x}I_x$-MgTe, J. Mater. Chem. C 2015, 3,
       10401–10408. Doi: 10.1039/C5TC01652E.

[62]   Ravich U I, Efimova B A, Smirnov I A. Semiconducting Lead Chalcogenides, New York, Plenum
       Press, 1970. Doi: 10.1007/978-1-4684-8607-0.

[63]   Crocker A J, Rogers L M. Valence band structure of PbTe, J. Phys. Colloq. 1968, 29, C4-129-C4-
       132. Doi: 10.1051/jphyscol:1968418.

[64]   Pei Y Z, Shi X Y, LaLonde A, Wang H, Chen L D, Snyder G J. Convergence of electronic bands
       for high performance bulk thermoelectrics, Nature 2011, 473, 66–69. Doi: 10.1038/
       nature09996.

[65]   Wu H J, Zhao L-D, Zheng F S, Wu D, Pei Y L, Tong X, Kanatzidis M G, He J Q. Broad temperature plateau for thermoelectric figure of merit ZT > 2 in phase-separated PbTe$_{0.7}$S$_{0.3}$, Nat. Comm. 2014, 5, 4515. Doi: 10.1038/ncomms5515.

[66]   Wu D, Zhao L-D, Tong X, Li W, Wu L, Tan Q, Pei Y, Huang L, Li J-F, Zhu Y. et al.. Superior thermoelectric performance in PbTe–PbS pseudo-binary: extremely low thermal conductivity and modulated carrier concentration, Energy Environ. Sci. 2015, 8, 2056–2068. Doi: 10.1039/C5EE01147G.

[67]   Bahk J-H, Shakouri A. Enhancing the thermoelectric figure of merit through the reduction of bipolar thermal conductivity with heterostructure barriers, Appl. Phys. Lett. 2014, 105, 52106:1–5. Doi: 10.1063/1.4892653.

[68]   Yang H R, Bahk J-H, Day T, Mohammed A M S, Snyder G J, Shakouri A, Wu Y. Enhanced thermoelectric properties in bulk nanowire heterostructure-based nanocomposites through minority carrier blocking, Nano Lett 2015, 15, 1349–1355. Doi: 10.1021/nl504624r.

[69]   Burke P G, Curtin B M, Bowers J E, Gossard A C. Minority carrier barrier heterojunctions for improved thermoelectric efficiency, Nano Energy 2015, 12, 735–741. Doi: 10.1016j.nanoen.2015.01.037.

[70]   Zhang C H, Ng H, Li Z, Khor K A, Xiong Q H. Minority carrier blocking to enhance the thermoelectric performance of solution-processed Bi$_x$Sb$_{2-x}$Te$_3$ nanocomposites via a liquid-phase sintering process, ACS Appl. Mater. Interfaces 2017, 9, 12501–12510. Doi: 10.1021/acsami.7b01473.

[71]   Chetty R, Kikuchi Y, Bouyrie Y, Suekuni K, Yamamoto A, Jood P, Ohta M. Power generation from the Cu$_{26}$Nb$_2$Ge$_6$S$_{32}$ -based single thermoelectric element with Au diffusion barrier, J. Mater. Chem. C. 2019, 7, 5184–5192. Doi: 10.1039/C9TC00868C.

[72]   Jood P, Ohta M. Hierarchical architecturing for layered thermoelectric sulfides and chalcogenides, Materials (Basel) 2015, 8, 1124–1149. Doi: 10.3390/ma8031124. Correction. (2015), Materials (Basel). 8, 6482–6483. doi:10.3390/ma8095315.

[73]   Koumoto K, Funahashi R, Guilmeau E, Miyazaki Y, Weidenkaff A, Wang Y F, Wan C L. Thermoelectric ceramics for energy harvesting, J. Am. Ceram. Soc. 2013, 96, 1–23. Doi: 10.1111/jace.12076.

[74]   Hébert S, Kobayashi W, Muguerra H, Bréard Y, Raghavendra N, Gascoin F, Guilmeau E, Maignan A. From oxides to selenides and sulfides: the richness of the CdI$_2$ type crystallographic structure for thermoelectric properties, Phys. status solidi 2013, 210, 69–81. Doi: 10.1002/pssa.201228505.

[75]   Hébert S, Berthebaud D, Daou R, Bréard Y, Pelloquin D, Guilmeau E, Gascoin F, Lebedev O, Maignan A. Searching for new thermoelectric materials: some examples among oxides, sulfides and selenides, J. Phys. Condens. Matter 2016, 28, 013001:1–23. Doi: 10.1088/0953-8984/28/1/013001.

[76]   Wiegers G A. Misfit layer compounds: structures and physical properties, Prog. Solid State Chem. 1996, 24, 1–139. Doi: 10.1016/0079-6786(95)00007-0.

[77]   Meerschaut A, Rabu P, Rouxel J. Preparation and characterization of new mixed sandwiched layered compounds Ln$_{32}$Nb$_{28}$S$_{88}$ (Ln = La, Ce), J. Solid State Chem. 1989, 78, 35–45. Doi: 10.1016/0022-4596(89)90125-4.

[78]   Imai H, Shimakawa Y, Kubo Y. Large thermoelectric power factor in TiS$_2$ crystal with nearly stoichiometric composition, Phys. Rev. B 2001, 64, 241104:1–4. Doi: 10.1103/PhysRevB.64.241104.

[79]   Ohta M, Satoh S, Kuzuya T, Hirai S, Kunii M, Yamamoto A. Thermoelectric properties of Ti$_{1+x}$S$_2$ prepared by CS$_2$ sulfurization, Acta Mater 2012, 60, 7232–7240. Doi: 10.1016/j.actamat.2012.09.035.

[80] Guilmeau E, Barbier T, Maignan A, Chateigner D. Thermoelectric anisotropy and texture of intercalated $TiS_2$, Appl. Phys. Lett. 2017, 111, 133903:1–4. Doi: 10.1063/1.4998952.

[81] Wan C L, Wang Y F, Wang N, Koumoto K. Low-thermal-conductivity $(MS)_{1+x}(TiS_2)_2$ (M = Pb, Bi, Sn) misfit layer compounds for bulk thermoelectric materials, Materials (Basel) 2010, 3, 2606–2617. Doi: 10.3390/ma3042606.

[82] Logothetis E M, Kaiser W J, Kukkonen C A, Faile S P, Colella R, Gambold J. Transport properties and the semiconducting nature of $TiS_2$, Phys. B+C 1980, 99, 193–198. Doi: 10.1016/0378-4363(80)90231-4.

[83] Kukkonen C A, Kaiser W J, Logothetis E M, Blumenstock B J, Schroeder P A, Faile S P, Colella R, Gambold J. Transport and optical properties of $Ti_{1+x}S$, Phys. Rev. B 1981, 24, 1691–1709. Doi: 10.1103/PhysRevB.24.1691.

[84] Inoue M, Koyano M, Negishi H, Ueda Y, Sato H. Localized impurity level and carrier concentration in self-intercalated $TiS_2$ crystals, Phys. status solidi 1985, 132, 295–303. Doi: 10.1002/pssb.2221320130.

[85] Koyano M, Negishi H, Ueda Y, Sasaki M, Inoue M. Electrical resistivity and thermoelectric power of intercalation compounds $M_xTiS_2$ (M = Mn, Fe, Co, and Ni), Phys. status solidi 1986, 138, 357–363. Doi: 10.1002/pssb.2221380137.

[86] Amara A, Frongillo Y, Aubin M J, Jandl S, Lopez-Castillo J M, Jay-Gerin J-P. Thermoelectric power of $TiS_2$, Phys. Rev. B 1987, 36, 6415–6419. Doi: 10.1103/PhysRevB.36.6415.

[87] Han S H, Cook B A (1994). An Experimental Search for High ZT Semiconductors: A Survey of the Preparation and Properties of Several Alloy Systems. In AIP Conference Proceedings (AIP), pp. 66–70. doi:10.1063/1.46837.

[88] Guilmeau E, Bréard Y, Maignan A. Transport and thermoelectric properties in copper intercalated $TiS_2$ chalcogenide, Appl. Phys. Lett. 2011, 99, 052107:1–3. Doi: 10.1063/1.3621834.

[89] Beaumale M, Barbier T, Bréard Y, Raveau B, Kinemuchi Y, Funahashi R, Guilmeau E. Mass fluctuation effect in $Ti_{1-x}Nb_xS_2$ bulk compounds, J. Electron. Mater. 2014, 43, 1590–1596. Doi: 10.1007/s11664-013-2802-x.

[90] Beaumale M, Barbier T, Bréard, Guelou G, Powell A V, Vaqueiro P, Guilmeau E. Electron doping and phonon scattering in $Ti_{1+x}S_2$ thermoelectric compounds, Acta Mater 2014, 78, 86–92. Doi: 10.1016/j.actamat.2014.06.032.

[91] Guélou G, Vaqueiro P, Prado-Gonjal J, Barbier T, Hébert S, Guilmeau E, Kockelmann W, Powell A V. The impact of charge transfer and structural disorder on the thermoelectric properties of cobalt intercalated $TiS_2$, J. Mater. Chem. C 2016, 4, 1871–1880. Doi: 10.1039/c5tc04217h.

[92] Wan C L, Wang Y F, Wang N, Norimatsu W, Kusunoki M, Koumoto K. Intercalation: building a natural superlattice for better thermoelectric performance in layered chalcogenides, J. Electron. Mater. 2011, 40, 1271–1280. Doi: 10.1007/s11664-011-1565-5.

[93] Wan C L, Wang Y F, Norimatsu W, Kusunoki M, Koumoto K. Nanoscale stacking faults induced low thermal conductivity in thermoelectric layered metal sulfides, Appl. Phys. Lett. 2012, 100, 101913:1–4. Doi: 10.1063/1.3691887.

[94] Putri Y E, Wan C L, Wang Y F, Norimatsu W, Kusunoki M, Koumoto K. Effects of alkaline earth doping on the thermoelectric properties of misfit layer sulfides, Scr. Mater. 2012, 66, 895–898. Doi: 10.1016/j.scriptamat.2012.02.010.

[95] Putri Y E, Wan C L, Dang F, Mori T, Ozawa Y, Norimatsu W, Kusunoki M, Koumoto K. Effects of transition metal substitution on the thermoelectric properties of metallic $(BiS)_{1.2}(TiS_2)_2$ misfit layer sulfide, J. Electron. Mater 2014, 43, 1870–1874. Doi: 10.1007/s11664-013-2894-3.

[96] Yin C, Hu Q, Tang M J, Liu, H T, Chen Z Y, Wang Z S, Ang R. Boosting the thermoelectric performance of misfit-layered $(SnS)_{1.2}(TiS_2)_2$ by a Co- and Cu-substituted alloying effect, J. Mater. Chem. 2018, A 6, 22909–22914. Doi: 10.1039/C8TA08426B.

[97]  Beaumale M, Barbier T, Bréard Y, Hébert S, Kinemuchi Y, Guilmeau E. Thermoelectric properties in the series $Ti_{1-x}Ta_xS_2$, J. Appl. Phys. 2014, 115, 043704:1–7. Doi: 10.1063/1.4863141.

[98]  Gascoin F, Raghavendra N, Guilmeau E, Bréard Y. $CdI_2$ structure type as potential thermoelectric materials: synthesis and high temperature thermoelectric properties of the solid solution $TiS_xSe_{2-x}$, J. Alloys Compd. 2012, 521, 121–125. Doi: 10.1016/j.jallcom.2012.01.067.

[99]  Wang Y F, Pan L, Li C, Tian R M, Huang R, Hu X H, Chen C, Bao N Z, Koumoto K, Lu C H. Doubling the ZT Record of $TiS_2$-based thermoelectrics by incorporation of ionized impurity scattering, J. Mater. Chem. C 2018, 6, 9345–9353. Doi: 10.1039/C8TC00914G.

[100] Wan C L, Gu X K, Dang F, Itoh T, Wang Y F, Sasaki H, Kondo M, Koga K, Yabuki K, Snyder G J, Yang R G, Koumoto K. Flexible n-type thermoelectric materials by organic intercalation of layered transition metal dichalcogenide $TiS_2$, Nat. Mater. 2015, 14, 622–627. Doi: 10.1038/nmat4251.

[101] Wan C L, Tian R M, Kondou, M, Yang R G, Zong P G, Koumoto K. Ultrahigh thermoelectric power factor in flexible hybrid inorganic-organic superlattice, Nat. Commun. 2017, 8, 1024:1–9. Doi: 10.1038/s41467-017-01149-4.

[102] Tian R M, Wan C L, Wang Y F, Wei Q S, Ishida T, Yamamoto A, Tsuruta A, Shin W S, Li S, Koumoto K. A solution-processed $TiS_2$/organic hybrid superlattice film towards flexible thermoelectric devices, J. Mater. Chem. A 2017, 5, 564–570. Doi: 10.1039/C6TA08838D.

[103] Wang L M, Zhang Z M, Geng L X, Yuan T Y, Liu Y C, Guo J C, Fang L, Qiu J J, Wang S R. Solution-printable fullerene/$TiS_2$ organic/inorganic hybrids for high-performance flexible n-type thermoelectrics, Energy Environ. Sci. 2018, 11, 1307–1317. Doi: 10.1039/C7EE03617E.

[104] Miyazaki Y, Ogawa H, Kajitani T. Preparation and thermoelectric properties of misfit-layered sulfide $[Yb_{1.90}S_2]_{0.62}NbS_2$, Jpn. J. Appl. Phys. 2004, 43, L1202–L1204. Doi: 10.1143/JJAP.43.L1202.

[105] Jood P, Ohta M, Nishiate H, Yamamoto A, Lebedev O I, Berthebaud D, Suekuni K, Kunii M. Microstructural control and thermoelectric properties of misfit layered sulfides $(LaS)_{1+m}TS_2$ (T = Cr, Nb): the natural superlattice systems, Chem. Mater. 2014, 26, 2684–2692. Doi: 10.1021/cm5004559.

[106] Jood P, Ohta M, Lebedev O I, Berthebaud D. Nanostructural and microstructural ordering and thermoelectric property tuning in misfit layered sulfide $[(LaS)_x]_{1.14}NbS_2$, Chem. Mater. 2015, 27, 7719–7728. Doi: 10.1021/acs.chemmater.5b03365.

[107] Miyazaki Y, Ogawa H, Nakajo T, Kikuchi Y, Hayashi K. Crystal Structure and Thermoelectric Properties of Misfit-Layered Sulfides $[Ln_2S_2]_pNbS_2$ (Ln = Lanthanides), J. Electron. Mater. 2013, 42, 1335–1339. Doi: 10.1007/s11664-012-2443-5.

[108] Suekuni K, Takabatake T. Research update: Cu–S based synthetic minerals as efficient thermoelectric materials at medium temperatures, APL Mater 2016, 4, 104503:1–11. Doi: 10.1063/1.4955398.

[109] Ohta M, Jood P, Murata M, Lee C-H, Yamamoto A, Obara H. An integrated approach to thermoelectrics: combining phonon dynamics, nanoengineering, novel materials development, module fabrication, and metrology, Adv. Energy Mater. 2018, 1801304:1–29. Doi: 10.1002/aenm.201801304.

[110] Bouyrie Y, Ohta M, Suekuni K, Jood P, Takabatake T. Addition of Co, Ni, Fe and their role in the thermoelectric properties of colusite $Cu_{26}Nb_2Ge_6S_{32}$, J. Alloys Compd. 2018, 735, 1838–1845. Doi: 10.1016/j.jallcom.2017.11.199.

[111] Suekuni K, Kim F S, Nishiate H, Ohta M, Tanaka H I, Takabatake T. High-performance thermoelectric minerals: colusites $Cu_{26}V_2M_6S_{32}$ (M = Ge, Sn), Appl. Phys. Lett. 2014, 105 (132107). 1–3. Doi: 10.1063/1.4896998.

[112] Kikuchi Y, Bouyrie Y, Ohta M, Suekuni K, Aihara M, Takabatake T. Vanadium-free colusites $Cu_{26}A_2Sn_6S_{32}$ (A = Nb, Ta) for environmentally friendly thermoelectrics, J. Mater. Chem. A 2016, 4, 15207–15214. Doi: 10.1039/C6TA05945G.

[113] Pavan Kumar V, Supka A R, Lemoine P, Lebedev O I, Raveau B, Suekuni K, Nassif V, Al Rahal Al Orabi R, Fornari M., R, Guilmeau E. High power factors of thermoelectric colusites $Cu_{26}T_2Ge_6S_{32}$ ( T = Cr, Mo, W): toward functionalization of the conductive "Cu–S" network, Adv. Energy Mater. 2019, 9, 1803249:1–11. Doi: 10.1002/aenm.201803249.

[114] Kim F S, Suekuni K, Nishiate H, Ohta M, Tanaka H I, Takabatake T. Tuning the charge carrier density in the thermoelectric colusite, J. Appl. Phys. 2016, 119, 175105:1–5. Doi: 10.1063/1.4948475.

[115] Bouyrie Y, Ohta M, Suekuni K, Kikuchi Y, Jood P, Yamamoto A, Takabatake T. Enhancement in the thermoelectric performance of colusites $Cu_{26}A_2E_6S_{32}$ (A = Nb, Ta; E = Sn, Ge) using E-site non-stoichiometry, J. Mater. Chem. C 2017, 5, 4174–4184. Doi: 10.1039/C7TC00762K.

[116] Suekuni K, Shimizu, Y, Nishibori E, Kasai H, Saito H, Yoshimoto D, Hashikuni K, Bouyrie Y, Chetty R, Ohta M. et al.. Atomic-scale phonon scatterers in thermoelectric colusites with a tetrahedral framework structure, J. Mater. Chem. A 2019, 7, 228–235. Doi: 10.1039/C8TA08248K.

[117] Bourgès C, Bouyrie Y, Supka A R, Al Rahal Al Orabi R, Lemoine P, Lebedev O I, Ohta M, Suekuni K, Nassif V, Hardy V. et al.. High-performance thermoelectric bulk colusite by process controlled structural disordering, J. Am. Chem. Soc. 2018, 140, 2186–2195. Doi: 10.1021/jacs.7b11224.

[118] Spry P G, Merlino S, Wang S, Zhang X M, Buseck P R. New occurrences and refined crystal chemistry of colusite, with comparisons to arsenosulvanite, Am. Mineral. 1994, 79, 750–762.

[119] Suekuni K, Tsuruta K, Ariga T, Koyano M. Thermoelectric properties of mineral tetrahedrites $Cu_{10}Tr_2Sb_4S_{13}$ with low thermal conductivity, Appl. Phys. Express 2012, 5, 051201:1–3. Doi: 10.1143/APEX.5.051201.

[120] Chetty R, Bali A, Mallik R C. Tetrahedrites as thermoelectric materials: an overview, J. Mater. Chem. C 2015, 3, 12364–12378. Doi: 10.1039/c5tc02537k.

[121] Lu X, Morelli D T, Xia Y X, Ozolins V. Increasing the thermoelectric figure of merit of tetrahedrites by co-doping with nickel and zinc, Chem. Mater. 2015, 27, 408–413. Doi: 10.1021/cm502570b.

[122] Kosaka Y, Suekuni K, Hashikuni K, Bouyrie Y, Ohta M, Takabatake T. Effects of Ge and Sn substitution on the metal–semiconductor transition and thermoelectric properties of $Cu_{12}Sb_4S_{13}$ tetrahedrite, Phys. Chem. Chem. Phys. 2017, 19, 8874–8879. Doi: 10.1039/C7CP00351J.

[123] Lu X, Morelli D T, Wang Y X, Lai W, Xia Y, Ozolins V. Phase stability, crystal structure, and thermoelectric properties of $Cu_{12}Sb_4S_{13-x}Se_x$ solid solutions, Chem. Mater. 2016, 28, 1781–1786. Doi: 10.1021/acs.chemmater.5b04796.

[124] Yan Y C, Wu H, Wang G Y, Lu X, Zhou X Y. High thermoelectric performance balanced by electrical and thermal transport in tetrahedrites $Cu_{12+x}Sb_4S_{12}Se$, Energy Storage Mater 2018, 13, 127–133. Doi: 10.1016/j.ensm.2018.01.006.

[125] Vaqueiro P, Guélou G, Kaltzoglou A, Smith R I, Barbier T, Guilmeau E, Powell A V. The influence of mobile copper ions on the glass-like thermal conductivity of copper-rich tetrahedrites, Chem. Mater. 2017, 29, 4080–4090. Doi: 10.1021/acs.chemmater.7b00891.

[126] Roche C, Pecheur P, Toussaint G, Jenny A, Scherrer H, Scherrer S. Study of Chevrel phases for thermoelectric applications: band structure calculations on compounds (M = metal), J. Phys. Condens. Matter 1998, 10, L333–L339. Doi: 10.1088/0953-8984/10/21/001.

[127] Caillat T, Fleurial J-P. Thermoelectric properties of the semiconducting Chevrel phase $Mo_2Re_4Se_8$, J. Phys. Chem. Solids 1998, 59, 1139–1144. Doi: 10.1016/S0022-3697(97)00183-2.

[128] Tsubota T, Ohtaki M, Eguchi K. Thermoelectric properties of Chevrel-type sulfides $AMo_6S_8$. (A = Fe, Ni, Ag, Zn, Sn, Pb, Cu), J. Ceram. Soc. Japan 1999, 107, 697–701. Doi: 10.2109/jcersj.107.697.

[129] Caillat T, Fleurial J-P, Snyder G J. Potential of Chevrel phases for thermoelectric applications, Solid State Sci 1999, 1, 535–544. Doi: 10.1016/S1293-2558(00)80105-3.

[130] Schmidt A M, McGuire M A, Gascoin F, Snyder G J, DiSalvo F J. Synthesis and thermoelectric properties of $(Cu_yMo_6Se_8)_{1-x}(Mo_4Ru_2Se_8)_x$ alloys, J. Alloys Compd. 2007, 431, 262–268. Doi: 10.1016/j.jallcom.2006.05.061.

[131] Ohta M, Yamamoto A, Obara H. Thermoelectric properties of Chevrel-phase sulfides $M_xMo_6S_8$ (M: Cr, Mn, Fe, Ni), J. Electron. Mater. 2010, 39, 2117–2121. Doi: 10.1007/s11664-009-0975-0.

[132] Chevrel R, Sergent M. Chemistry and Structure of Ternary Molybdenum Chalcogenides, In: Superconductivity in Ternary Compounds I: Structural, Electronic, and Lattice Properties, Fischer Ø and Maple M B, eds., Springer-Verlag Berlin Heidelberg, 1982, 25–86. Doi: 10.1007/978-3-642-81868-4.

[133] Ohta M, Obara H, Yamamoto A. Preparation and thermoelectric properties of Chevrel-phase $Cu_xMo_6S_8$ (2.0≤x≤4.0), Mater. Trans 2009, 50, 2129–2133. Doi: 10.2320/matertrans. MAW200918.

[134] Zhou T, Lenoir B, Colin M, Dauscher A, Al Orabi R A R, Gougeon P, Potel M, Guilmeau E. Promising thermoelectric properties in $Ag_xMo_9Se_{11}$ compounds (3.4 ≤ x ≤ 3.9), Appl. Phys. Lett 2011, 98(162106). 1–3. Doi: 10.1063/1.3579261.

[135] Gougeon P, Gall P, Al Rahal Al Orabi R, Fontaine B, Gautier R, Potel M, Zhou T, Lenoir B, Colin M, Candolfi C. Al Rahal Al Orabi, R et al. Synthesis, crystal and electronic structures, and thermoelectric properties of the novel cluster compound $Ag_3In_2Mo_{15}Se_{19}$, Chem. Mater 2012, 24, 2899–2908. Doi: 10.1021/cm3009557.

[136] Ohta M, Chung D Y, Kunii M, Kanatzidis M G. Low lattice thermal conductivity in $Pb_5Bi_6Se_{14}$, $Pb_3Bi_2S_6$, and $PbBi_2S_4$: promising thermoelectric materials in the cannizzarite, lillianite, and galenobismuthite homologous series, J. Mater. Chem. A 2014, 2, 20048–20058. Doi: 10.1039/C4TA05135A.

[137] Olvera A, Shi G S, Djieutedjeu H, Page A, Uher C, Kioupakis E, Poudeu P F P. $Pb_7Bi_4Se_{13}$: a lillianite homologue with promising thermoelectric properties, Inorg. Chem. 2015, 54, 746–755. Doi: 10.1021/ic501327u.

[138] Sassi S, Candolfi C, Delaizir G, Migot S, Ghanbaja J, Gendarme C, Dauscher A, Malaman B, Lenoir B. Crystal structure and transport properties of the homologous compounds $(PbSe)_5$ $(Bi_2Se_3)_{3m}$(m = 2, 3), Inorg. Chem. 2018, 57, 422–434. Doi: 10.1021/acs.inorgchem.7b02656.

# Index

https://doi.org/10.1515/9783110596526-010

www.ingramcontent.com/pod-product-compliance
Lightning Source LLC
Chambersburg PA
CBHW061415210326
41598CB00035B/6224

* 9 7 8 3 1 1 0 5 9 6 4 8 9 *